OPTICAL DEVICES IN COMMUNICATION AND COMPUTATION

Edited by **Peng Xi**

—

Optical Devices in Communication and Computation
http://dx.doi.org/10.5772/2859
Edited by Peng Xi

Contributors
Mohammad Syuhaimi Ab-Rahman, V. Aboites, Y. Barmenkov, A. Kir'yanov, M. Wilson, Wei Li, Xunya Jiang, Po-Chang Wu, Wei Lee, Tetsuzo Yoshimura, Feng Liu, Biqin Dong, Xiaohan Liu, Lin Chen, Xu Guang Huang, Jin Tao, Duc Doan Hong, Fushinobu Kazuyoshi, Kyung M. Choi

Published by InTech
Janeza Trdine 9, 51000 Rijeka, Croatia

Publishing Process Manager Dragana Manestar
Technical Editor InTech Prepress, Novi Sad
Cover InTech Design team

Additional hard copies can be obtained from orders@intechopen.com

Optical Devices in Communication and Computation, Edited by Peng Xi
p. cm.
ISBN 978-953-51-0763-7

For EU product safety concerns:
IN TECH d.o.o., Prolaz Marije Krucifikse Kozulić 3, 51000 Rijeka, Croatia,
info@intechopen.com or visit our website at intechopen.com.

Meet the editor

Dr. Peng Xi obtained his PhD in optical engineering from Shanghai Institute of Optics and Fine Mechanics, Chinese Academy of Sciences. He then worked as a post-doctoral Research Associate in Hong Kong University of Science and Technology (2003-2004), Purdue University (2005-2006), and Michigan State University (2006-2007). From 2008 to 2009 he worked at Shanghai Jiao Tong University as an Associate Professor. He is currently working at the Department of Biomedical Engineering, Peking University as an Associate Professor. He has been invited to give several keynote speeches and serve on the scientific committee board of Focus On Microscopy conference. His research interests include optical nanoscopy, multiphoton microscopy, as well as novel biomedical optical imaging techniques.
Optical devices in co

Contents

Section 1

Advances in Theoretical Analysis

Chapter 1

Zero Loss Condition Analysis on Optical Cross Add and Drop Multiplexer (OXADM) Operational Scheme in Point-to-Point Network

Mohammad Syuhaimi Ab-Rahman

Additional information is available at the end of the chapter

http://dx.doi.org/10.5772/51006

1. Introduction

OXADMs are element which provide the capabilities of add and drop function and cross connecting traffic in the network, similar to OADM and OXC. OXADM consists of three main subsystem; a wavelength selective demultiplexer, a switching subsystem and a wavelength multiplexer. Each OXADM is expected to handle at least two distinct wavelength channels each with a coarse granularity of 2.5 Gbps of higher (signals with finer granularities are handled by logical switch node such as SDH/SONET digital cross connects or ATM switches). There are eight ports for add and drop functions, which are controlled by four lines of MEMs optical switch. The other four lines of MEMs switches are used to control the wavelength routing function between two different paths. The functions of OXADM include node termination, drop and add, routing, multiplexing and also providing mechanism of restoration for point-to-point, ring and mesh metropolitan and also customer access network in FTTH. The asymmetrical architecture of OXADM consists of 3 parts; selective port, add/drop operation, and path routing. Selective port permits only the interest wavelength pass through and acts as a filter. While add and drop function can be implemented in second part of OXADM architecture. The signals can then be re-routed to any port of output or/and perform an accumulation function which multiplex all signals onto one path and then exit to any interest output port. This will be done by the third part. OXADM can also perform 'U' turn to enable the line protection (Ring Protection) in the event of breakdown condition. This will be done by the first and third part. These two features have differed OXADM with the other existing device such as OADM and OXC. The purpose of this study was to obtain the maximum allowed loss for the device OXADM and input power required to maintain the satisfactory performance of the BER (BER $<10^{-9}$) in the specific loss value. Ideal situation is a situation where all the devices that form the optical device were considered to have zero loss. However, this loss is replaced in the BER measurement with the use of optical attenuator is set at 25 dB. The value of 25 dB will represent the total loss in the OXADM device. In zero loss condition, the only contributor to the system loss is the non-linear effect of power penalty. The decrement of data transmission rate with the increment of loss and maximum loss for each operation in the network OXADM point is also studied. The relationship between allowable power loss and the magnitude of input signal is shown in proposed equation. Optical fiber with nonlinear dispersion (attenuation constant, α = 0.25 dB/km) used for connecting two nodes OXADM at a distance of 60 km.

This paper also measured the operational loss value for three main operation of OXADM such as pass through, dropping and adding signal. The relationship between minimum input power and attenuation given by the linear equation, $y = x + 25$ to intercept the y axis is 25 dB (maximum loss in the input power 0 dBm). Gradient, $m = 1$ shows no change at 1 dBm of input power will change the power loss of 1 dB. The restoration scheme offers by OXADM is also been investigated. We examine the relationship between the attenuation/loss at optical node on output power and the BER performance of the ring protection mechanism is activated. The simulation study also seeks to obtain the magnitude of the attenuation is allowed during the operation of this ring of protection (if attenuation increases due to inclusion of other optical devices and connectors). Rate of decreasing of output power due to attenuation increased will also be studied and based on the value of the internal amplifier gain can be determined relatively. Finally, the proposed value of the internal amplifier which is suitable for miniaturization compensate signal to a directional orientation to the West and East to have the same attenuation as a ring of protection is turned on.

2. OXADM device

Optical switch based devices is one of the most promising element that is used in optical communication network. Starting with Modulator at the receiver site, then moving to Optical Add and Drop Multiplexing (OADM) and Optical Cross Connect (OXC) at the distribution site and finally ending with Receiver (demodulator) at recovery site have shown the significant useful of the device. However, the rapid change and evolvement in optical network and service today has required the new type of optical switching device to be developed. Optical Cross Node, Tuneable Ring Node, Customer Access Protection Switch (CAPU), Arrayed Waveguide Grating Multiplexing are amongst the new generation of optical switch device [Mutafungwa 2000][Eldada & Nunen 2000][Aziz et al. 2009]. In this paper we introducing of new architecture of switch device that is designed to overcome drawbacks that occur in wavelength management in expected. The device is called optical cross add and drop multiplexing (OXADM) which use combination concept of OXC and OADM. Its enable the operating wavelength on two different optical trunks to be switched to each other and implementing accumulating function simultaneously. Here, the operating wavelengths can be multiplexed together and exit to any interested output port. The wavelength transfer between two different cores of fiber will increase the flexibility, survivability and also efficiency of the network structure. To make device operational more efficient by reducing the power penalty, zero leakage MEMs switches are used to control the mechanism of operation such as wavelength add/drop and wavelength routing operation. As a result, the switching performed within the optical layer will be able to achieve high speed restoration against failure/degradation of cables, fibers and optical amplifiers which had been proposed in [Rahman et al. 2006a][Rahman et al. 2006b]. We had proposed previously the migration of topology will be easier and reduce the restructuring process by eliminating the installation of new nodes because OXADMs are applicable for both types of topologies beside provide efficiency, reliability and survivability to the network [Rahman et al. 2006c][Rahman & Shaari 2007].

OXADMs are element which provide the capabilities of add and drop function and cross connecting traffic in the network, similar to OADM and OXC. OXADM consists of three main subsystem; a wavelength selective demultiplexer, a switching subsystem and a wavelength multiplexer. Each OXADM is expected to handle at least two distinct wavelength channels each with a coarse granularity of 2.5 Gbps of higher (signals with finer granularities are handled by logical switch node such as SDH/SONET digital cross connects or ATM switches. There are eight ports for add and drop functions, which are controlled by four lines of MEMs optical switch. The other four lines of MEMs switches are used to control the wavelength routing function between two different paths. The functions of OXADM include node termination, drop and add, routing, multiplexing and also providing mechanism of restoration for point-to-point, ring and mesh metropolitan and also customer access network in FTTH. With the setting of the MEMs optical switch configuration, the device can be programmed to function as another optical devices such as multiplexer, demultiplexer, coupler, wavelength converter (with fiber grating filter configuration), OADM, wavelength round about an etc for the single application. The designed 4-channel OXADM device is expected to have maximum operational loss of 0.06 dB for each channel when device components are in ideal/zero loss condition. The maximum insertion loss when considering the component loss at every channel is less than 6 dB [Rahman et al. 2006a]-[Rahman 2008].

In this paper we analyze the performance of OXADM in zero loss condition to obtain the achievable loss of point-to-point network at a specific receiver sensitivity value. Finally to address the operational loss and can be called as power penalty to each function or operation performed by this device.

2.1. Insertion loss calculation

Table 1 shows the modulated launched power to characterize the insertion loss of OXADM operation. Since the launch of the four modulated wavelength operation is almost similar, therefore the process of leveling (equating the amplitude of) the wavelength is not necessary. Specification for the characterization of the insertion loss calculation is as follows:

Attenuation = 25 dB (representing insertion loss)
Photodetector sensitivity = -28.4 dBm at1550 nm
Data transmission rate = 2.5 Gps (OC-48)
WDM analyzer resolution bandwidth = 0.1 nm
Photodetector thermal noise = $1x10^{-23}$ W/Hz
Launched power (before modulation) = 0 dBm

The word 'sensitivity' is used in this paper are based on the simulation using optisystem tool. 'Sensitivity' is actually referring to the power allocation or budget power in actual application. The actual sensitivity in photodetector is defined as

$$\text{Photodetector Sensitivity}(\text{dBm}) = -(\text{Power Budget} + \text{Safety Margin}) \tag{1}$$

Wavelength	Launched Power (Watt)	Launched Power (dBm)
1510 nm	4.680×10^{-4}	-3.297
1530 nm	4.808×10^{-4}	-3.180
1550 nm	4.872×10^{-4}	-3.123
1570 nm	4.808×10^{-4}	-3.180

Table 1. Modulated launched power which injected to OXADM device.

2.2. Attenuation representing network total loss

The purpose of this simulation study was to determine allowable loss of OXADM to maintain the network performance in point to point network and be tested under ideal condition. The decrement of data transmission rate with the increment of loss and maximum loss for each operation in the network OXADM point is also studied. The relationship between allowable power loss and the magnitude of input signal is shown in equation (1). Optical fiber with nonlinear dispersion (attenuation constant, α = 0.25 dB/km) used for connecting two nodes OXADM at a distance of 60 km (Figure 1).

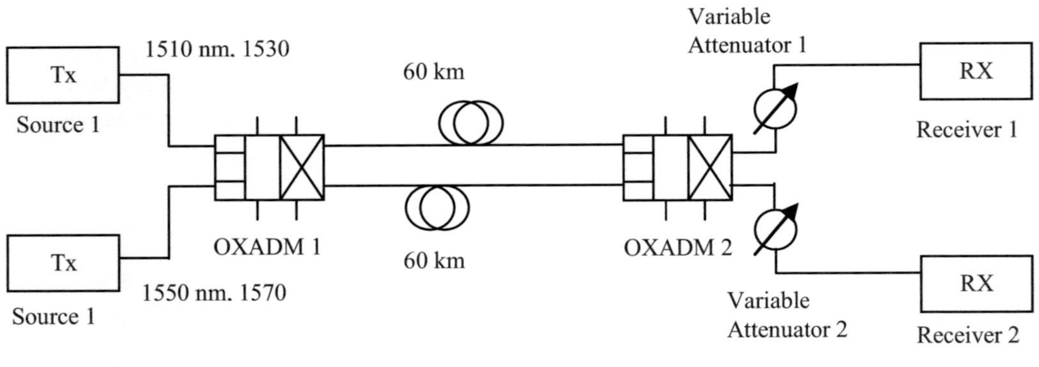

Figure 1. Experimental set up of point-to-point network which uses OXADM as an optical node. The value of OXADM insertion loss is determined by adjusting the attenuator.

2.3. The effect of attenuation to the BER

a) Addition Signal to the Output Signal

Figure 2 shows the effects of attenuation on the BER for the operation of additional signals into the device OXADM. Attenuation value is set starting at 20 dB to 29 dB. The purpose of this characterization was to obtain the actual value of the total insertion loss is acceptable to maintain the BER measurement of $1\text{x}10^{-9}$.

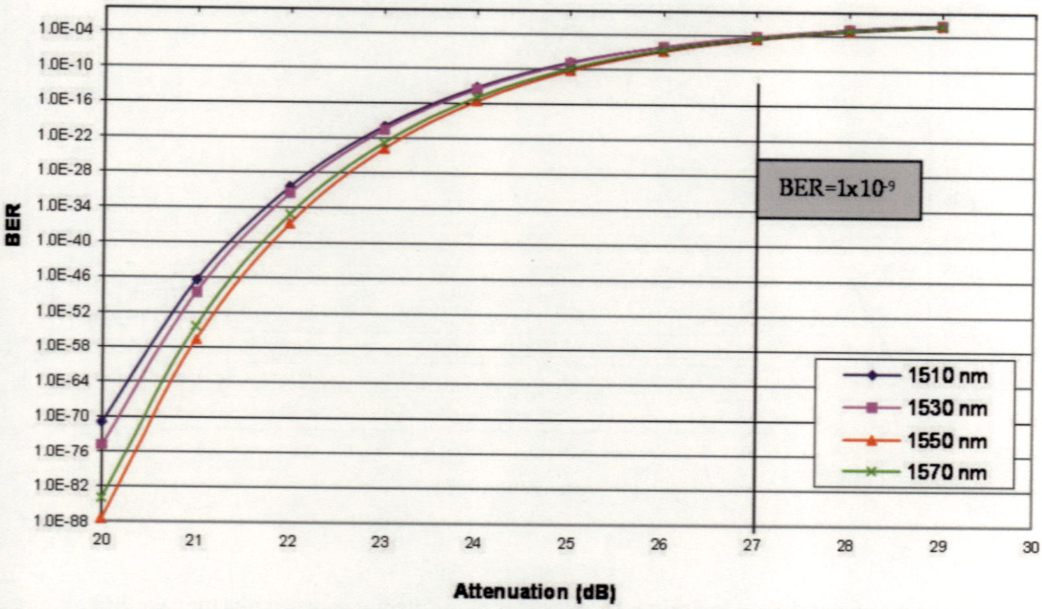

Figure 2. Effect of attenuation to BER performance for four different wavelengths for new additional signal.

From the graph, the value of the attenuation that gave readings equivalent to the BER is $1\text{x}10^{-9}$ at 25 dB. This means that the maximum acceptable amount of insertion loss in the OXADM device is 25 dB. However, this value can be increased by increasing the sensitivity of the system depends on the receiver system used.

b) Launched Signal to Output Signal

Figure 3 shows the effect of attenuation on the BER measurements for the operation of pass through to the signal. At 25 dB attenuation values give the same BER measurement readings $1\text{x}10^{-9}$. This value is equal to the value obtained for the operation of adding a new signal of OXADM. This shows OXADM single unit provides good performance in the value of the maximum insertion loss of 25 dB to the sensitivity of -28.4 dBm.

Conclusions from the studies on this part of the overall estimated value of OXADM device is 25 dB. Studies in the next section (the theory of product) will have an estimated value of real power for the provision of optical networks based on different values OXADM 25 dB with the insertion loss of OXADM which measured in the product theory.

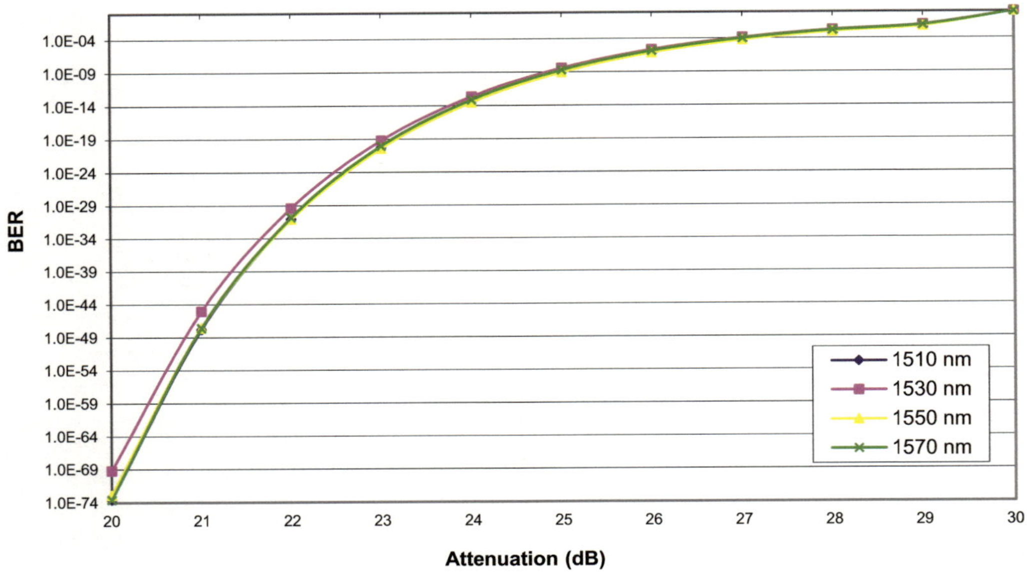

Figure 3. Effect of attenuation to BER performance for four different wavelengths for pass through operation.

2.4. Input signal to BER performance

Figure 4 shows the effect of input power diode laser (before the signal is modulated with data) on BER performance in a variety of attenuation. Attenuation value is set between 20 dB to 26 dB. The purpose of this characterization is to obtain the minimum power required by the device OXADM to operate in a satisfactory condition. The relationship between minimum input power and attenuation given by the linear equation, $y = x + 25$ to intercept the y axis is 25 dB (maximum loss in the input power 0 dBm). Gradient, $m = 1$ shows no change at 1 dBm of input power will change the power loss of 1 dB. The changes are shown in Figure 5.

The insertion loss under ideal condition is called as operational loss. The magnitude is rely on the operation functioned by OXADM. This term can also be used as power penalty. Power penalty is the other loss need to be compensated instead of insertion loss. Power penalty is the loss due to the non-linear effect such as SRS, FWM and others.

The loss under zero loss condition is also measured for each operation of OXADM. Table 2 listing the operational loss or power penalty for three OXADM operations; pass through, dropping, adding signal. The values is range from 0.05 to 0.18 depend to the number of switch device involve of each operation.

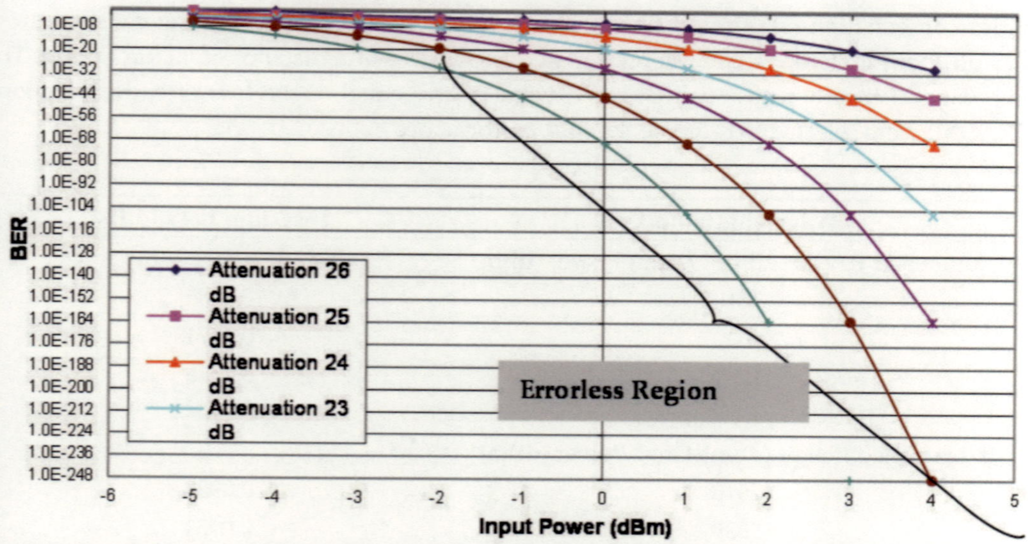

Figure 4. Effect of Input Power to the BER performance at different attenuation values (1530 nm)

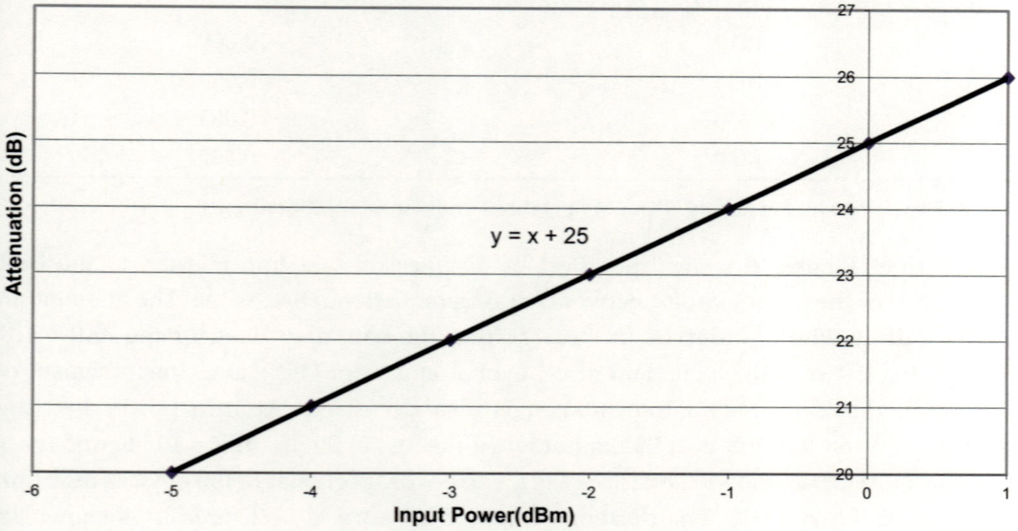

Figure 5. The required input power by OXADM to maintain the performance at different attenuation (λ=1530 nm, BER=$3.98x10^{-9}$).

2.5. Attenuation over distance

The purpose of this simulation study is to determine the performance of the OXADM in point to point network under the certain loss value. The decrement of achievable distance due to the increment of loss value is also studied. As a result, the relationship between the achievable distances for point to point network to the OXADM insertion loss has been defined in equation (2). Optical fiber with nonlinear dispersion (attenuation constant, α = 0.25 dB / km) is used for connecting two nodes OXADM at a distance of 60 km (Figure 1). Five value if insertion loss (which has a value nearly equal to the loss of each operation OXADM) were selected to estimate the BER performance in this network.

Wavelength (nm)	Insertion Loss (dB)
i. Launched Power (dBm)-Output Power (dBm)	
1510	0.1185
1530	0.1171
1550	0.1128
1570	0.1144
ii. Launched Power (dBm)-Drop Power (dBm)	
1510	0.0938
1530	0.0931
1550	0.0891
1570	0.0903
iii. Add Power (dBm)-Output Power (dBm)	
1510	0.0600
1530	0.0584
1550	0.0599
1570	0.0572

Table 2. Power Penalty for several OXADM operations under ideal condition.

Figure 6 until Figure 10 shows the effect of distance of data transmission to the BER performance of the point to point networks at different attenuation value. The attenuation is set at 0 dB to 20 dB. Observed in these graphs, the boundary lines for the BER = 10^{-9} shifted to the left with the increment of value of attenuation. This shows the increment of device loss, distance of data transmission is also decreased. At zero power loss the boundary lines on the BER is at 95 km but when the loss at 20 dB, BER = 10^{-9} boundary is located at 14 km. This shows the distance is inversely proportional to the devices insertion loss (Saleh & Teich 1991). The decrement rate of distance is 3.92 km/dB, as shown in Figure 11 and equations (2).

$$y = -3.9151x + 94.434 \tag{2}$$

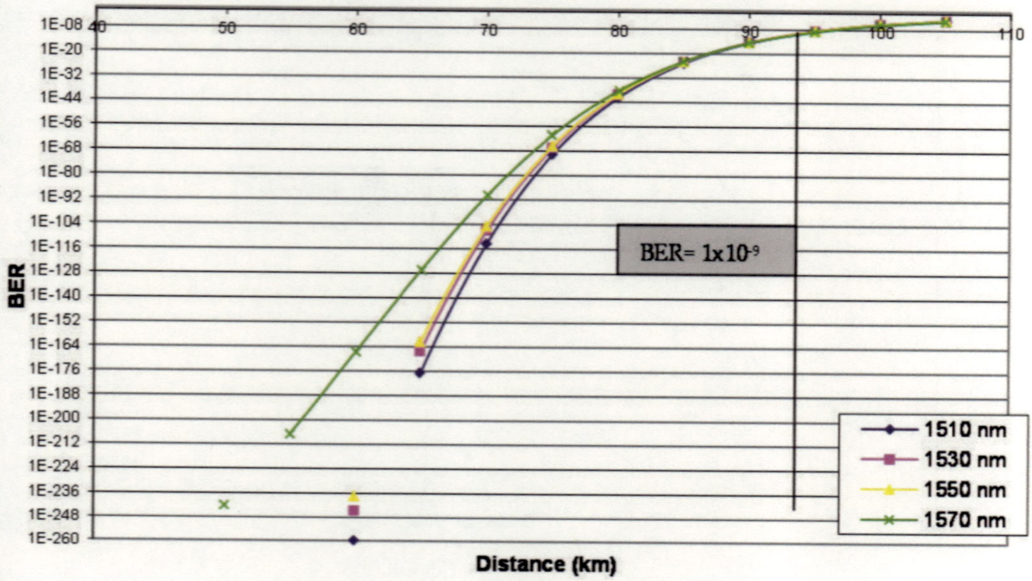

Figure 6. Effect of distance to the BER performance in npoint-to-point network at zero attenuation.

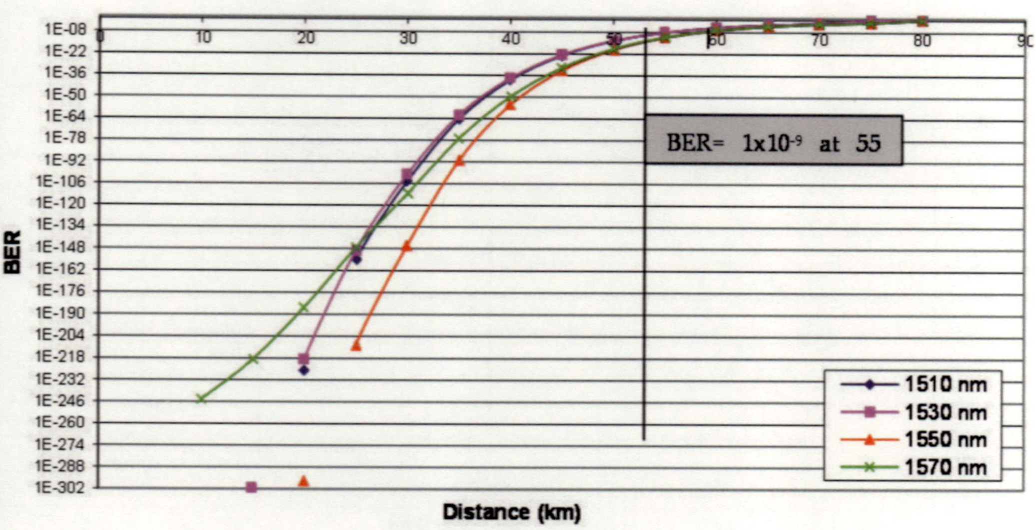

Figure 7. Effect of distance to the BER performance in point-to-point network at attenuation of 10 dB.

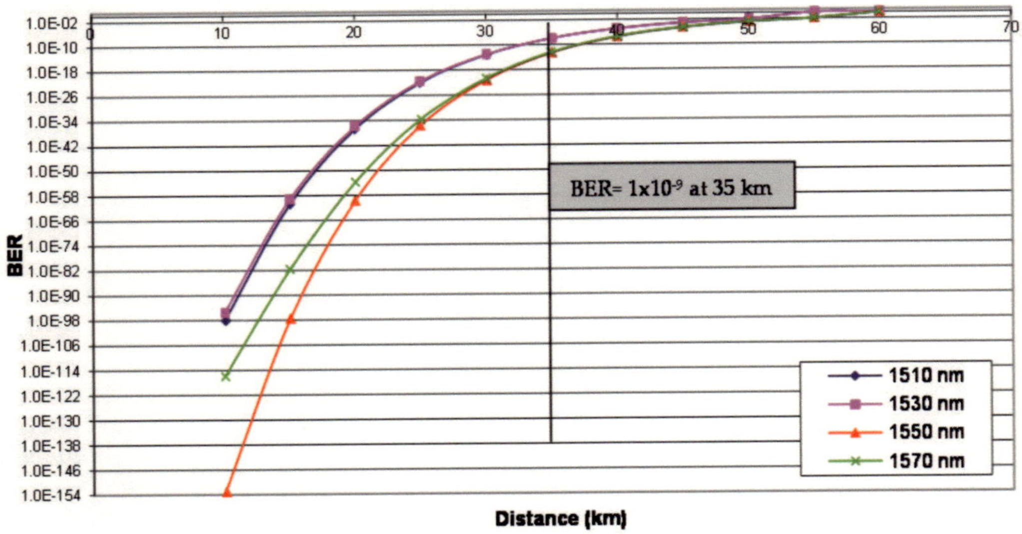

Figure 8. Effect of distance to the BER performance in npoint-to-point network at attenuation of 15 dB.

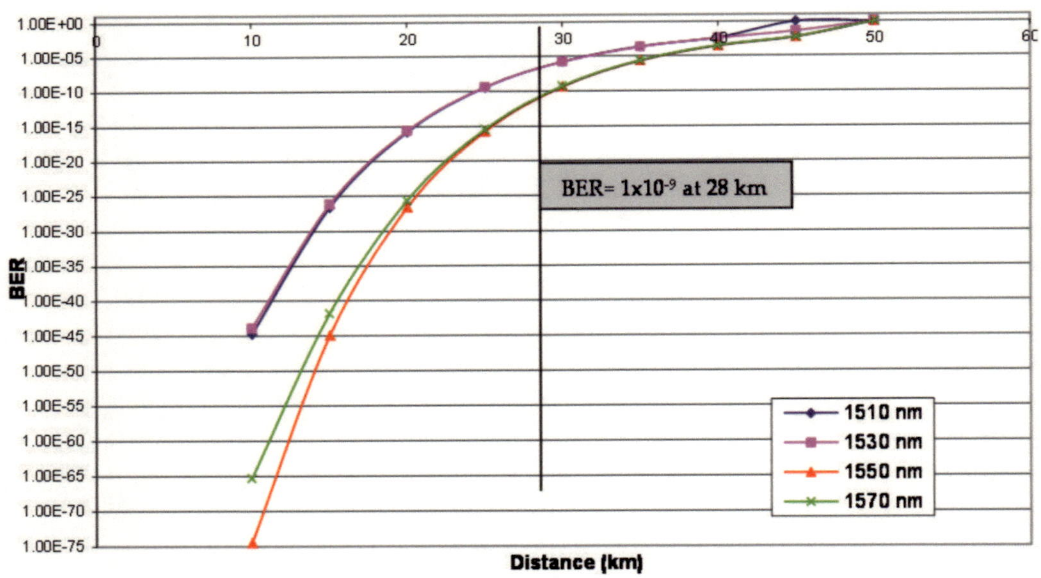

Figure 9. Effect of distance to the BER performance in npoint-to-point network at attenuation of 17dB.

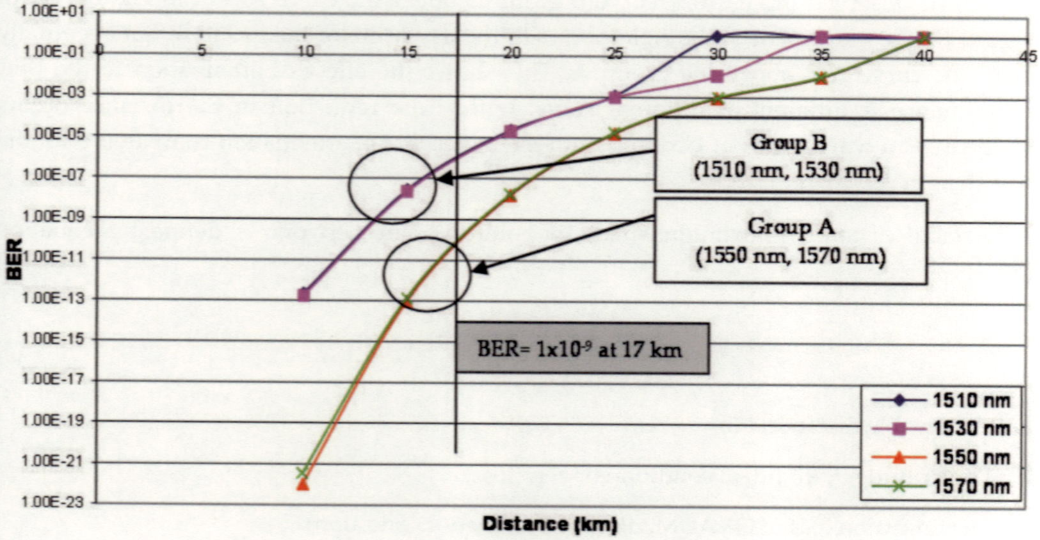

Figure 10. Effect of distance to the BER performance in npoint-to-point network at attenuation of 20 dB.

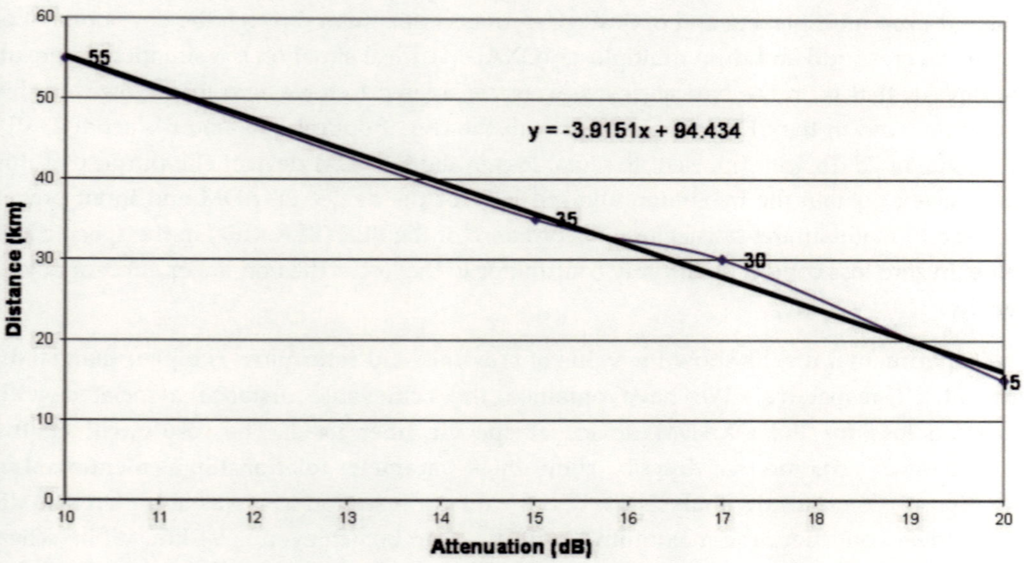

Figure 11. Achievable distance at specific attenuation values in point-to-point network at sensitivity -28.4 dBm (1550 nm at OC-48).

Wavelength of operation of the network is point to point can be divided into two groups: the group A (wavelength 1550 nm and 1570 nm) and group B (wavelength 1510 nm and 1530 nm). Observed in Figure 6, curve B group is under the curve A. But in Figure 7 until Figure 10, the movement to the right occurs at the curvature of the group B that eventually bend the curve B is above the group A. This shows the effect of attenuation to the BER performance is different at different wavelengths. The reduction in the distance occurs suddenly at a wavelength of Group B with the increasing of attenuation compared with the wavelength A.

Achievable distance (maximum span) in point-to-point network is define bu equation (3)

$$L = \frac{P - l_{OXADM}}{\alpha} \tag{3}$$

L = Achievable distance, km

P = Power Budget, dB (ideal condition or zero loss)

l_{OXADM} = Insertion loss of OXADM, dB (product theory condition)

α = fiber constant, dB/km

3. Conclusion

We have introduced a new switching device which utilizes the combined concepts of optical add and drop multiplexing and optical cross connect operation through the development of an optical cross add and drop multiplexer (OXADM). Ideal situation is a situation where all the devices that form the optical device were considered to have zero loss. However, this loss is replaced in the BER measurement with the use of optical attenuator is set at 25 dB. The value of 25 dB will represent the total loss in the OXADM device. The purpose of this study was to obtain the maximum allowed loss for the device OXADM and input power required to maintain the satisfactory performance of the BER (BER $<10^{-9}$) in the specific loss value. In zero loss condition, the only contributor to the loss is the non-linear effect of power penalty.

The experimental results show the value of crosstalk and return loss is bigger than 60 dB and 40 dB respectively.We have obtained the achievable distance associated with insertion loss for the OXADM device at specific fiber used. The result will be the mathematical equation that describe about these parameter relationship as mentioned in equation (3). As a result, analysis using the value of insertion loss was less than 0.06 dB under ideal condition, the maximum length that can be achieved is 94 km. While when considering the loss, with the transmitter power of 0 dBm and sensitivity –22.8 dBm at a point-to-point configuration with safety margin, the required transmission is 71 km with OXADM.

Author details

Mohammad Syuhaimi Ab-Rahman
Universiti Kebangsaan Malaysia (UKM),
Malaysia

Acknowledgement

This project is supported by Ministry of Science, Technology and Innovation (MOSTI), Government of Malaysia, through the National Top-Down Project fund and National Science Fund (NSF). The authors would like to thank the Photonic Technology Laboratory in Institute of Micro Engineering and Nanoelectronics (IMEN), Universiti Kebangsaan Malaysia (UKM), Malaysia, for providing the facilities to conduct the experiments. The OXADM had firstly been exhibited in 19th International Invention, Innovation and Technology Exhibition (ITEX 2008), Malaysia, and was awarded with Bronze medal in telecommunication category.

4. References

Mutafungwa, E. An improved wavelength-selective all fiber cross-connect node, *IEEE Journal of Applied Optics. pp 63-69. 2000.*

Eldada, L. & Nunen, J.v. Architecture and performance requirements of optical metro ring nodes in implementing optical add/drop and protection functions, *Telephotonics Review.* 2000

Rahman, M.S.A.; Husin, H.; Ehsan A.A., & Shaari, S. Analytical modeling of optical cross add and drop multiplexing switch", *Proceeding 2006 IEEE International Conference on Semiconductor* Electronics, pub. IEEE Malaysia Section. pp 290-293. 2006a.

Rahman, M.S.A.; Shaari, S. OXADM restoration scheme: Approach to optical ring network protection, *IEEE International Conference on Networks*. pp 371-376. 2006b.

Rahman, M.S.A.; Ehsan, A.A. & Shaari, S. Mesh upgraded ring in metropolitan network using OXADM, *Proceeding of the 5th International Conference on Optical Communications and Networks & the 2nd International Symposium on Advances and Trends in Fiber. 2006c.*

Rahman, M.S.A. & Shaari, S. 2007. *Survivable Mesh Upgraded Ring in Metropolitan Network,* Journal of Optical Communication, JOC (German). 28(2007)3, pp. 206-211

Rahman, M.S.A. 2008. *First Experimental on OXADM restoration scheme Using Point-to-Point Configuration.* Journal of Optical Communication, JOC (German). 29(2008)3. Pp. 174-177.

Aziz, S.A.C.; Ab-Rahman, M.S. & Jumari,K., 2009 Customer access protection unit for survivable FTTH network. Proceedings of International Conference on Space Science and Communication, 2009. 26-27 Oct, pp: 71 – 73, Negeri Sembilan, 10.1109/ICONSPACE.2009.5352668

Saleh, B. E.A.& Teich, M.C. Fundamentals of photonics. Wiley (New York). 1991

Chapter 2

Optical Resonators and Dynamic Maps

V. Aboites, Y. Barmenkov, A. Kir'yanov and M. Wilson

Additional information is available at the end of the chapter

http://dx.doi.org/10.5772/51854

1. Introduction

In recent years, optical phase conjugation (OPC) has been an important research subject in the field of lasers and nonlinear optics. OPC defines a link between two coherent optical beams propagating in opposite directions with reversed wave front and identical transverse amplitude distributions. The distinctive characteristic of a pair of phase-conjugate beams is that the aberration influence imposed on the forward beam passed through an inhomogeneous or disturbing medium can be automatically removed for the backward beam passed through the same disturbing medium. There are three main approaches that are efficiently able to produce the backward phase-conjugate beam. The first one is based on the degenerate (or partially degenerate) four-wave mixing processes (FWM), the second is based on a variety of backward simulated (e.g. Brillouin, Raman or Kerr) scattering processes, and the third is based on one-photon or multi-photon pumped backward stimulated emission (lasing) processes. Among these different approaches, there is a common physical mechanism in generating a backward phase-conjugate beam, which is the formation of the induced holographic grating and the subsequent wave-front restoration via a backward reading beam. In most experimental studies, certain types of resonance enhancements of induced refractive-index changes are desirable for obtaining higher grating-refraction efficiency. OPC-associated techniques can be effectively utilized in many different application areas: such as high-brightness laser oscillator/amplifier systems, cavity-less lasing devices, laser target-aiming systems, aberration correction for coherent-light transmission and reflection through disturbing media, long distance optical fiber communications with ultra-high bit-rate, optical phase locking and coupling systems, and novel optical data storage and processing systems (see Ref. [1] and references therein).

The power performance of a phase conjugated laser oscillator can be significantly improved introducing intracavity nonlinear elements, e.g. Eichler et al. [2] and O'Connor et al. [3] showed that a stimulated-Brillouin-scattering (SBS) phase conjugating cell placed inside the resonator of a solid-state laser reduces its optical coherence length, because each axial mode of the phase conjugated oscillator experiences a frequency shift at every reflection by the SBS cell resulting in a multi-frequency lasing spectrum, that makes the laser insensitive to changing operating conditions such as pulse repetition frequency, pump energy, etc. This ability is very important for many laser applications including ranging and remote sensing. The intracavity cell is also able to compensate optical aberrations from the resonator and from thermal effects in the active medium, resulting in near diffraction limited output [4], and eliminate the need for a conventional Q-switch as well, because its intensity-dependent reflectivity acts as a passive Q-switch, typically producing a train of nanosecond pulses of diffraction limited beam quality. One more significant use of OPC is a so-called short hologram, which does not exhibit in-depth diffraction deformation of the fine speckle pattern of the recording fields [5]. A thermal hologram in the output mirror was recorded by two speckle waves produced as a result of this recording a ring Nd:YAG laser [6]. Phase conjugation by SBS represents a fundamentally promising approach for achieving power scaling of solid-state lasers [7, 8] and optical fibers [9].

There are several theoretical models to describe OPC in resonators and lasers. One of them is to use the SBS reflection as one of the cavity mirrors of a laser resonator to form a so-called linear phase conjugate resonator [10], however ring-phase conjugate resonators are also possible [11]. The theoretical model of an OPC laser in transient operation [12] considers the temporal and spatial dynamic of the input field the Stokes field and the acoustic-wave amplitude in the SBS cell. On the other hand the spatial mode analysis of a laser may be carried out using transfer matrices, also know as ABCD matrices, which are a useful mathematical tool when studying the propagation of light rays through complex optical systems. They provide a simple way to obtain the final key characteristics (position and angle) of the ray. As an important example we could mention that transfer matrices have been used to study self-adaptive laser resonators where the laser oscillator is made out of a plane output coupler and an infinite nonlinear FWM medium in a self-intersecting loop geometry [13].

In this chapter we put forward an approach where the intracavity element is presented in the context of an iterative map (e.g. Tinkerbell, Duffing and Hénon) whose state is determined by its previous state. It is shown that the behavior of a beam within a ring optical resonator may be well described by a particular iterative map and the necessary conditions for its occurrence are discussed. In particular, it is shown that the introduction of a specific element within a ring phase-conjugated resonator may produce beams described by a Duffing, Tinkerbell or Hénon map, which we call "Tinkerbell, Duffing or Hénon beams". The idea of introducing map generating elements in optical resonators from a mathematical viewpoint was originally explored in [14-16] and this chapter is mainly based on those results.

This chapter is organized as follows: Section 2 discusses the matrix optics elements on which this work is based. Section 3 presents as an illustration some basic features of Tinkerbell, Duffing and Hénon maps, Sections 4,5 and 6 show, each one of them, the main characteristics of the map generation matrix and Tinkerbell, Duffing and Hénon Beams, as well as the general case for each beams in a ring phase conjugated resonator. Finally Section 6 presents the conclusions.

2. ABCD matrix optics

Any optical element may be described by a 2×2 matrix in paraxial optics. Assuming cylindrical symmetry around the optical axis, and defining at a given position z both the perpendicular distance of any ray to the optical axis and its angle with the same axis as $y(z)$ and $\theta(z)$, when the ray undergoes a transformation as it travels through an optical system represented by the matrix $[A,B,C,D]$, the resultant values of y and θ are given by [17]:

$$\begin{pmatrix} y_{n+1} \\ \theta_{n+1} \end{pmatrix} = \begin{pmatrix} A & B \\ C & D \end{pmatrix} \begin{pmatrix} y_n \\ \theta_n \end{pmatrix}. \tag{1}$$

For any optical system, one may obtain the total $[A,B,C,D]$ matrix, by carrying out the matrix product of the matrices describing each one of the optical elements in the system.

2.1. Constant ABCD elements

For passive optical elements such as lenses, interfaces between two media, reflections, propagation, and many others, the elements A, B, C, D are constants and the determinant $\mathrm{Det}[A,B,C,D] = n_n/n_{n+1}$, where n_n and n_{n+1} are the refraction index before and after the optical element described by the matrix. Since typically n_n and n_{n+1} are the same, it holds that $\mathrm{Det}[A,B,C,D] = 1$.

2.2. Non constant ABCD elements

However, for active or non-linear optical elements the A, B, C, D matrix elements are not constant but may be functions of various parameters. The following three examples are worth mentioning.

2.2.1. *Curved interface with a Kerr electro-optic material*

Due to the electro-optic Kerr effect the refraction index of an optical media n is a function of the electric field strength E [18]. The change of the refraction index is given by $\Delta n = \lambda K E^2$, where λ is the wavelength and K is the Kerr constant of the media. For example, the $[A,B,C,D]$ matrix of a curved surface of radius of curvature r separating two regions of refractive index n_1 and n_2 (taking the center of the radius of curvature positive to the right in the zone of refractive index n_2) is given as:

$$\begin{pmatrix} 1 & 0 \\ -\dfrac{(n_2 - n_1)}{r} & 1 \end{pmatrix}. \tag{2}$$

Having vacuum ($n_1 = 1$) on the left of the interface and a Kerr electro-optic material on the right, the above [*ABCD*] matrix becomes

$$\begin{pmatrix} 1 & 0 \\ -\dfrac{\left(n_2(E)-1\right)}{r} & 1 \end{pmatrix}. \tag{3}$$

Clearly the elements *A*, *B*, *D* are constants but element *C* is a function of the electric field *E*.

2.2.2. *Phase conjugate mirror*

A second example is a phase conjugate mirror. The process of phase conjugation has the property of retracing an incoming ray along the same incident path [7]. The ideal ABCD phase conjugate matrix is

$$\begin{pmatrix} 1 & 0 \\ 0 & -1 \end{pmatrix}. \tag{4}$$

One may notice that the determinant of this particular matrix is not 1 but -1. The ABCD matrix of a real phase conjugated mirror must take into account the specific process to produce the phase conjugation. As already mentioned, typically phase conjugation is achieved in two ways; Four Wave Mixing or using a stimulated scattering process such as Brillouin, i.e. SBS. However upon reflection on a stimulated SBS phase conjugated mirror, the reflected wave has its frequency ω downshifted to $\omega - \delta = \omega(1 - \delta/\omega)$ where δ is the characteristic Brillouin downshift frequency of the mirror material (typically $\delta/\omega \ll 1$). In a non-ideal (i.e. real) case one must take the downshifting frequency into account and the ABCD matrix reads

$$\begin{pmatrix} 1-\dfrac{\delta}{\omega} & 0 \\ 0 & -1 \end{pmatrix}. \tag{5}$$

Furthermore, since in phase conjugation by SBS a light intensity threshold must be reached in order to have an exponential amplification of the scattered light, the above ideal matrix (4) must be modified. The scattered light intensity at position *z* in the medium is given as

$$I_S(z) = I_S(0)\exp\left(g_B I_L l\right), \tag{6}$$

where $I_S(0)$ is the initial level of scattering, g_B denotes the characteristic exponential gain coefficient of the scattering process, I_L is the intensity of the incident light beam, and l is the interaction length over which amplification takes place. Given the amplification $G = \exp(g_B(\nu) I_L l)$ the threshold gain factor is commonly taken as $G \sim \exp(30) \approx 10^{13}$ which corresponds to a threshold intensity

$$I_{L,th} = \frac{30}{g_B l}. \tag{7}$$

The modeling of a real stimulated Brillouin scattering phase conjugate mirror usually takes into account a Gaussian aperture of radius a at intensity $1/e^2$ placed before an ideal phase conjugator. In this way the reflected beam is Gaussian and only the parts of the Gaussian incident beam with intensity above threshold are phase conjugate reflected. The matrix of this aperture is given by:

$$\begin{pmatrix} 1 & 0 \\ -\frac{i\lambda}{\pi a^2} & 1 \end{pmatrix}, \tag{8}$$

where the aperture a is a function of the incident light intensity $a(I_L)$ (I_L must reach threshold to initiate the scattering process). As we can see, depending on the model, the *ABCD* matrix elements of a phase conjugated mirror may depend on several parameters such as the Brillouin downshifting frequency, the Gaussian aperture radius and the incident light intensity [19].

2.3. Systems with hysteresis

At last, as third example we may consider a system with hysteresis. It is well known that such systems exhibit memory. There are many examples of materials with electric, magnetic and elastic hysteresis, as well as systems in neuroscience, biology, electronics, energy and even economics which show hysteresis. As it is known in a system with no hysteresis, it is possible to predict the system's output at an instant in time given only its input at that instant in time. However in a system with hysteresis, this is not possible; there is no way to predict the output without knowing the system's previous state and there is no way to know the system's state without looking at the history of the input. This means that it is necessary to know the path that the input followed before it reached its current value. For an optical element with hysteresis the *ABCD* matrix elements are function of the $y_n, y_{n-1}, \ldots y_{n-i}$ and $\theta_n, \theta_{n-1}, \ldots, \theta_{n-i}$ and its knowledge is necessary in order to find the state y_{n+1}, θ_{n+1}. In general, taking into account hysteresis, the $[A,B,C,D]$ matrix of Eq. (1) may be written as:

$$\begin{pmatrix} A & B \\ C & D \end{pmatrix} = \begin{pmatrix} A(y_n, y_{n-1}, \ldots y_{n-i}, \theta_n, \theta_{n-1}, \ldots \theta_{n-i}) & B(y_n, y_{n-1}, \ldots y_{n-i}, \theta_n, \theta_{n-1}, \ldots \theta_{n-i}) \\ C(y_n, y_{n-1}, \ldots y_{n-i}, \theta_n, \theta_{n-1}, \ldots \theta_{n-i}) & D(y_n, y_{n-1}, \ldots y_{n-i}, \theta_n, \theta_{n-1}, \ldots \theta_{n-i}) \end{pmatrix}. \tag{9}$$

3. Dynamic maps

An extensive list of two-dimensional maps may be found in Ref. [20]. A few examples are Tinkerbell, Duffing and Hénon maps. As will be shown next they may be written as a matrix dynamical system such as the one described by Eq. (1) or equivalently as

$$
\begin{aligned}
y_{n+1} &= Ay_n + B\theta_n, \quad \text{(a)} \\
\theta_{n+1} &= Cy_n + D\theta_n. \quad \text{(b)}
\end{aligned}
\tag{10}
$$

3.1. Tinkerbell map

The Tinkerbell map [21, 22] is a discrete-time dynamical system given by the equations:

$$
\begin{aligned}
y_{n+1} &= y_n^2 - \theta_n^2 + \alpha\, y_n + \beta\, \theta_n, \quad \text{(a)} \\
\theta_{n+1} &= 2y_n\theta_n + \gamma\, y_n + \delta\, \theta_n. \quad \text{(b)}
\end{aligned}
\tag{11}
$$

where y_n and θ_n are the scalar state variables and α, β, γ, and δ the map parameters. In order to write the Tinkerbell map as a matrix system such as Eq. (1) the following values for the coefficients *A, B, C* and *D* must hold:

$$A(y_n, \alpha) = y_n + \alpha, \tag{12}$$

$$B(\theta_n, \beta) = -\theta_n + \beta, \tag{13}$$

$$C(\theta_n, \gamma) = 2\theta_n + \gamma, \tag{14}$$

$$D(\delta) = \delta. \tag{15}$$

It should be noted that these coefficients are not constants but depend on the state variables y_n and θ_n and the Tinkerbell map parameters α, β, γ, and δ. Therefore as an *ABCD* matrix system the Tinkerbell map may be written as:

$$
\begin{pmatrix} y_{n+1} \\ \theta_{n+1} \end{pmatrix} = \begin{pmatrix} y_n + \alpha & -\theta_n + \beta \\ 2\theta_n + \gamma & \delta \end{pmatrix} \begin{pmatrix} y_n \\ \theta_n \end{pmatrix}.
\tag{16}
$$

3.2. Hénon map

The Hénon map has been widely studied due to its nonlinear chaotic dynamics. Hénon map is a popular example of a two-dimensional quadratic mapping which produces a discrete-time system with chaotic behavior. The Hénon map is described by the following two difference equations [23, 24]:

$$
\begin{aligned}
y_{n+1} &= 1 - \alpha y_n^2 + \theta_n, \quad \text{(a)} \\
\theta_{n+1} &= \beta y_n. \quad \text{(b)}
\end{aligned}
\tag{17}
$$

Following similar steps as those of the Tinkerbell map, this map may be written as a dynamic matrix system:

$$\begin{pmatrix} y_{n+1} \\ \theta_{n+1} \end{pmatrix} = \begin{pmatrix} \frac{1}{y_n} - \alpha y_n & 1 \\ \beta & 0 \end{pmatrix} \begin{pmatrix} y_n \\ \theta_n \end{pmatrix} \tag{18}$$

where y_n and θ_n are the scalar state variables which can be measured as time series and α and β the map parameters. In many control systems α is a control parameter. The Jacobian β $(0 \le \beta \le 1)$ is related to dissipation. The dynamics of the Hénon map is well studied (see, for instance, Ref. [25]) and its fixed points are given by:

$$(y_1, \theta_1) = \left(\frac{-\beta - 1 - \sqrt{(\beta+1)^2 + 4\alpha}}{2\alpha}, -\beta y_1 \right) \tag{19}$$

$$(y_2, \theta_2) = \left(\frac{-\beta - 1 + \sqrt{(\beta+1)^2 + 4\alpha}}{2\alpha}, -\beta y_2 \right). \tag{20}$$

And the corresponding eigenvalues are

$$\lambda_{1,2} = -\alpha y \pm \sqrt{(\alpha y)^2 - \beta}. \tag{21}$$

3.3. Duffing map

The study of the stability and chaos of the Duffing map has been the topic of many articles [26-27]. The Duffing map is a dynamical system which may be written as follows:

$$\begin{aligned} y_{n+1} &= \theta_n, && \text{(a)} \\ \theta_{n+1} &= -\beta y_n + \alpha \theta_n - \theta_n^3, && \text{(b)} \end{aligned} \tag{22}$$

where y_n and θ_n are the scalar state variables and α and β the map parameters. In order to write the Duffing map equations as a matrix system Eq. (1) the following values for the coefficients A, B, C and D must hold. It should be noted that these coefficients are not constants but depend on θ_n and the Duffing map parameters are as follows:

$$A = 0, \tag{23}$$

$$B = 1, \tag{24}$$

$$C(\beta) = -\beta, \tag{25}$$

$$D(\theta_n, \alpha) = \alpha - \theta_n^2. \tag{26}$$

Therefore as an ABCD matrix system the Duffing map may be written as:

$$\begin{pmatrix} y_{n+1} \\ \theta_{n+1} \end{pmatrix} = \begin{pmatrix} 0 & 1 \\ -\beta & \alpha - \theta_n^2 \end{pmatrix} \begin{pmatrix} y_n \\ \theta_n \end{pmatrix} \tag{27}$$

4. Maps in a ring phase-conjugated resonator

In this section an optical resonator with a specific map behavior for the variables y and θ is presented. Figure 1 shows a ring phase-conjugated resonator consisting of two ideal mirrors, an ideal phase conjugate mirror and a yet unknown optical element described by a matrix $[a,b,c,e]$. The two perfect plain mirrors [M] and the ideal phase conjugated mirror [PM] are separated by a distance d. The matrices involved in this resonator are: the identity matrix: $\begin{pmatrix} 1 & 0 \\ 0 & 1 \end{pmatrix}$ for the plane mirrors [M], $\begin{pmatrix} 1 & 0 \\ 0 & -1 \end{pmatrix}$ for the ideal phase conjugated mirror [PM], $\begin{pmatrix} 1 & d \\ 0 & 1 \end{pmatrix}$ for a distance d translation and, in addition, the unknown map generating device matrix represented by $\begin{pmatrix} a & b \\ c & e \end{pmatrix}$, is located between the plain mirrors [M] at distance $d/2$ from each one.

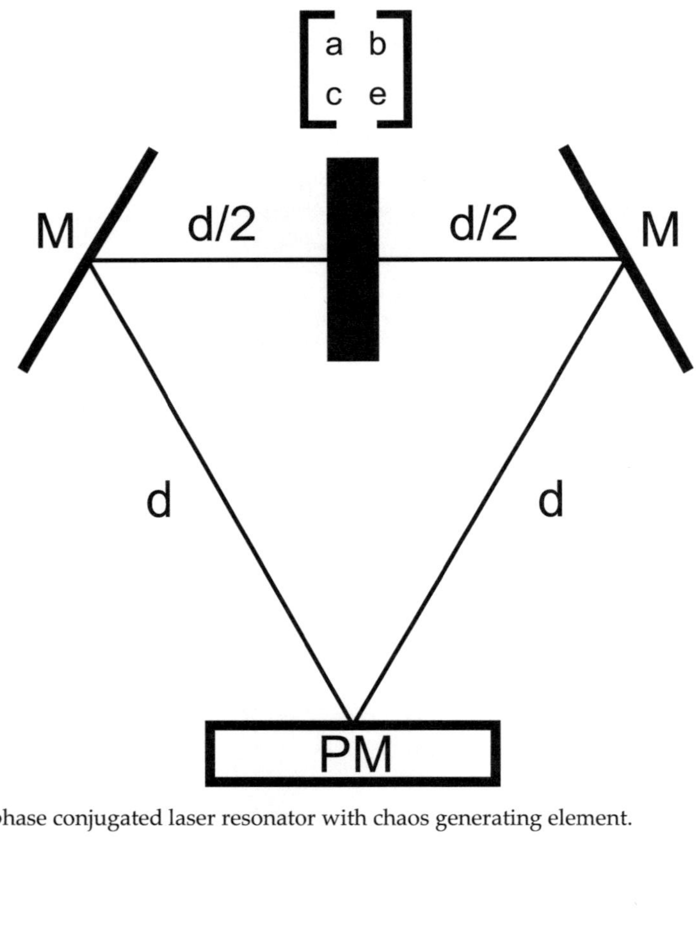

Figure 1. Ring phase conjugated laser resonator with chaos generating element.

For this system, the total transformation matrix [*A,B,C,D*] for a complete round trip is:

$$\begin{pmatrix} A & B \\ C & D \end{pmatrix} = \begin{pmatrix} 1 & 0 \\ 0 & -1 \end{pmatrix}\begin{pmatrix} 1 & d \\ 0 & 1 \end{pmatrix}\begin{pmatrix} 1 & 0 \\ 0 & 1 \end{pmatrix}\begin{pmatrix} 1 & d/2 \\ 0 & 1 \end{pmatrix}\begin{pmatrix} a & b \\ c & e \end{pmatrix}\begin{pmatrix} 1 & d/2 \\ 0 & 1 \end{pmatrix}\begin{pmatrix} 1 & 0 \\ 0 & 1 \end{pmatrix}\begin{pmatrix} 1 & d \\ 0 & 1 \end{pmatrix}. \quad (28)$$

The above one round trip total transformation matrix is

$$\begin{pmatrix} a + \frac{3cd}{2} & b + \frac{3d}{4}(2a + 3cd + 2e) \\ -c & -\frac{3cd}{2} - e \end{pmatrix}. \quad (29)$$

As can be seen, the elements of this matrix depend on the elements of the map generating matrix device [*a,b,c,e*]. If one does want a specific map to be reproduced by a ray in the ring optical resonator, then each round trip a ray described by (y_n,θ_n) has to be considered as an iteration of the desired map. Then, the *ABCD* matrix of the map system (16), (18), (27) must be equated to the total *ABCD* matrix of the resonator (29), this in order to generate a specific map dynamics for (y_n,θ_n).

It should be noticed that the results given by equations (28) and (29) are only valid for b small ($b \approx 0$). This is due to the fact that before and after the matrix element [*a,b,c,e*] we have a propagation of *d*/2. For a general case, expression (29) has to be substituted by:

$$\begin{pmatrix} A & B \\ C & D \end{pmatrix} = \begin{pmatrix} 1 & 0 \\ 0 & -1 \end{pmatrix}\begin{pmatrix} 1 & d \\ 0 & 1 \end{pmatrix}\begin{pmatrix} 1 & 0 \\ 0 & 1 \end{pmatrix}\begin{pmatrix} 1 & \frac{d-b}{2} \\ 0 & 1 \end{pmatrix}\begin{pmatrix} a & b \\ c & e \end{pmatrix}\begin{pmatrix} 1 & \frac{d-b}{2} \\ 0 & 1 \end{pmatrix}\begin{pmatrix} 1 & 0 \\ 0 & 1 \end{pmatrix}\begin{pmatrix} 1 & d \\ 0 & 1 \end{pmatrix} \quad (30)$$

Therefore the round trip total transformation matrix is:

$$\begin{pmatrix} a - \frac{c}{2}(b - 3d) & \frac{1}{4}\left[b^2c - 2b(-2 + a + 3cd + e) + 3d(2a + 3cd + 2e)\right] \\ -c & \frac{1}{2}(bc - 3cd - 2e) \end{pmatrix}. \quad (31)$$

Matrix (29) describes a simplified ideal case whereas matrix (31) describes a general more complex and realistic case. These results will be widely used in the next three sections.

5. Tinkerbell beams

This section presents an optical resonator that produces beams following the Tinkerbell map dynamics; these beams will be called "Tinkerbell beams". Equation (29) is the one round trip total transformation matrix of the resonator. If one does want a particular map to be reproduced by a ray in the optical resonator, each round trip described by (y_n, θ_n), has to be considered as an iteration of the selected map. In order to obtain Tinkerbell beams, Eqs. (12) to (15) must be equated to Eq. (29), that is:

$$a+\frac{3cd}{2}=\alpha+y_n, \tag{32}$$

$$b+\frac{3d}{4}\left(2a+3cd+2e\right)=\beta-\theta_n, \tag{33}$$

$$c=-\gamma-2\theta_n, \tag{34}$$

$$e+\frac{3cd}{2}=-\delta. \tag{35}$$

Equations (32-35) define a system for the matrix elements *a, b, c, e,* that guarantees a Tinkerbell map behaviour for a given ray (y_n, θ_n). These elements can be written in terms of the map parameters (α, β, γ and δ), the resonator's main parameter *d* and the ray state variables y_n and θ_n as:

$$a=\alpha+\frac{3}{2}\gamma d+3d\theta_n+y_n, \tag{36}$$

$$b=\frac{1}{4}\left(4\beta-6\alpha d+6\delta d-9\gamma d^2-4\theta_n-18d^2\theta_n-6dy_n\right), \tag{37}$$

$$c=-2\theta_n-\gamma, \tag{38}$$

$$e=-\delta+\frac{3}{2}d\left(\gamma+2\theta_n\right). \tag{39}$$

The introduction of the above values for the $\begin{pmatrix} a & b \\ c & e \end{pmatrix}$ matrix in Eq. (28) enables us to obtain Eq. (16). For any transfer matrix elements *A* and *D* describe the lateral magnification while *C* describe the focal length, whereas the device's optical thickness is given by $B = L/n$, where *L* is its length and *n* its refractive index. From Eqs. (36-39) it must be noted that the upper elements (*a* and *b*) of the device matrix depend on both state variables (y_n and θ_n) while the lower elements (*c* and *e*) only on the state variable θ_n. The study of the stability and chaos of the Tinkerbell map in terms of its parameters is a well-known topic [21,22]. The behaviour of element *b* is quite interesting; figure 2 shows a computer calculation for the first 100 round trips of matrix element *b* of the Tinkerbell map generating device for a resonator of unitary length (d = 1) and map parameters α = 0, β = -0.6, γ = 0 and δ = -1, these parameters were found using brute force calculations and they were selected due to the matrix-element *b* behaviour (i.e. we were looking for behaviour able to be achievable in experiments). As can be seen, the optical length of the map generating device varies on each round trip in a periodic form, this would require that the physical length of the device, its refractive index - or a combination of both- change in time. The actual design of a physical Tinkerbell map generating device for a unitary ring resonator must satisfy Eqs. (36-39), to do so its elements (*a*, *b*, *c* and *e*) must vary accordingly.

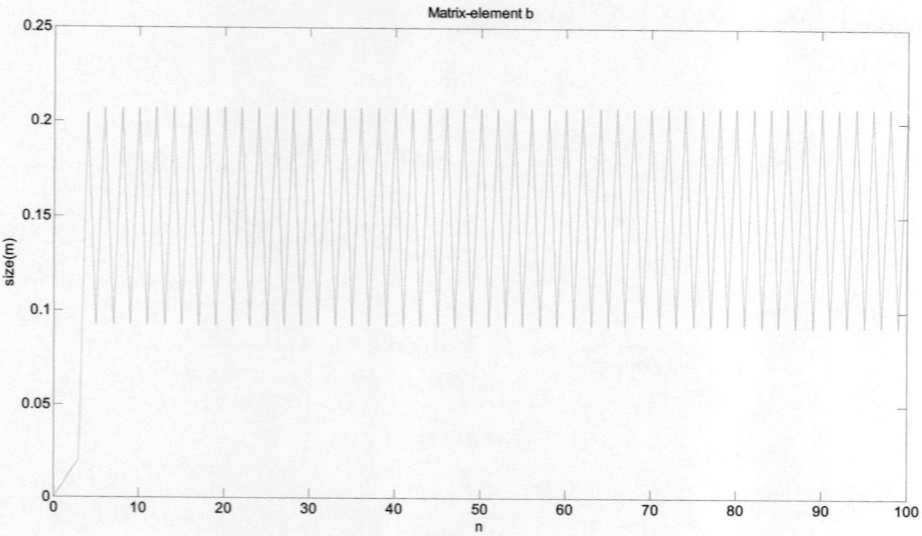

Figure 2. Computer calculation of the magnitude of matrix element b of the Tinkerbell map generating device for a resonator with $d = 1$ and Tinkerbell parameters $\alpha = 0$, $\beta = -0.6$, $\gamma = 0$ and $\delta = -1$ for the first 100 round trips.

5.1. Tinkerbell beams: General case

To obtain the Eqs. (36-39) b, the thickness of the Tinkerbell generating device, has to be very small (close to zero), so the translations before and after the device can be over the same distance $d/2$. In the previous numeric simulation b takes values up to 0.2, so the general case where the map generating element b does not have to be small must be studied. As previously explained Eq. (28) must be substituted by Eq. (30).

From Eqs. (16) and (31) we obtain the following system of equations for the matrix elements a, b, c and e:

$$a - \frac{c}{2}(b - 3d) = \alpha + y_n, \tag{40}$$

$$\frac{1}{4}(b^2 c - 2b(-2 + a + 3cd + e) + 3d(2a + 3cd + 2e)) = \beta - \theta_n, \tag{41}$$

$$-c = \gamma + 2\theta_n, \tag{42}$$

$$\frac{bc - 3cd - 2e}{2} = \delta. \tag{43}$$

The solution to this new system is written as:

$$a = \alpha + \frac{3}{2}\gamma d + 3d\theta_n + y_n + \frac{1}{2\gamma + 4\theta_n}\left(\begin{array}{l} \gamma(2 - \alpha + \delta - 3\gamma d - 12d\theta_n - y_n) \\ +\theta_n(4 - 2\alpha + 2\delta - 12d\theta_n - 2y_n) \\ -\left(-\frac{\gamma}{2} - \theta_n\right)\sqrt{P^2 - Q} \end{array}\right), \tag{44}$$

$$b = \frac{1}{\gamma + 2\theta_n}\left(-2 + \alpha - \delta + 3\gamma d + 6d\theta_n + y_n + \frac{\sqrt{P^2 - Q}}{2}\right), \tag{45}$$

$$c = -\gamma - 2\theta_n, \tag{46}$$

$$e = \delta + \frac{3}{2}\gamma d + 3d\theta_n + \frac{1}{2\gamma + 4\theta_n}\left(\begin{array}{l} \gamma(2 - \alpha + \delta - 3\gamma d - 12d\theta_n - y_n) \\ +\theta_n(4 - 2\alpha + 2\delta - 12d\theta_n - 2y_n) \\ -\left(-\frac{\gamma}{2} - \theta_n\right)\sqrt{P^2 - Q} \end{array}\right), \tag{47}$$

where:

$$P = 4 - 2\alpha + 2\delta - 6\gamma d - 12d\theta_n - 2y_n$$

and

$$Q = (4\gamma + 8\theta_n)\left(-4\beta + 6\gamma d - 6\delta d + 9\gamma d^2 + 4\theta_n + 18d^2\theta_n + 6dy_n\right).$$

It should be noted that if one takes into account the thickness of the map generating element, the equations complexity is substantially increased. Now only c has a simple relation with θ_n and γ, on the other hand a, b and e are dependent on both state variables, on all Tinkerbell parameters, as well as on the resonator length. When the calculation is performed for this new matrix with the following map parameters: $\alpha = 0.4$, $\beta = -0.4$, $\gamma = -0.3$ and $\delta = 0.225$, figure 3 is obtained. The behaviour observed in figure 3 for the matrix-element b can be obtained for several different parameters' combinations, as well as other dynamical regimes with a lack of relevance to our work. One can note that after a few iterations the device's optical thickness is small and constant, this should make easier a physical implementation of this device.

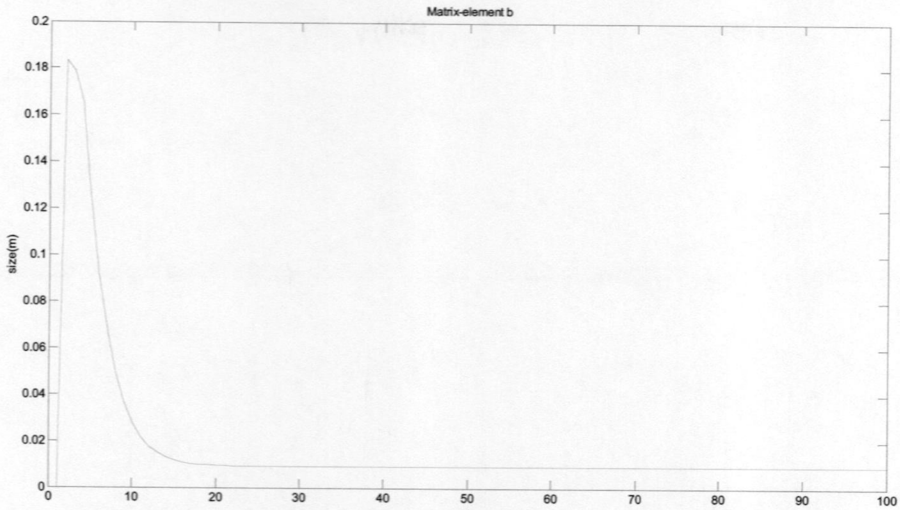

Figure 3. Computer calculation of the magnitude of matrix element b of the Tinkerbell map generating device for a resonator with $d = 1$ and Tinkerbell parameters $\alpha = 0.4$, $\beta = -0.4$, $\gamma = -0.3$ and $\delta = 0.225$ for the first 100 round trips.

6. Duffing beams

This section presents an optical resonator that produces beams following the Duffing map dynamics; these beams will be called "Duffing beams". Equation (29) is the one round trip total transformation matrix of the resonator. If one does want a particular map to be reproduced by a ray in the optical resonator, each round trip described by (y_n, θ_n), has to be considered as an iteration of the selected map. In order to obtain Duffing beams, Eqs. (23) to (26) must be equated to Eq. (29), that is:

$$a + \frac{3cd}{2} = 0, \tag{48}$$

$$b + \frac{3d}{4}\left(2a + 3cd + 2e\right) = 1, \tag{49}$$

$$-c = -\beta, \tag{50}$$

$$-\frac{3cd}{2} - e = \alpha - \theta_n^2. \tag{51}$$

Equations (48-51) define a system for the matrix elements of a, b, c, e, enabling the generation of a Duffing map for the y_n and θ_n state variables. Its solution is:

$$a = -\frac{3\beta d}{2}, \tag{52}$$

$$b = \frac{1}{4}\left(4 + 6\alpha d + 9\beta d^2 - 6d\theta_n^2\right), \tag{53}$$

$$c = \beta, \tag{54}$$

$$e = -\alpha - \frac{3d\beta}{2} + \theta_n^2. \tag{55}$$

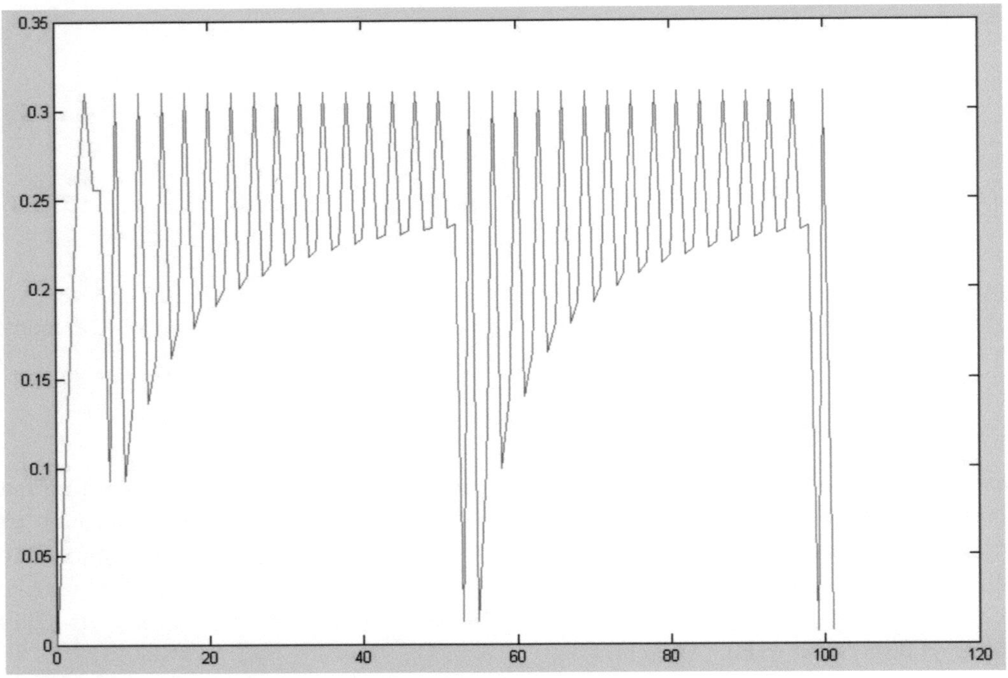

Figure 4. Computer calculation of the magnitude of matrix element b of the Duffing map generating device for a resonator with $d = 1$ and Duffing parameters $\alpha = 1.04$ and $\beta = -1$ for the first 100 round trips.

As can be seen these matrix elements depend on the Duffing parameters α and β as well as on the resonator main parameter d and on the state variable θ_n. These are the values which must be substituted for the [a,b,c,e] matrix in equation (28) for the round trip matrix. As expected, the introduction of the above [a,b,c,e] matrix elements in Eq. (29) produces the $ABCD$ matrix of the Duffing Map, Eq. (27). For a general $ABCD$ transfer matrix, elements A and D are related to the lateral magnification and element C to the focal length, whereas element B gives the optical length of the device. The optical thickness of the $ABCD$ is; $B = L/n$, where L is the physical length of the device and n its refractive index. From Eqs. (52-55) we may see that the A and C elements of the matrix [a,b,c,e] are constants depending only on the resonator parameter d and the Duffing parameters α and β. However matrix elements B and D are dynamic ones and depend on the state variable θ_n. Of special interest is element B of the map generating matrix [a,b,c,e]. Figure 4 shows a computer calculation of matrix element B of the Duffing map generating device for a resonator with $d = 1$ and Duffing parameters $\alpha = 1.04$ and $\beta = -1$ for the first 100 round trips. As it is well known, depending on the α and β map parameters different dynamic states may be obtained including chaos. As can be seen the optical length of the map generating device given by the B matrix element varies on each round trip. This requires that either the physical length of the device or its refractive index, or a combination of both, changes as shown in Figure 4. The design of a physical Duffing map generating device for this resonator must satisfy Eqs. (52-55). A physical implementation of this device is possible as long as its *ABCD* elements vary according to these equations.

6.1. Duffing beams: General case

The results given by Eqs. (52-55) are valid only when the b element of the $[a,b,c,e]$ matrix is small. As can be seen from Eq. (28), the thickness of the Duffing map generating element (described by matrix $[a,b,c,e]$) must be close to zero. This because in Eq. (28) the matrix before and after the $[a,b,c,e]$ is a matrix for a $d/2$ translation which is possible only if $b = 0$ or very small. The previous numeric simulation shows that the b element takes the values of up to 0.3. Therefore one must consider a general case where the map generating element b has not the limitation of being asked to be small. For a general case, Eq. (28) must be substituted by Eq. (30) and (31). From expressions (27) and (31) we obtain the following system of equations for the matrix elements a, b, c, e:

$$a - \frac{c}{2}(b - 3d) = 0, \tag{56}$$

$$\frac{1}{4}(b^2 c - 2b(-2 + a + 3cd + e) + 3d(2a + 3cd + 2e)) = 1, \tag{57}$$

$$-c = -\beta, \tag{58}$$

$$\frac{bc - 3cd - 2e}{2} = \alpha - \theta_n^2. \tag{59}$$

The solution to this system is given by:

$$a = \frac{2 + \alpha - \theta_n^2 + \sqrt{\alpha^2 + 4\beta(-1 + 3d) - 2\alpha(-2 + \theta_n^2) + (-2 + \theta_n^2)^2}}{2}, \tag{60}$$

$$b = \frac{2 + \alpha + 3\beta d - \theta_n^2 + \sqrt{\alpha^2 + 4\beta(-1 + 3d) - 2\alpha(-2 + \theta_n^2) + (-2 + \theta_n^2)^2}}{\beta}, \tag{61}$$

$$c = \beta, \tag{62}$$

$$e = \frac{2 - \alpha + \theta_n^2 + \sqrt{\alpha^2 + 4\beta(-1 + 3d) - 2\alpha(-2 + \theta_n^2) + (-2 + \theta_n^2)^2}}{2}. \tag{63}$$

As we may see, taking into account the thickness of the map generating element device described by matrix [a,b,c,e] greatly increases its complexity. Now only the C matrix element is constant, being elements A, B and D dependent on the state variable θ_n and on the Duffing parameters α and β as well as on the resonator main parameter d. Figure 5 shows a computer calculation of the matrix element B of the Duffing map generating device for a resonator with $d = 1$ and Duffing parameters $\alpha = 1.04$ and $\beta = -0.6$ for the first 100 round trips. As can be seen, the optical thickness variation of the map generating device now is rather small, which means that the length and/or refractive index variation of the map generating element is also small and favors a physical realization of this device.

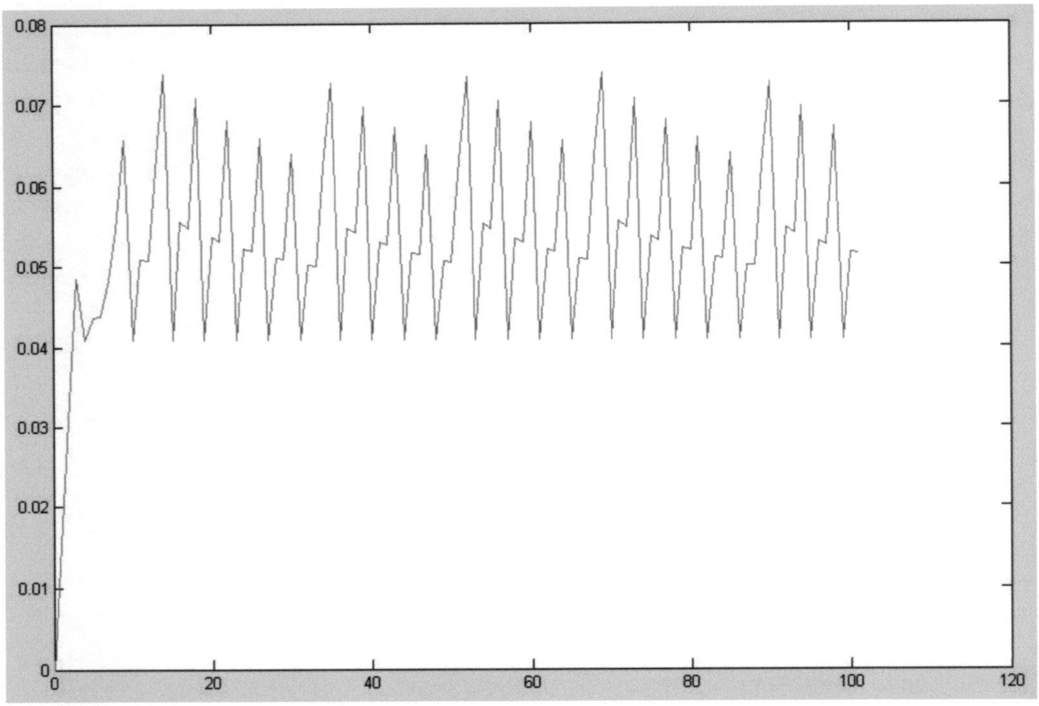

Figure 5. Computer calculation of the magnitude of matrix element b of the Duffing map generating device for a resonator with $d = 1$ and Duffing parameters $\alpha = 1.04$ and $\beta = -0.6$ for the first 100 round trips.

7. Hénon beams

This section presents an optical resonator that produces beams following the Hénon map dynamics; these beams will be called "Hénon beams". Equation (29) is the one round trip total transformation matrix of the resonator. If one does want a particular map to be reproduced by a ray in the optical resonator, each round trip described by (y_n, θ_n), has to be considered as an iteration of the selected map. In order to obtain Hénon beams, the [A, B, C, D] elements of Eq. (18) must be equated to Eq. (29), that is:

$$a+\frac{3cd}{2}=\frac{1}{y_n}-\alpha y_n, \tag{64}$$

$$b+\frac{3d}{4}\left(2a+3cd+2e\right)=1, \tag{65}$$

$$-c=\beta, \tag{66}$$

$$-\frac{3cd}{2}-e=0. \tag{67}$$

The solution for the Hénon chaos matrix elements [*a,b,c,e*], able to produce Hénon beams in terms of the Hénon Map are the following:

$$a=\frac{3\beta d}{2}+\frac{1}{y_n}-\alpha y_n, \tag{68}$$

$$b=1+\frac{3}{2}d(-\frac{1}{y_n}+\alpha y_n-\frac{3d\beta}{2}), \tag{69}$$

$$c=-\beta, \tag{70}$$

$$e=\frac{3d\beta}{2}. \tag{71}$$

As can be seen the matrix elements depend on the Hénon parameters α and β as well as on the resonator main parameter d and on the state variable y_n. However when analyzing the behavior of element "*b*" (Eq. (69)) we may see that there is a problem caused by the term $1/y_n$. While for the case of Tinkerbell and Duffing beams we were able to look at the behavior of the obtained "*b*" element for small values of y_n, as it is shown in figures (2-5), this is not possible for the Henon case because small values of y_n will produce very large values for "*b*", therefore making very difficult to obtain solutions with practical value.

7.1. Hénon beams: General case

In an analogous way to the two previous cases, using expression (18) and (31) we obtain for the general Hénon chaos matrix elements [*a,b,c,e*]:

$$a=\frac{-1-2y_n+\alpha y_n^2+\sqrt{1-4y_n-2\left(-2+\alpha-2\beta+6\beta d\right)y_n^2+4\alpha y_n^3+\alpha^2 y_n^4}}{2y_n}, \tag{72}$$

$$b=\frac{1+\left(-2+3\beta d\right)y_n-\alpha y_n^2+\sqrt{1-4y_n-2\left(-2+\alpha-2\beta+6\beta d\right)y_n^2+4\alpha y_n^3+\alpha^2 y_n^4}}{2y_n}, \tag{73}$$

$$c = -\beta, \tag{74}$$

$$e = \frac{-1 + 2y_n + \alpha y_n^2 - \sqrt{1 - 4y_n - 2(-2 + \alpha - 2\beta + 6\beta d)y_n^2 + 4\alpha y_n^3 + \alpha^2 + y_n^4}}{2y_n}. \tag{75}$$

8. Conclusions

This chapter presents a description of the application of non-constant ABCD matrix in the description of ring optical phase conjugated resonators. It is shown how the introduction of a particular map generating device in a ring optical phase-conjugated resonator can generate beams with the behavior of a specific two dimensional map. In this way beams that behave according to the Tinkerbell, Duffing or Henon Maps which we call "Tinkerbell, Duffing or Henon Beams", are obtained.

In particular, this chapter shows how Tinkerbell beams can be produced if a particular device is introduced in a ring optical phase-conjugated resonator. The difference equations of the Tinkerbell map are explicitly introduced in an *ABCD* transfer matrix to control the beams behaviour. The matrix elements *a, b, c* and *e* of a map generating device are found in terms of the map parameters (α, β, γ and δ), the state variables (y_n and θ_n) and the resonator length. The mathematical characteristics of an optical device inside an optical resonator capable to produce Tinkerbell beams are found. In the general case a device with fixed size was obtained, opening the possibility of a continuance of this work; that is the actual building of an optical device with these *a, b, c* and *d* matrix elements according to the description given and the experimental observation of Tinkerbell beams.

Also, it is explicitly shown how the difference equations of the Duffing map can be used to describe the dynamic behavior of what we call Duffing beams i.e. beams that behave according to the Duffing map. The matrix elements *a, b, c, e* of a map generating device are found in terms of α and β, the Duffing parameters, the state variable θ_n and the resonator parameter *d*.

Finally it is shown that the difference equations of the Hénon map can be used to describe the dynamical behavior of Hénon beams. The matrix elements *a, b, c, e* of a chaos generating device are found in terms of α and β the Hénon parameters, and *d* the resonator parameter.

Author details

V. Aboites*, Y. Barmenkov and A. Kir'yanov
Centro de Investigaciones en Óptica, México

M. Wilson
Université des Sciences et Technologies de Lille, France

* Corresponding Author

9. References

[1] G.S. He, Optical Phase Conjugation: Principles, Techniques and Applications, Progress in Quantum Electronics 26, No, 3, (2002), 61p.

[2] H.-J. Eichler, R. Menzel, and D. Schumann, Appl. Opt., 31 No. 24 (1992) 5038-5043

[3] M. O'Connor, V. Devrelis, and J. Munch, in Proc. Int. Conf. on Lasers'95 (1995) pp. 500-504

[4] M. Ostermeyer, A. Heuer, V. Watermann, and R. Menzel in Int. Quantum Electronics Conf., 1996 OSA Technical Digest Series (Optical Society of America, Washington, DC, 1996), p. 259

[5] Bel'dyugin I.M., Galushkin M.G., and Zemskov E.M. Kvantovaya Elektron., 11, 887 (1984) [Sov. J. Quantum Electron., 14, 602 (1984); Bespalov V.I. and Betin A.A. Izv. Akad. Nauk SSSR., Ser. Fiz., 53 1496 (1989)

[6] V.V. Yarovoi, A.V. Kirsanov, Phase conjugation of speckle-inhomogeneous radiation in a holographic Nd:YAG laser with a short thermal hologram, Quantum Electronics 32(8) 697-702 (2002)

[7] M.J. Damzen, V.I. Vlad, V. Babin, and A. Mocofanescu, Stimulated Brillouin Scattering: Fundamentals and Applications, Institute of Physics, Bristol (2003)

[8] D. A. Rockwell, A review of phase-conjugate solid-state lasers, IEEE Journal of Quantum Electronics 24, No. 6, (1988) 1124-1140

[9] Dämmig, M., Zinner, G., Mitschke, F., Welling, H. Stimulated Brillouin scattering in fibers with and without external feedback (1993) Physical Review A, 48 (4), pp. 3301-3309

[10] P.J. Soan, M.J. Damzen, V. Aboites and M.H.R. Hutchinson, Opt. Lett. 19 (1994), 783

[11] A.D. Case, P.J. Soan, M.J. Damzen and M.H.R. Hutchinson, J. Opt. Soc. Am. B 9 (1992) 374

[12] B. Barrientos, V. Aboites, and M. Damzen, Temporal dynamics of a ring dye laser with a stimulated Brillouin scattering mirror, Applied Optics, 35 (27) 5386-5391 (1996)

[13] E. Rosas, V. Aboites, M.J. Damzen, FWM interaction transfer matrix for self-adaptive laser oscillators, Optics Communications 174 243-247 (2000)

[14] V. Aboites, Int. J. of Pure and Applied Mathematics, 36, No. 4 (2007) 345-352.

[15] V. Aboites and M. Wilson, Int. J. of Pure and Applied Mathematics, 54 No. 3 (2009) 429-435.

[16] V. Aboites, A.N. Pisarchik, A. Kiryanov, X. Gomez-Mont, Opt. Comm., 283, (2010) 3328-3333

[17] A. Gerrard and J.M. Burch, Introduction to Matrix Methods in Optics, Dover Publications Inc., New York (1994).

[18] Y. Hisakado, H. Kikuchi, T. Nagamura, and T. Kajiyama, Advanced Materials, 17 No.1 (2005) 96-97

[19] A.V. Kir'yanov, V. Aboites and N.N. Il'ichev, JOSA B, 17 (2000) 11-17

[20] http://en.wikipedia.org/wiki/List_of_chaotic_maps

[21] R.L. Davidchack, Y.C. Lai, A.Klebanoff, E.M. Bollt, Physics Letters A, 287 (2001) 99-104

[22] P.E. McSharry, P.R.C. Ruffino, Dynamical Systems, 18, No. 3 (2003) 191-200

[23] E. Eschenazi, H.G. Solari, and R. Gilmore, Phys. Rev. A 39 (1989) 2609.
[24] M. Hénon, Commun. Math. Phys. 50 (1976) 69.
[25] R.L. Devaney, An Introduction to Chaotic Dynamical Systems, Addison-Wesley, Redwood City (1989).
[26] L.M. Saha and R. Tehri, Int. J. of Appl. Math and Mech., 6 (1) (2010) 86-93
[27] C. Murakami, W. Murakami, K. Hirose and W.H. Ichikawa, Chaos, Solitons & Fractals, 16 (2) (2003) 233-244

Chapter 3

Electrodynamics of Evanescent Wave in Negative Refractive Index Superlens

Wei Li and Xunya Jiang

Additional information is available at the end of the chapter

http://dx.doi.org/10.5772/50025

1. Introduction

Early in 1968, Veselago[1] predicted that a new type of artificial metamaterial, which possesses simultaneously negative permittivity and permeability, could function as a lens to focus electromagnetic waves. These research direction was promoted by Pendry's work[2, 3] and other latter works [4–26]. They show that with such a metamaterial lens, not only the radiative waves but also the evanescent waves, can be collected at its image, so the lens could be a superlens which can break through or overcome the diffraction limit of the conventional imaging system. This beyond-limit property gives us a new window to design devices.

It is well-known that evanescent wave plays an important role in the beyond-limit property of the metamaterial superlens. Furthermore, evanescent waves become more and more important when the metamaterial devices enter sub-wavelength scales[27, 28]. Therefore, the quantitatively study of *pure* evanescent waves in the metamaterial superlens is very significant. However, the quantitatively effects of *pure* evanescent wave in the metamaterial superlens have not been intensively studied, since so far almost all studies were only interested in the image properties with global field[8, 9], which is the summation of radiative wave and evanescent wave. On the other hand, many theoretical works were performed to study the metamaterial superlens, employing either finite-difference-time domain (FDTD) simulations[16] or some approximate approaches[29, 30]. However, one cannot obtain the rigorous *pure* evanescent wave by these numerical methods, since the image field of the metamaterial superlens obtained by FDTD is global field, and other approximate approaches cannot be rigorous.

In reviewing these existing efforts, we feel desirable to develop a rigorous method that can be used to study quantitatively the transient phenomena of the evanescent wave in the image of the metamaterial superlens. In this paper, we will present a new method based on the Green's function[7, 8] to serve this purpose. Our method can be successfully used to calculate the evanescent wave, as well as the radiative wave and the global field. The main idea of our method can be briefly illustrated as follows. Since the metamaterial superlens is a linear system, so all dynamical processes can be solved by sum of multi-frequency components. And each frequency component can be solved by sum of multi-wavevector components. So we can use Green's function of multi-frequency components to obtain the strict numerical results. Therefore, our method based on the Green's function is strict, and it is quite a universe method in any linear system, for example, it can be used to study the two dimensional (2D) and three dimensional (3D) metamaterial superlens.

The content of the chapter is organized as follows. We mainly focus on the details of our theory and method in Section 2. After that, in Section 3, we will calculate the field of the image of a NIR superlens by using our method, including radiative waves, evanescent waves, SEWs, and global field. The method will be confirmed by using FDTD simulations. In Section 4, we will present our study on the group delay of evanescent wave in the superlens by using our method. Finally, conclusions are presented in Section 5.

2. Theoretical method

In this section, we will focus on the theoretical details of our method. First, a time-dependent Green's function will be introduced. Then, based on the Green's function, the method to obtain evanescent waves as well as radiative waves will be presented.

2.1. A time-dependent Green's function

First, we would like to introduce a very useful time-dependent Green's function for the solution of inhomogeneous media. The time-dependent Green's function can be applied to study the dynamical scattering processes[7, 22]. In the inhomogeneous media, the problem we study can be solved by Maxwell equations:

$$\begin{aligned} \nabla \times \vec{E}(r,t) &= -\mu(r)\mu_0 \frac{\partial}{\partial t}\vec{H}(r,t), \\ \nabla \times \vec{H}(r,t) &= \epsilon(r)\epsilon_0 \frac{\partial}{\partial t}\vec{E}(r,t) + \vec{J}(r,t), \end{aligned} \tag{1}$$

where $\vec{J}(r,t) = \sigma(r,t)\vec{E}(r,t)$ is the current density and $\sigma(r,t)$ is the conductivity. We rewrite Eq.(1) as:

$$\nabla \times \nabla \times \vec{E}(r,t) + \epsilon(r)\mu(r)\epsilon_0\mu_0 \frac{\partial^2}{\partial t^2}\vec{E}(r,t) = -\mu(r)\mu_0 \frac{\partial}{\partial t}[\sigma(r,t)\vec{E}(r,t)] \tag{2}$$

To solve Eq.(2), we introduce a dynamic Green's function $\overset{\rightarrow\rightarrow}{G}(r,r';t,t')$, which satisfies:

$$\left(\nabla \times \nabla \times + \epsilon(r)\mu(r)\epsilon_0\mu_0 \frac{\partial^2}{\partial t^2}\right) \overset{\rightarrow\rightarrow}{G}(r,r';t,t') = \delta(r-r')\delta(t-t')\overset{\rightarrow\rightarrow}{I} \tag{3}$$

where $\overset{\rightarrow\rightarrow}{I}$ is a unit dyad. In this system, when r and r' is given, $\overset{\rightarrow\rightarrow}{G}(r,r';t,t')$ is just a function of $(t-t')$ in time domain, so it yields

$$\overset{\rightarrow\rightarrow}{G}(r,r';t,t') = \overset{\rightarrow\rightarrow}{G}(r,r';t-t'). \tag{4}$$

Therefore, Eq.(3) becomes:

$$\left(\nabla \times \nabla \times +\epsilon(r)\mu(r)\epsilon_0\mu_0\frac{\partial^2}{\partial t^2}\right)\overleftrightarrow{G}(r,r';t-t') = \delta(r-r')\delta(t-t')\overleftrightarrow{I} \tag{5}$$

If $\overleftrightarrow{G}(r,r';t-t')$ is known, then the $\vec{E}$ field can be found as

$$\vec{E}(r,t) = -\mu_0 \int \overleftrightarrow{G}(r,r';t-t') \cdot \vec{J}(r',t')dr'dt' \tag{6}$$

By Fourier transforming Eq.(5) from time domain to frequency domain, we obtain

$$\left(\nabla \times \nabla \times -\mu(r,\omega)\epsilon(r,\omega)\epsilon_0\mu_0\omega^2\right)\overleftrightarrow{G}(r,r';\omega) = \delta(r-r')\overleftrightarrow{I} \tag{7}$$

where $\epsilon(r,\omega)$ and $\mu(r,\omega)$ are the relative permittivity and the relative permeability of the dispersive material at a frequency ω, respectively, and $\overleftrightarrow{G}(r,r';\omega)$ is the Green's Function in the frequency domain, which satisfies

$$\overleftrightarrow{G}(r,r';t-t') = \frac{1}{2\pi}\int d\omega \overleftrightarrow{G}(r,r';\omega)e^{-i\omega(t-t')} \tag{8}$$

By Fourier transforming the electric field $\vec{E}(r,t)$ in time domain to the frequency domain, we obtain $\vec{E}(r,\omega)$, satisfying

$$\vec{E}(r,t) = \frac{1}{2\pi}\int d\omega e^{-i\omega t}\vec{E}(r,\omega) \tag{9}$$

Considering the general properties:

$$\begin{aligned} \epsilon(r,-\omega) &= \epsilon(r,\omega)^*, \\ \mu(r,-\omega) &= \mu(r,\omega)^*, \end{aligned} \tag{10}$$

and

$$\overleftrightarrow{G}(r,r';-\omega) = \overleftrightarrow{G}(r,r';\omega)^*. \tag{11}$$

Substituting Eq.(8), Eq.(9), Eq(10) and Eq.(11) into Eq.(6), we can obtain

$$\vec{E}(r,\omega) = \overleftrightarrow{G}(r,r';\omega) \cdot \vec{E}(r',\omega). \tag{12}$$

Eq.(12) exhibits the role of $\overleftrightarrow{G}(r,r';\omega)$ that $\overleftrightarrow{G}(r,r';\omega)$ is a propagator for the field in the frequency domain between r and r'. Here we have supposed that $\sigma(r',t) = 1$ in the source region for simplicity. If $\vec{E}(r',\omega)$ is the exciting source's electric field in the frequency domain, i.e., $\vec{E}_s(r',\omega) = \vec{E}(r',\omega)$, then Eq.(12) can be rewritten as:

$$\vec{E}(r,\omega) = \overleftrightarrow{G}(r,r';\omega) \cdot \vec{E}_s(r',\omega) \tag{13}$$

where $\vec{E}_s(r',\omega)$ is the spectrum of the exciting source.

So the field $\vec{E}(r,t)$ can be obtained by the inverse Fourier transformation:

$$\vec{E}(r,t) = \frac{1}{2\pi}\int d\omega e^{-i\omega t}\vec{E}(r,\omega), \tag{14}$$

where ω_0 is the working frequency of the exciting source.

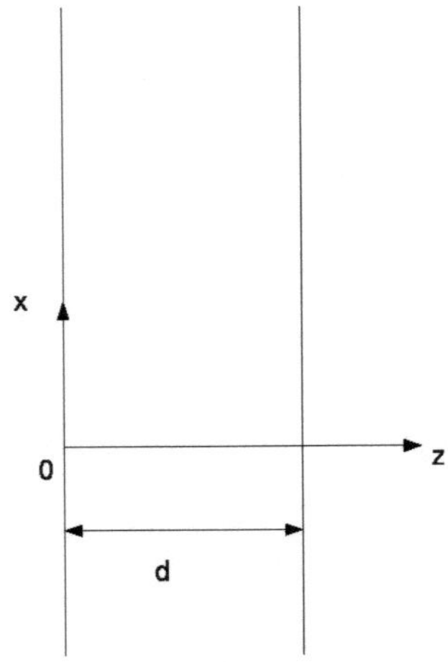

Figure 1. The shematic of the three-layer inhomogeneous medium.

Now the remaining problem is to solve Eq(7) to get $\overrightarrow{\overrightarrow{G}}(r,r';\omega)$. With the exciting source $E_s(r',t)$ (whose spectrum is $E_s(r',\omega)$), after obtain $\overrightarrow{\overrightarrow{G}}(r,r';\omega)$, then we can calculate $\vec{E}(r,\omega)$ directly by Eq.(12).

As a typical example, the time-dependent Green's function for the three-layer inhomogeneous media is presented, which is shown in Fig.1. The inhomogeneous media of the system can be described by:

$$\epsilon(r)\epsilon_0, \mu(r)\mu_0 = \begin{cases} \epsilon_0, \mu_0 & z > d \\ \epsilon_l^r\epsilon_0, \mu_l^r\mu_0 & 0 < z < d \\ \epsilon_0, \mu_0 & z < 0 \end{cases} \tag{15}$$

$\overrightarrow{\overrightarrow{G}}(r,r';\omega)$ is related to the transmission coefficient and/or reflection coefficient which are dependent on the boundary conditions of each layer. Here, we only consider the Green's function in the region that $z > d$. Under these conditions, $\overrightarrow{\overrightarrow{G}}(r,r';\omega)$ has been obtained in Ref.[7]. $\overrightarrow{\overrightarrow{G}}(r,r';\omega)$ is different in different spatial dimensions. For three-dimension, we can rewrite $\overrightarrow{\overrightarrow{G}}(r,r';\omega)$ as follows:

$$\begin{aligned} \overrightarrow{\overrightarrow{G}}(r,r';\omega) = -\frac{i\sigma(r',\omega)}{8\pi}\int \frac{1}{k_z}e^{-ik_z z}[T^{TE}(k_\parallel)(J_0(k_\parallel x) - J_2(k_\parallel x)) \\ + \frac{k_z^2}{k^2}T^{TM}(k_\parallel)(J_0(k_\parallel x) + J_2(k_\parallel x))k_\parallel dk_\parallel], \end{aligned} \tag{16}$$

where $J_0(z)$ and $J_2(z)$ are the usual zeroth-order Bessel function and second-order Bessel function, respectively, and $k_{\|}^2 + k_{lz}^2 = \epsilon_l^r \mu_l^r (\omega/c)^2$, $k_{\|}^2 + k_z^2 = (\omega/c)^2$ are the dispersion relation in the region $(0 < z < d)$ and the region $(z < 0, z > d)$ respectively. Here, T^{TE} and T^{TM} are the transmission coefficients for TE wave ($\vec{H}$ polarized) and TM wave ($\vec{E}$ polarized), respectively, which are respectively given by

$$T^{TE}(k_{\|}) = \frac{4\triangle e^{-ik_z d}}{(\triangle + 1)^2 e^{-ik_{lz} d} - (\triangle - 1)^2 e^{ik_{lz} d}}, \tag{17}$$

and

$$T^{TM}(k_{\|}) = \frac{4\triangle' e^{-ik_z d}}{(\triangle' + 1)^2 e^{-ik_{lz} d} - (\triangle' - 1)^2 e^{ik_{lz} d}}, \tag{18}$$

where $\triangle = \frac{k_{lz}}{k_z \mu_l^r}$ and $\triangle' = \frac{k_{lz}}{k_z \epsilon_l^r}$.

$\sigma(r', \omega)$ is the conductivity in source region, i.e., $\vec{J}(r', \omega) = \sigma(r', \omega) \vec{E}_s(r', \omega)$. $\sigma(r', \omega)$ has been assumed to be $\sigma(r', \omega) = 1$. So Eq(16) becomes:

$$\begin{aligned} \overset{\rightarrow\rightarrow}{G}(r, r'; \omega) = &- \frac{i}{8\pi} \int \frac{1}{k_z} e^{-ik_z z} [T^{TE}(k_{\|})(J_0(k_{\|} x) - J_2(k_{\|} x)) \\ &+ \frac{k_z^2}{k^2} T^{TM}(k_{\|})(J_0(k_{\|} x) + J_2(k_{\|} x)) k_{\|} dk_{\|}], \end{aligned} \tag{19}$$

For the 2D case, the wave vector $k_{\|} = k_x$, we have

$$\overset{\rightarrow\rightarrow}{G}(r, r'; \omega) = -\frac{i}{4\pi} \int \frac{e^{ik_x x}}{k_z} T^{TE}(k_x) e^{ik_z z} dk_x. \tag{20}$$

And for the 3D case, we have

$$\overset{\rightarrow\rightarrow}{G}(r, r'; \omega) = -\frac{i}{2k} T^{TE}(0) e^{-ikz}. \tag{21}$$

After $\overset{\rightarrow\rightarrow}{G}(r, r'; \omega)$ is obtained, we can obtain the field in the frequency domain in the region $z > d$ by Eq.(12). And then by the inverse Fourier transformation, the field in time domain can be obtained.

2.2. The Green's function for radiative waves and evanescent waves

Now, we will apply a time-dependent Green's function for a radiative wave and an evanescent wave. This Green's function can be directly developed from the Green's function introduced in the Sec.2.1. The schematic model is shown in Fig.1. As we know, the plane solution wave for the electric field in vacuum is of the form $E_z(r_{||}, z, t) = E_{z0} \exp(i(k_{||} r_{||} + k_z z - \omega t))$, where $k_{||}$ and k_z are wave numbers along the xy plane and z directions respectively, and they satisfy the dispersion relation as follows: $k_{||}^2 + k_z^2 = \omega^2/c^2$, where c is the light velocity in the vacuum. In the case of $k_{||}^2 < \omega^2/c^2$, k_z is real, corresponding to the radiative waves along the z direction. While if $k_{||}^2 > \omega^2/c^2$, k_z is imaginary, corresponding to the evanescent waves along the z

direction. Similarly, in the slab region, the real or imaginary k_{lz} corresponds to the radiative wave or the evanescent wave along z direction, respectively.

For simplicity, we first consider the 2D system, in which $k_{||} = k_x$. The global field of image region is the superposition of radiative waves and evanescent waves. We can rewrite Eq.(20),Eq.(13) and Eq(14) as follows:

$$\overleftrightarrow{G}(r,r';\omega;k_{min},k_{max}) = -\frac{i}{4\pi}\int_{k_{min}}^{k_{max}} \frac{e^{ik_x x}}{k_z} T^{TE}(k_x)e^{ik_z z}dk_x, \tag{22}$$

$$\vec{E}(r;\omega;k_{min},k_{max}) = \overleftrightarrow{G}(r,r';\omega;k_{min},k_{max}) \cdot \vec{E}_s(r',\omega), \tag{23}$$

and

$$\vec{E}(r;t;k_{min},k_{max}) = \frac{1}{2\pi}\int d\omega e^{-i\omega t}\vec{E}(r;\omega;k_{min},k_{max}), \tag{24}$$

where $k_{min} < k_x < k_{max}$ is the integral range. The integral range is of great significance, since different integral range corresponds to different wave. For example, the integral range $[k_{min} \to -\infty, k_{max} \to \infty]$ is for global field, and the range $[k_{min} = -\omega/c, k_{max} = \omega/c]$ is for radiative wave. Obviously, for linear system, the integral range can be chosen arbitrarily.

From Eq.(22) to Eq.(24), we can directly calculate the radiative wave and evanescent wave. In the case of radiative wave ($k_x^2 < \omega^2/c^2$), the integral range is $[k_{min} = -\omega/c, k_{max} = \omega/c]$, so the radiative wave Green's function $\overleftrightarrow{G}_{rad}(r,r';\omega)$ satisfies

$$\begin{aligned}\overleftrightarrow{G}_{rad}(r,r';\omega) &= \overleftrightarrow{G}(r,r';\omega;k_{min} = -\omega/c, k_{max} = \omega/c) \\ &= -\frac{i}{4\pi}\int_{-\omega/c}^{\omega/c} \frac{e^{ik_x x}}{k_z} T^{TE}(k_x)e^{ik_z z}dk_x\end{aligned} \tag{25}$$

for radiative waves.

In the case of evanescent wave ($k_x^2 > \omega^2/c^2$), the integral range is $k_x > |\omega/c|$, so the evanescent wave Green's function $\overleftrightarrow{G}_{eva}(r,r';\omega)$ satisfies:

$$\begin{aligned}\overleftrightarrow{G}_{eva}(r,r';\omega) &= 2\lim_{k_{max}\to\infty} \overleftrightarrow{G}(r,r';\omega;k_{min} = \omega/c, k_{max}) \\ &= -\frac{i}{2\pi}\lim_{k_{max}\to\infty}\int_{\omega/c}^{k_{max}} \frac{e^{ik_x x}}{k_z} T^{TE}(k_x)e^{ik_z z}dk_x\end{aligned} \tag{26}$$

for evanescent waves.

And for the global field, the global field Green's function $\overleftrightarrow{G}_{glob}(r,r';\omega)$ satisfies:

$$\overleftrightarrow{G}_{glob}(r,r';\omega) = \overleftrightarrow{G}_{rad}(r,r';\omega) + \overleftrightarrow{G}_{eva}(r,r';\omega) \tag{27}$$

Additionally, we can also focus our observation on the *subdivided evanescent wave* (SEW), with a certain integral range $[k_{min} = k_x^a, k_{max} = k_x^b]$. The SEW, with a certain integral range, can also be obtained from Eq.(22) to Eq.(24) such that

$$\overleftrightarrow{G}_{SEW}(r,r';\omega)|_{k_{min}=k_x^a}^{k_{max}=k_x^b} = \overleftrightarrow{G}(r,r';k_x^a,k_x^b) \\ = -\frac{i}{4\pi}\int_{k_x^a}^{k_x^b}\frac{e^{ik_x x}}{k_z}T^{TE}(k_x)e^{ik_z z}dk_x. \tag{28}$$

As an typical example, a SEW with an integral range $[k_{min} = 1.1\omega/c, k_{max} = 1.2\omega/c]$, whose Green's function can be obtained easily from Eq.(28) as:

$$\overleftrightarrow{G}_{SEW}(r,r';\omega)|_{k_{min}=1.1\omega/c}^{k_{max}=1.2\omega/c} = -\frac{i}{4\pi}\int_{1.1\omega/c}^{1.2\omega/c}\frac{e^{ik_x x}}{k_z}T^{TE}(k_x)e^{ik_z z}dk_x.$$

In this way, we can obtain the Green's function for the SEW with any integral range. Obviously, evanescent wave could be regarded as the superposition of SEWs. Therefore, we have

$$\overleftrightarrow{G}_{eva}(r,r';\omega) = \sum_{SEWs}\overleftrightarrow{G}_{SEW}(r,r';\omega). \tag{29}$$

From Eq.(28), Eq.(26), Eq.(25), and Eq.(27), one can obtain the SEW Green's function $\overleftrightarrow{G}_{SEW}(r,r';\omega)$, the evanescent wave Green's function $\overleftrightarrow{G}_{eva}(r,r';\omega)$, the radiative wave Green's function $\overleftrightarrow{G}_{rad}(r,r';\omega)$, and the global field Green's function $\overleftrightarrow{G}_{glob}(r,r';\omega)$, respectively. Substituting them to Eq.(23) and Eq.(24) respectively, we can obtain the field of the SEW, the evanescent wave, the radiative wave, and the global field, respectively.

For the 3D system, obviously, the methods to get Green's function for the SEW, the evanescent wave, the radiative wave, and the global field are respectively very similar with the above discussion, i.e., just replace k_x by $k_{||}$ and let $k_{||}^2 = k_x^2 + k_y^2$ in Eq.(28), Eq.(26), Eq.(25), and Eq.(27), respectively.

3. Electromagnetic waves in the image of the superlens

In this section, we will discuss the image's field of a 2D metamaterial superlens, which is shown in Fig.2. The thickness of the metamaterial slab is d, which is placed at the *xy-plane* between $z = d/2$ and $z = 3d/2$. The source is set in the object plane at $(z = 0)$. Obviously, the image will be formed in the image plane at $z = 2d$. The source is the quasi-monochromatic random source with the field expressed as $E_s(r,t)$=$U_s(r,t)exp(-i\omega_0 t)$, where $U_s(r,t)$ is a slow-varying random function, $\omega_0 = 1.33 \times 10^{15}/s$ is the central frequency of our random source (the details of the random source can be seen in Ref.[29]). The exciting source $E_s(r,t)$ is a quasi-monochromatic field with the central frequency ω_0, whose electric field $\vec{E}_s(t)$ and frequency spectrum $\vec{E}_s(\omega)$ are shown in Fig.3(a) and Fig.3(b), respectively. In this paper, only the TM modes are investigated (the TM modes have the electric field perpendicular to the two-dimensional plane of our model).

The inhomogeneous media of the metamaterial superlens system are described by:

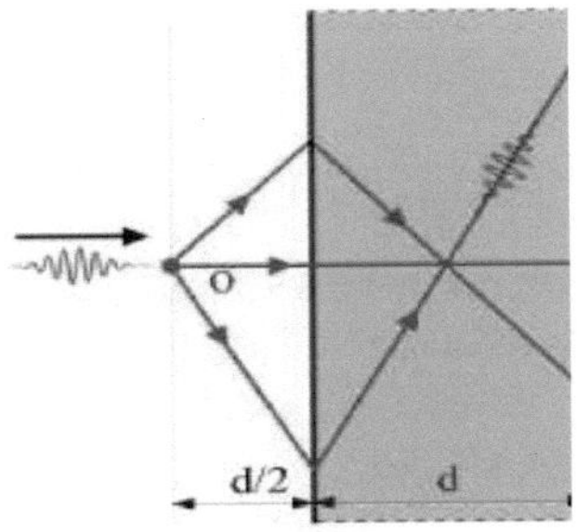

Figure 2. The schematic diagram of our model.

$$\epsilon(r),\mu(r)=\begin{cases}\epsilon_0,\mu_0 & z>3/2d\\ \epsilon_0\epsilon_l^r,\mu_0\mu_l^r & 1/2d<z<3/2d\\ \epsilon_0,\mu_0 & z<1/2d\end{cases} \tag{30}$$

The negative relative permittivity ϵ_l^r and the negative relative permeability μ_l^r of metamaterial are phenomenologically introduced by the Lorenz model. The negative relative permittivity ϵ_l^r and the negative relative permeability μ_l^r are satisfied as follows:

$$\epsilon_l^r(\omega)=\mu_l^r(\omega)=1+\omega_p^2/(\omega_a^2-\omega^2-i\Delta\omega\cdot\omega) \tag{31}$$

where $\omega_a = 1.884\times10^{15}/s$ and $\Delta\omega = 1.88\times10^{14}/s$ are the resonant frequency and the resonant line-width of the "resonators" in the metamaterial respectively, and $\omega_p = 10\times\omega_a$ is the plasma frequency. At $\omega=\omega_0$, we have $\epsilon_l^r=\mu_l^r=-1.0+i0.0029$.

In order to excite the evanescent wave strong enough in the image of the metamaterial superlens, the distance $d/2$ between the source and the superlens should be small enough. Here we choose $d=\lambda_0/2$, where $\lambda_0=1.42\mu m$ is the wavelength corresponding to the central frequency ω_0.

For this metamaterial superlens system, it is very easy to obtain $\overset{\rightarrow\rightarrow}{G}_{glob}(r,r';\omega)$, $\overset{\rightarrow\rightarrow}{G}_{eva}(r,r';\omega)$, and $\overset{\rightarrow\rightarrow}{G}_{SEW}(r,r';\omega)$ from Eq.(26) to Eq.(28), respectively. Let $r=2d$ and $r'=0$, and thus the Green's functions for the image will be obtained. After that, we can obtain the global field, the evanescent wave and the SEW of the image via:

$$\begin{aligned}\vec{E}_{glob}(2d,t)&=\frac{1}{2\pi}\int d\omega e^{-i\omega t}\vec{E}_{glob}(2d,\omega)\\ \vec{E}_{eva}(2d,t)&=\frac{1}{2\pi}\int d\omega e^{-i\omega t}\vec{E}_{eva}(2d,\omega)\\ \vec{E}_{SEW}(2d,t)&=\frac{1}{2\pi}\int d\omega e^{-i\omega t}\vec{E}_{SEW}(2d,\omega)\end{aligned} \tag{32}$$

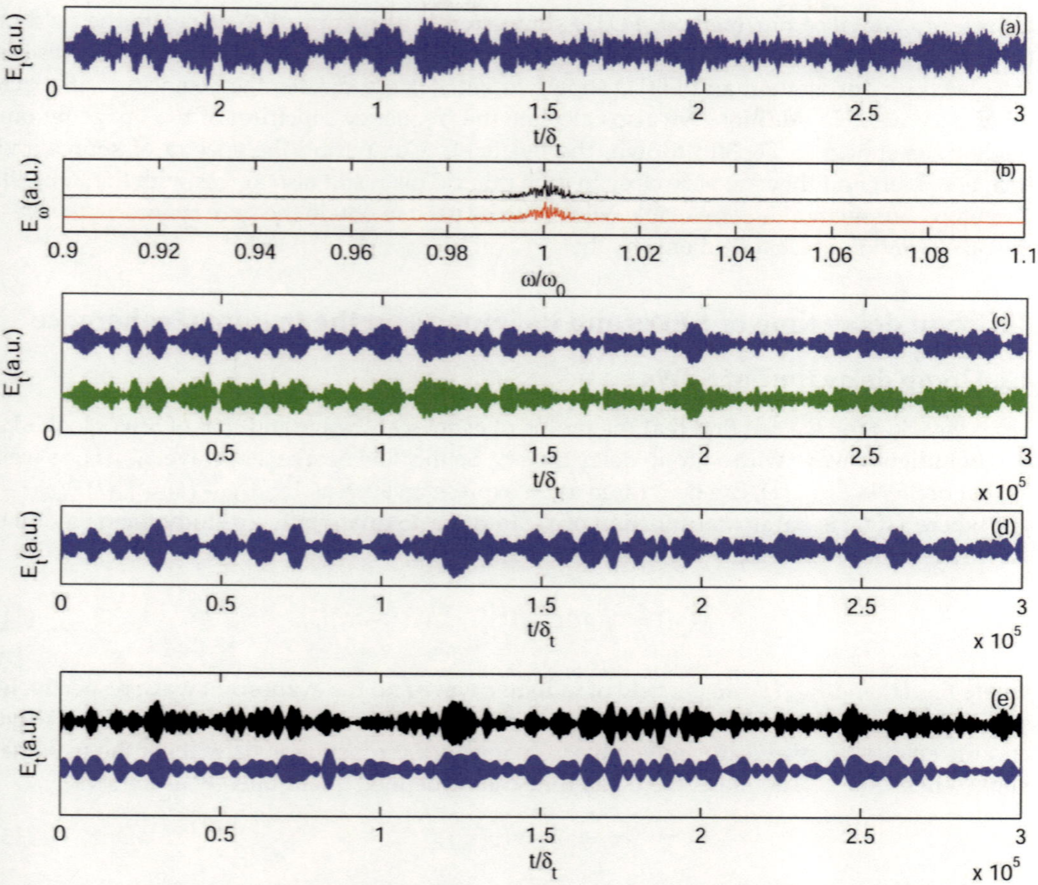

Figure 3. (a) Electric field of the source. (b) Spectrum of the source (top) and the image obtained by using our method (bottom) in units of ω_0. (c) Electric field of the image vs time. The global field calculated by using our method (top) and by using FDTD (bottom). (d) The evanescent wave of the image calculated by using our method. (e) Two typical SEWs of the image.

respectively, where

$$\begin{aligned}\vec{E}_{glob}(2d,\omega) &= \overrightarrow{\overrightarrow{G}}_{glob}(2d,0;\omega)\cdot\vec{E}_s(\omega)\\ \vec{E}_{eva}(2d,\omega) &= \overrightarrow{\overrightarrow{G}}_{eva}(2d,0;\omega)\cdot\vec{E}_s(\omega)\\ \vec{E}_{SEW}(2d,\omega) &= \overrightarrow{\overrightarrow{G}}_{SEW}(2d,0;\omega)\cdot\vec{E}_s(\omega).\end{aligned} \tag{33}$$

The numerical results calculated by our method are shown in Fig.3. Fig.3(c)(up, the blue one), (d) and (e) show the global field, the evanescent wave, and two typical SEWs respectively. The integral k_x range of the two SEWs are $[k_{min} = 1.1\omega/c, k_{max} = 1.2\omega/c]$ (shown in Fig.3(e)(up, the black one)), and $[k_{min} = 1.3\omega/c, k_{max} = 1.4\omega/c]$ (shown in Fig.3(e)(down, the blue one)), respectively.

In order to convince our method, FDTD simulation is also applied to calculate the field of the image, which is shown in Fig.3(c) (down, the green one). Comparison with the results calculated by our method and FDTD shown in Fig.3(c), we can see they coincide with each other very well. In addition, we also calculate the frequency sepctrum of the image by our method, as shown in Fig.3(b) (down, the red one). Comparing the spectra of source and image, we can find they are very close to each other. This result also agrees with the Ref.[29]. Therefore, our method is convincible, which can be used to obtain the pure evanescent waves, the SEWs, and the global field effectively.

4. Group delay time of SEWs and its impacts on the temporal coherence

4.1. Group delay time of SEWs

From Figs.3(c)-(e), we can find that the profile of evanescent wave and that of SEWs look like that of radiative wave with a group delay time τ_r. So the field evanescent wave $\vec{E}_{eva}(t)$, as well as that of SEWs $\vec{E}_{SEW}(t)$, can be written as an expression such as $\vec{E}_{eva(SEW)}(t)= f_a(t)\vec{E}_{rad}(t-\tau_r)$, where $f_a(\tau_r)$ is parameter function of τ_r. In order to quantitatively study the delay time τ_r, we introduce a function $y(\tau_i)$ which satisfies:

$$y(\tau_i) = \int_t dt|\vec{E}_{rad}(t)| \cdot |\vec{E}_{eva}(t-\tau_i)|, \tag{34}$$

where $\vec{E}_{rad}(t)$ and $\vec{E}_{eva}(t)$ are the field of radiative wave and the evanescent wave respectively, τ_i is an independent variable with the time dimension. Since the profile of the radiative wave and the evanescent wave are very similar, obviously, the function $y(\tau_i)$ will get the maximal value when $\tau_i = \tau_r$. Therefore, the delay time can be defined quantitatively as follows:

$$\tau_r = [Max(y(\tau_i))]^{-1}. \tag{35}$$

Here $[\cdots]^{-1}$ means the inverse function.

Similarly, we can also study the group delay time of the SEWs. We rewrite Eq.(34) as follows:

$$y(\tau_i) = \int_t dt|\vec{E}_{rad}(t)| \cdot |\vec{E}_{SEW}(t-\tau_i)|, \tag{36}$$

where $E_{SEW}(t)$ is the field of the SEW with a certain integral k_x range. So we can calculate the delay time of the SEWs from Eq.(36) and Eq.(35).

In our numerical experiment, in order to calculate the group delay time of the SEWs, we choose 70 SEWs with integral k_x range as $[k_{mx} - 0.01\omega_0/c,\ k_{mx} + 0.01\omega_0/c]$, where k_{mx}=$1.01\omega_0/c, 1.02\omega_0/c, \cdots, 1.69\omega_0/c, 1.7\omega_0/c$, respectively. The delay time of the 70 SEWs is shown in Fig.4(a). In this figure, we can see that the SEW with larger integral variable k_x will have a larger delay time. Obviously, the function $\tau_r = \tau_r(k_x)$ is a continuous-monotone increasing function, when the integral range $[k_x - dk_x, k_x + dk_x]$ is infinitesimal ($dk_x \to 0$). The result can be obtained by using a polynomial fitting, as shown in Fig.4(a). Therefore, from Fig.4(a), we can find that the SEWs with larger integral variable k_x corresponds to a larger delay time, which means the group velocity of SEWs with larger integral variable is smaller in the superlens system.

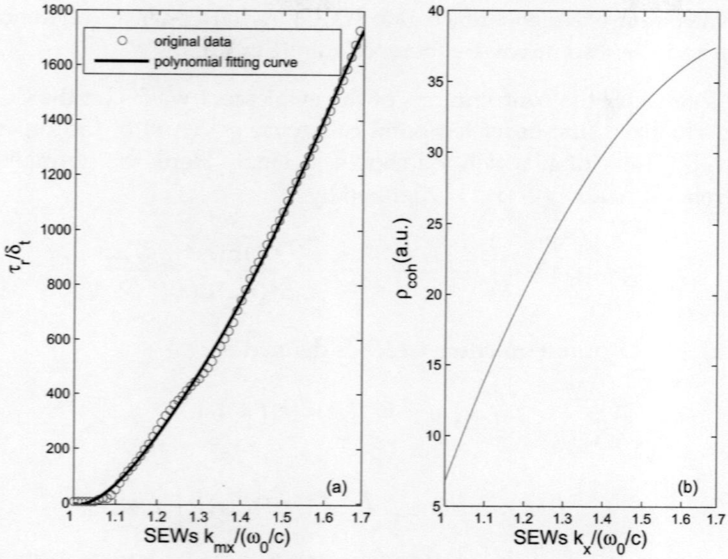

Figure 4. (a) The group delay time τ_r of SEWs. (b) ρ_{coh} of SEWs.

Since the field profile of SEWs looks like that of a radiative wave, so we can write the field of the SEW with the integral range $[k_x - \delta k_x, k_x + \delta k_x]$ as follows:

$$\vec{E}_{SEW}(t) = A(\tau_r(k_x))\vec{E}_{rad}(t - \tau_r(k_x)), \tag{37}$$

where $A(\tau_r(k_x)) = B(\tau_r(k_x))exp(-2|k_z(k_x)|d)$, $B(\tau_r)$ is a slowly-verifying function of τ_r, and $exp(-2|k_z|d)$ is an exponentially-decreasing function of k_z and a general-function of τ_r. Obviously, when τ_r(or k_x) becomes larger, $A(\tau_r)$ will trends to become smaller. Then the field of evanescent wave $\vec{E}_{eva}(t)$ can be obtained by:

$$\begin{aligned} \vec{E}_{eva}(t) &= \int \vec{E}_{SEW}(t)dk_x \\ &= \int A(\tau_r(k_x))\vec{E}_{rad}(t - \tau_r(k_x))dk_x. \end{aligned} \tag{38}$$

The physical meaning of $A(\tau_r)$ and its impacts will be discussed in the following.

4.2. Impacts of the group delay of SEWs on temporal coherence gain in the image of the superlens

One of the most interesting impacts of the group delay of SEWs is related to the first-order temporal coherence gain (CG). Here, we would like to discuss the CG caused by the SEWs in the image of the superlens. In our previous work [29, 30], we have investigated a prominent CG of the image by the radiative waves even when the frequency-filtering effects are very weak. Then, a natural question is what about the role of the evanescent waves play in the CG of a superlens? In this section, we will show that not only the radiative waves but also the evanescent waves, and the SEWs that can be responsible for the CG. Furthermore, we will show that the total CG in the image of a superlens is the weighted averaged of evanescent-wave coherence gain (ECG), radiative-wave coherence gain (RCG), radiative-wave and evanescent-wave coherence gain (RECG).

First of all, let's consider the contributions of the evanescent waves on the CG. For this, we calculate the normalized first-order temporal coherence $g^{(1)}(r,\tau)$ of the superlens with the random source $E_s(t)$ exciting, which are shown in Fig.5. Here, the normalized first-order temporal coherence function $g^{(1)}(r,\tau)$ is defined by

$$g^{(1)}(r,\tau) = \frac{G^{(1)}(r,\tau)}{G^{(1)}(r,0)} = \frac{< \vec{E}^*(r,t)\vec{E}(r,t+\tau_r) >}{< \vec{E}^*(r,t)\vec{E}(r,t) >} \tag{39}$$

where $G^{(1)}(r,\tau)$ is a coherence function, which is defined by

$$G^{(1)}(r,\tau) = < \vec{E}^*(r,t)\vec{E}(r,t+\tau_r) > \tag{40}$$

or

$$G^{(1)}(r,\tau) = \lim_{T\to\infty} \frac{1}{2T} \int_{-T}^{T} \vec{E}^*(r,t)\vec{E}(r,t+\tau)dt, \tag{41}$$

here $< \cdots >$ means the statistic average (ensemble average) and τ is time delay. From Fig.5, we can see that the temporal coherence of image of evanescent wave, radiative wave and global field, are all obviously better than that of source. Comparing with the temporal coherence of source and image, there are three kinds of coherence gain as follows. The first one is the radiative-wave coherence gain (RCG, this mechanism has been discussed in our previous work[29, 30]), which is determined by the radiative waves. The second one is the evanescent-wave coherence gain (ECG), which is determined by the evanescent waves. The last one is the global field coherence gain (GCG), which is from the global field with the co-effect of RCG and ECG.

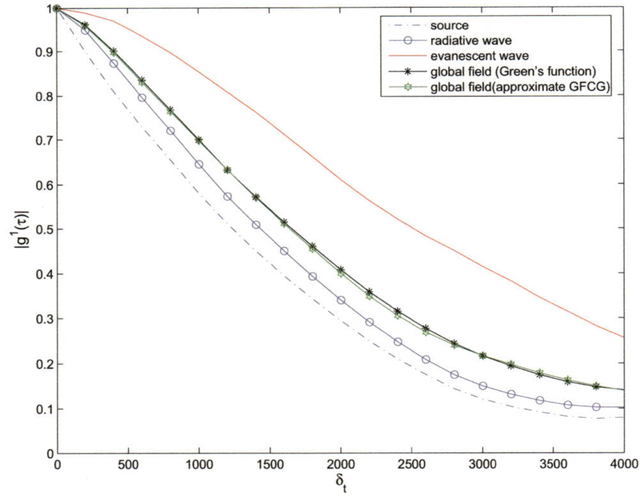

Figure 5. The first temporal coherence of image and source. The dashed-dotted one is the $g^{(1)}(r,\tau)$ of the source. Others are for the image, i.e., the red one, the black one (and also the green one), and the blue one are the $g^{(1)}(r,\tau)$ of the evanescent waves, the global field, and the radiative waves, respectively.

To show how the ECG is produced from the interference of the SEWs with different group delay time, we assume there are only two SEWs, such as $\vec{E}^{\alpha}_{SEW}(t)$ and $\vec{E}^{\beta}_{SEW}(t)$ with the range $[k^{\alpha}_{mx} - \delta k^{\alpha}_{mx}, k^{\alpha}_{mx} + \delta k^{\alpha}_{mx}]$ and $[k^{\beta}_{mx} - \delta k^{\beta}_{mx}, k^{\beta}_{mx} + \delta k^{\beta}_{mx}]$ respectively, which correspond to the delay time τ^{α}_r and τ^{β}_r, respectively. The two SEWs can be expressed by Eq.(37). We assume the two SEWs have an integral range very close to each other, i.e., $k^{\alpha}_{mx} \simeq k^{\beta}_{mx}$, and so we have $A(\tau_r(k^{\alpha}_{mx})) \simeq A(\tau_r(k^{\beta}_{mx}))$ and both of them are close to a certain constant A.

From Eq.(38), we have the evanescent wave field:

$$\begin{aligned} \vec{E}_{eva}(t) &= \vec{E}^{\alpha}_{SEW}(t) + \vec{E}^{\beta}_{SEW}(t) \\ &= A(\tau_r(k^{\alpha}_{mx}))\vec{E}_{rad}(t - \tau^{\alpha}_r) + A(\tau_r(k^{\beta}_{mx}))\vec{E}_{rad}(t - \tau^{\beta}_r) \\ &\simeq A \cdot (\vec{E}_{rad}(t - \tau^{\alpha}_r) + \vec{E}_{rad}(t - \tau^{\beta}_r)), \end{aligned} \tag{42}$$

then the temporal coherence of the evanescent wave in the image is given by

$$\begin{aligned} G_{eva}(\tau) &=< \vec{E}^{*}_{eva}(t)\vec{E}_{eva}(t+\tau) > \\ &=< \vec{E}^{*}_{rad}(t - \tau^{\alpha}_r)\vec{E}_{rad}(t - \tau^{\alpha}_r + \tau) + \vec{E}^{*}_{rad}(t - \tau^{\beta}_r)\vec{E}_{rad}(t - \tau^{\beta}_r + \tau) + \\ &\quad \vec{E}^{*}_{rad}(t - \tau^{\alpha}_r)\vec{E}_{rad}(t - \tau^{\beta}_r + \tau) + \vec{E}^{*}_{rad}(t - \tau^{\beta}_r)\vec{E}_{rad}(t - \tau^{\alpha}_r + \tau) >, \end{aligned} \tag{43}$$

The first two terms are the same as the coherence function of the radiative wave, so they do not contribute to ECG. The last two terms are from the interference between SEWs, they can be very large at the condition $\tau \simeq |\tau^{\alpha}_r - \tau^{\beta}_r|$. This condition can always be satisfied between SEWs since τ_r is a continuous variable. So the relative delay time $|\tau^{\alpha}_r - \tau^{\beta}_r|$ of SEWs are responsible for ECG .

Therefore, ECG can always exist in the superlens when two conditions are satisfied: (1) $A(\tau_r(k^{\alpha}_{mx})) \simeq A(\tau_r(k^{\beta}_{mx}))$; and (2) $\tau \simeq |\tau^{\alpha}_r - \tau^{\beta}_r|$. Unfortunately, the two conditions could not always be satisfied at the same time. When $\tau \simeq |\tau^{\alpha}_r - \tau^{\beta}_r|$ is large, which also means the integral variable k^{α}_{mx} and k^{β}_{mx} are far from each other, so the value of $|A(\tau^{\alpha}_r) - A(\tau^{\beta}_r)|$ will be very large, and thus the former condition could not be satisfied. The condition that $A(\tau_r(k_x)) \simeq A(\tau_r(k_{mx}))$ is the direct reflection that ***only** the interference of those SEWs with **close** integral variable k_x can produce the ECG.*

Here the integral variable "k_x *close* to k_{mx}" means when k_{mx} is given, for any k_x satisfying $|k_x - k_{mx}| \to \delta k_{mx}$, the condition $A(\tau_r(k_x)) \simeq A(\tau_r(k_{mx}))$ is always satisfied, where δk_{mx} is a threshold value with small positive value near zero(for example δk_{mx}=0.01). For two SEWs with the integral range $[k_{mx} - \delta k_{mx}, k_{mx} + \delta k_{mx}]$ and $[k_x - \delta k_x, k_x + \delta k_x]$ respectively, we can obtain the relative delay time $\Delta\tau_r(k_x) = |\tau_r(k_x) - \tau_r(k_{mx})|$. As the discussion above, the relative delay time $\Delta\tau_r$ is responsible to the ECG. When k_x is *close* to k_{mx}, $\Delta\tau_r(k_x)$ is a monotonic increasing function of k_x in the range $[k_{mx}, k_{mx} + \delta k_{mx}]$, which gives a threshold value τ_d as:

$$\tau_d(k_x) = \lim_{k_x \to k_{mx} + \delta.k_{mx}} \Delta\tau_r(k_x) \tag{44}$$

τ_d shows the upper-limit of the effective coherent relative delay time that only the $\Delta\tau_r \leq \tau_d$ is the effective responsible to ECG. As $\Delta\tau_r$ increasing, when $\Delta\tau_r > \tau_d$, which means $|k_x - k_{mx}| > \delta k_{mx}$ (i.e. k_x and k_{mx} is not close to each other), the difference between $A(\tau_r(k_x))$ and $A(\tau_r(k_{mx}))$ becomes greater, and so their interference becomes weaker and their contribution to the coherence gain will decrease rapidly. While when $\Delta\tau_r \gg \tau_d$, which means the integral variable k_x and k_{mx} are very far from each other, then the SEW with the much larger integral variable one is too weak to have any effective interference, so their contribution to the coherence gain trends to be zero.

Therefore, the coherence gain from the interference of SEWs is limited by τ_d. The SEW with larger τ_d corresponds larger coherence gain. In order to study the temperas coherence gain of SEWs, we introduce a parameter function as: $\rho_{coh}(k_x) = \dfrac{d\tau_r(k_x)}{dk_x}$, which is shown in Fig.4(b). The physical meaning of $\rho_{coh}(k_x)$ is very clear, which gives a relation

$$\tau_d(k_x) \simeq \rho_{coh}(k_x) \cdot \delta k_x \tag{45}$$

when $\delta k_x \rightarrow 0$, we can get $\tau_d = \rho_{coh}\delta k_x$. Thus, τ_d is proportional to ρ_{coh}. From Fig.4(b) we can see that ρ_{coh} is an increasing function of k_x, so the SEWs with larger integral variable k_x correspond larger τ_d and stronger ECG, and so the field of the SEWs with larger integral various k_x will have better temporal-coherence $g^{(1)}(\tau)$.

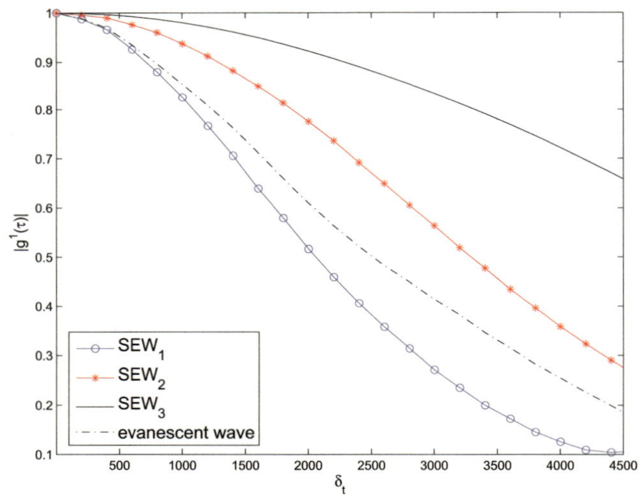

Figure 6. $g^{(1)}(\tau)$ of SEW_1, SEW_2, and SEW_3, comparing with $g^{(1)}(\tau)$ of evanescent wave.

To convince it, three SEWs (SWE_1,SWE_2,SWE_3)with the different integral k_x range $[k_{min} = 1.0\omega/c, k_{max} = 1.15\omega/c]$, $[k_{min} = 1.15\omega/c, k_{max} = 1.3\omega/c]$ and $[k_{min} = 1.3\omega/c, k_{max} = 1.45\omega/c]$ respectively are chosen to calculate the normalized first-order temporal-coherence function $g^{(1)}(\tau)$, which are shown in Fig.6. In this figure, we can see the temporal-coherence of SWE_3 with the largest integral variable is the best (black), the temporal coherence of SWE_2 with the second largest integral variable(red) is the second best, and the temporal coherence of SWE_1 with the smallest integral variable is the worst, which is expected.

5. Conclusion

In conclusion, based on the Green's function, we have numerically and theoretically obtained the evanescent wave, as well as the SEWs, separating from the global field. This study could help us to investigate the effect of an evanescent wave on a metamaterial superlens directly and give us a new way to design new devices.

Author details

Wei Li and Xunya Jiang
State Key Laboratory of Functional Materials for Informatics, Shanghai Institute of Microsystem and Information Technology, Chinese Academy of Sicences, Shanghai 200050, China

6. References

[1] V. C. Veselago, Sov. Phys. Usp. 10, 509 (1968).
[2] J. B. Pendry, Phys. Rev. Lett. 85, 3966 (2000).
[3] J. B. Pendry, Phys. Rev. Lett. 91, 099701 (2003).
[4] N. Garcia and M. Nieto-Vesperinas, Phys. Rev. Lett. 88, 207403 (2002).
[5] G. Gomez-Santos, Phys. Rev. Lett. 90, 077401 (2003).
[6] X. S. Rao and C. K. Ong, Phys. Rev. B 68, 113103 (2003).
[7] Y. Zhang, T. M. Grzegorczyk, and J. A. Kong, Prog. Electromagn. Res. 35, 271 (2002).
[8] L. Zhou and C. T. Chan, Appl. Phys. Lett. 86, 101104 (2005).
[9] Lei Zhou, Xueqin Huang and C.T. Chan, Photonics and Nanostructures 3, 100 (2005).
[10] Nader Engheta, IEEE Antennas and Wireless Propagation Lett. 1,10 (2002).
[11] Ilya V.Shadrivov, Andrey A.sukhorukov and Yuri S.Kivshar, Phys.Rev.E 67,057602 (2003).
[12] A.C.Peacock and N.G.R.Broderick,11,2502 (2003).
[13] R.A.Shelby,D.R.Smith,and S.Schultz, Science 292,77 (2001).
[14] E.Cubukcu,K.Aydin,E.Ozbay,S.Foteinopoulou,and C.M.Soukoulis,Nature 423,604 (2003)
[15] D. R. Smith, D. Schurig, M. Rosenbluth, S. Schultz, S.A.Ramakrishna, and J.B. Pendry, Appl. Phys. Lett. 82, 1506 (2003).
[16] X. S. Rao and C. K. Ong, Phys. Rev. E 68, 067601 (2003).
[17] S. A. Cummer, Appl. Phys. Lett. 82, 1503 (2003).
[18] P. F. Loschialpo, D. L. Smith, D. W. Forester, F. J. Rachford, and J. Schelleng, Phys. Rev. E 67, 025602(R) (2003).
[19] S.Foteinopoulou, E.N.Economou, and C.M.Soukoulis, Phys,Rev.Lett.90,107402 (2003).
[20] R. W. Ziolkowski and E. Heyman, Phys. Rev. E 64, 056625 (2001).
[21] L. Chen, S. L. He and L. F. Shen, Phys. Rev. Lett. 92, 107404 (2004).
[22] L. Zhou and C. T. Chan, Appl. Phys. Lett. 84, 1444 (2004).
[23] Michael W.Feise,Yuri S.Kivshar,Phys.Lett.A 334,326(2005).
[24] R. Merlin, Appl. Phys. Lett. 84, 1290 (2004).
[25] B.I.Wu, T.M.Grzegorczyk, Y.Zhang and J.A.Kong, J.Appl.Phys 93,9386 (2003).
[26] R.Ruppin, Phys.Lett.A 277,61(2000); R.Ruppin, J.Phys.:Condens.Matter 13,1811 (2001).
[27] W. Li, J. Chen, G. Nouet, L. Chen, and X. Jiang, Appl. Phys. Lett. 99, 051112 (2011).

[28] R. G. Hunsperger, Integrated Optics: Theory and Technology (Springer, New York, 1985).
[29] Peijun Yao, Wei Li, Songlin Feng and Xunya Jiang, Opt. Express 14, 12295 (2005).
[30] Xunya Jiang, Wenda Han, Peijun Yao and Wei Li, Appl. Phys. Lett. 89, 221102 (2006).

Section 2

Novel Structures in Optical Devices

Chapter 4

Tunable and Memorable Optical Devices with One-Dimensional Photonic-Crystal/Liquid-Crystal Hybrid Structures

Po-Chang Wu and Wei Lee

Additional information is available at the end of the chapter

http://dx.doi.org/10.5772/48127

1. Introduction

1.1. Concept of defect-mode tunability in a one-dimensional photonic-crystal/liquid-crystal hybrid cell

Photonic crystals (PCs) are a special class of artificial structures with spatially periodic dielectric permittivity and their investigations stem from 1987 when both Yablonovitch and John independently demonstrated their own findings (Yablonovitch, 1987; John, 1987). The most attractive feature of PCs is the existence of photonic bandgap (PBG), characterized by the spatial distribution of refractive index or dielectric constant. The PBG of PCs is analogous to the electronic bandgap of semiconductors, meaning that certain photons will be localized and forbidden in propagation through PCs. According to this characteristic, PCs can be of wide use; thus, various types of PCs have successively been proposed and devised for photonic applications. (Fleming & Lin, 1999; Imada et al., 1999; Knight, 2003; Krauss et al., 2000; Nelson et al., 2000; Park, 1999). If a defect layer is infiltrated in a PC to disrupt its periodicity, partial defect modes that allow the transmission of photons at specific wavelengths will be generated within the PBG. Based on this design of PCs with defect layers, several photonic devices made of two- or three-dimensional PCs have been suggested for lasers (Painter et al., 1999), optical fibers (Knight et al., 1998), and some other applications (Blanco et al., 2000; Chow et al., 2000). Notably, the spectral profile of a PC can be more flexible when the PBG is intrinsically tunable or when the PC is composed of a tunable defect layer.

Liquid crystals (LCs) are anisotropic materials whose physical properties such as electrical and optical anisotropy can be tuned by the electric field, magnetic field, temperature, and the like in that the LC molecules are susceptible to external stimuli. Depending on the molecular orientation, LCs can serve as a phase retardation medium and an optical polarization rotator to manipulate the incoming light via the electrically controlled birefringence and polarization rotation, respectively. As a result, LCs are widely applied to various types of currently used electro-optical devices, especially the information display. As a matter of fact, certain LCs such as the cholesteric LC (CLC), blue phase LC (BP LC) and ferroelectric LC (FLC) can be regarded as PCs due to their periodical orientation of molecular helix. Among them, the PBG of CLCs depends on the LC and chiral parameters. The tunable CLC PCs can thus be realized by adjusting the external factors such as temperature (Morris et al., 2005; Hung et al., 2000), light irradiation (Bobrovsky et al., 2003; Lin et al., 2005), or electric field (Choi et al., 2009; Lin et al., 2005; Yu et al., 2005). Some other tunable PCs based on FLC (Kasano et al., 2003) or BP LC (Yokoyama et al., 2006) have also been demonstrated.

By inserting a LC as defect layer in the PC, tunabilities in the profiles of defect modes are expected. The first tunable PC/LC hybrid structure was developed by Ozaki et al. when they used a planar aligned nematic LC as a defect layer sandwiched between two one-dimensional (1D) periodical multilayers (i.e., dielectric mirrors) and successfully demonstrated the electrically tunable wavelength of defect modes (Ozaki et al., 2002). Their concept of tunability in the 1D PC/LC cell can briefly be described as follows: Figure 1 illustrates the setup for investigation of the transmission spectra of the PC/LC cell. Note that the configuration includes a single linear polarizer in front of the PC/LC cell. The LC with positive dielectric anisotropy is aligned homogeneously along the x-axis whereas the incoming light propagates in the z-axis. The multilayer consists of two dielectric materials, with high- and low-refractive-index layers stacked alternatively. When the incoming light is normally incident to the PC/LC cell with its polarization direction parallel to the LC molecular axis, the optical path length (OPL) is contributed by the sole extraordinary refractive index (n_e). The appearance of peaks in the PBG of the transmission spectrum thus represents extraordinary defect modes. Once an electric field is applied across the cell to align the molecule along the z-axis, the contribution of the ordinary refractive index (n_o) to the OPL leads to the shift of defect modes to the shorter wavelength in the spectrum. On the contrary, the wavelength positions of all defect modes remain unchanged under the field-on and field-off conditions when the polarization direction of the impinging light becomes parallel to the y-axis due to the equal contribution of the OPL. It is concluded that the tunability of defect modes is attributed to the change in refractive index as well as the OPL manipulated by the electric field. After the publication of the milestone paper by Ozaki et al. (Ozaki et al., 2002), the idea has been extended to the 1D PC/CLC color-tunable lasing (Matsuhisa et al., 2006, 2007; Ozaki et al., 2003a, 2005; Park et al., 2009) and fast-response PC/LC structure (Ozaki et al., 2003b). Furthermore, based on this mechanism, several approaches, such as varying the incident angle of light (Arkhipkin et al., 2007) and temperature (Arkhipkin et al., 2008) or using magnetic field (Zyryanov et al., 2008) to adjust the optical anisotropy of LCs, have successively been proposed for realization of tunable 1D PC/LC cells. Particularly, Zyryanov and colleagues further took the investigation of a 1D PC/LC cell to a new stage by setting the hybrid cell between crossed polarizers (Zyryanov et al., 2010). Their results conclude some attractive features for the tunable mechanism of defect modes within the PBG. First, the shift of defect-mode wavelengths depends on the change in effective refractive index (n_{eff}) (Zyryanov et al., 2008, 2010). The decrease in n_{eff} gives rise to the blueshift of defect-mode wavelengths whereas redshift occurs as n_{eff} increases. In addition, a larger number of defect modes can be obtained by using LC with high refractive index or by increasing the thickness of the defect layer, enabling the 1D PC/LC structure to be applicable for the use of a narrow band filter. Moreover, they experimentally and theoretically demonstrated that the transmission tunability of defect modes can be achieved by placing the cell between crossed polarizers to allow "interference" between two orthogonal polarization components through vector sum in the projection direction along the transmission axis of the analyzer. As a result, the transmittance is increased when the defect-mode wavelength of an extraordinary component overlaps that of an ordinary one (Zyryanov et al., 2008).

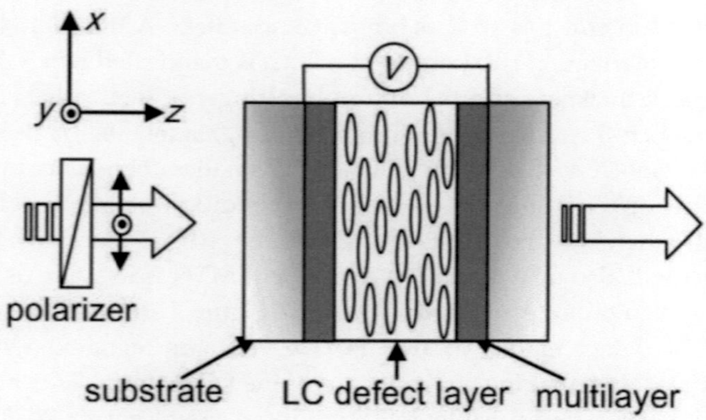

Figure 1. Schematic of a 1D PC/LC hybrid structure in a typical experimental setup.

1.2. Overview of the development of one-dimensional photonic crystals with a dynamic-mode liquid crystal

Current available LC devices can be classified into two categories, dynamic-mode (DM) and memory-mode (MM)LCs, according to their operation as a function of applied electric field. The LC molecules in the DMLC have only one stable state, determined by the condition of the alignment film, and they are continuously reoriented with applied voltage. On the contrary, there are two or multistable states in the MMLC and the stable states can be switched from one to another by the applied voltage. In the DMLC, it can serve as a phase retarder or optical rotator and the light passing them is modified by electrically controlled birefringence (ECB) and polarization rotation effect, respectively. For the 1D PC/LC cells mentioned in Section 1.1, the LC defect layers used are DMLCs acting as phase retarders. The mechanism of the tunable defect modes in these cells can thus be explained in terms of the ordinary and extraordinary refractive indices (Ozaki et al., 2002, 2004b; Zyryanov et al., 2008). On the other hand, the twisted-nematic (TN) LC, in which the molecular orientation exhibits 90° twist, acts as an optical polarization rotator so that the light passing through the TN LC is characterized by the rotation of polarization. The hybrid structure configured by a 1D PC and a TN LC was demonstrated in 2010 (Lin et al., 2010). Several phenomena attributable to the adiabatic following are quite different from the tunability mechanism mentioned in the preceding section. To realize how the adiabatic following enables optical tunability, described here are the spectral characteristics of an electrically controlled 1D PC/TN cell acquired in both the single-polarizer (SP) and crossed-polarizer (CP) schemes.

In the study reported by Lin and coworkers (Lin et al., 2010), the 90° TN LC modes are divided into three groups, depending on the polarization angle β between the transmission axis of the first polarizer (i.e., input polarization) and the director axis (i.e., averaged molecular axis) lying in the front substrate. They are the ordinary-mode (O-mode), extraordinary-mode (E-mode), and mixed-mode (M-mode) TN satisfying the conditions of β = 90°, 0°, and 45°, respectively. It is worth mentioning that the M-mode TN considered by Lin et al. is only a specific one. Generally, the M-mode TN, abbreviated as MTN, combines the polarization-rotation effect as well as birefringence effect. A traditional TN is primarily characterized by its polarization-rotation effect, which is manifested when the multiplication of the mesogenic bulk thickness and the optical birefringence (i.e., optical anisotropy), $d\Delta n$, is larger than the Gooch–Tarry first minimum condition, namely, $(\sqrt{3}/2)\lambda \approx 0.866\lambda$, where λ is the wavelength (Gooch & Tarry, 1975). If $d\Delta n$ is smaller than that condition, then the polarization-rotation effect is incomplete. Another condition of MTN is that $0° < \beta < 90°$ without the requirement of the twisted angle to be 90°. Under such a circumstance, birefringence effect will also take place. The triumph of MTN is its wide use in reflective LC displays, including pico projectors (Wu & Wu, 1996). Figure 2 shows the phenomenon of the wavelength shift of defect modes in two PC/TN cells impregnated with two different nematic LC materials. Note that the birefringence of the LC material E7 is higher than that of CYLC43. Unlike the ECB-based tunable defect modes in some specific PC/LC cells, in which the defect modes for the ordinary-ray are independent of the applied voltage, the defect-mode wavelengths in both the E-mode and O-mode PC/TN cell show blueshift when the applied voltage increases. Compared with the O-mode configuration, the E-mode has a more perceptible shift in wavelength (blueshift) due to significant change (decrease) in effective refractive index, dedicated by the unwinding process of the molecular helix. It is also clear from Fig. 2 that the extent of buleshift for E7 is greater than that for CYLC43. This result can be explained by their Mauguin parameters (Gooch & Tarry, 1975)

$$u = \frac{2d\Delta n}{\lambda}, \tag{1}$$

giving superior adiabatic following capability of E7 (Lin et al., 2010). A perfect adiabatic following in the TN LC cell would enable the linearly polarized light to traverse the LC layer with the rotation of the molecular twist, which makes n_{eff} for the incident beam nearly equal to n_e in E-mode and n_o in O-mode. In fact, the elliptic polarization can hardly be avoided in the TN LC (Yeh & Gu, 1999a); thus, n_{eff} is no longer a constant in the O-mode TN LC but becomes a weak function of applied voltage. As a result, small shifts for the defect modes are observed in the O-mode PC/TN cells.

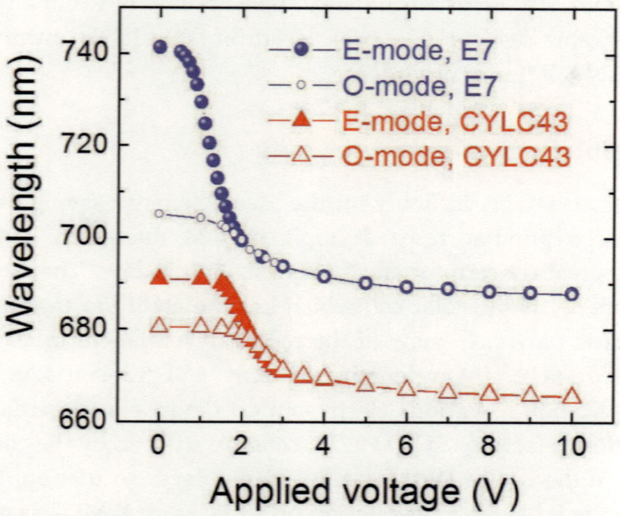

Figure 2. Voltage-induced blueshift of defect modes for two comparative 1D PC/LC cells with distinct defect materials (adapted from Lin et al., 2010).

Figure 3 demonstrates an integrated effect on the transmission of defect modes in the M-mode PC/TN cell. One can obviously identify that the peaks of the M-mode spectrum are located at the exactly same positions of the E-mode and O-mode peaks in wavelength. This result indicates that each defect mode in the M-mode PC/TN is contributed by both the adiabatic following and birefringence effects. Moreover, the intensity of the transmitted light in either E- or O-mode spreads to the other, making the integrated intensity of the peaks in the M-mode almost the same as that in either E- or M-mode. This implies that the M-mode spectrum is a superposition of those of both E-mode and O-mode. Interestingly, while looking carefully into the spectra of E- and O-mode, small satellite peaks accompanying the defect modes are observed, presumably due to the unavoidable elliptic polarization of light in a conventional TN cell. This phenomenon has left untouched in other types of 1D PC/LC cells. Recently, it has been explained by the Mauguin condition violation and the coupling between the slow semi-longitudinal mode (i.e., twisted extraordinary mode) and fast semi-transverse mode (i.e., twisted ordinary mode) PC/TN cell (Timofeev et al., 2012).

In regard to a typical 90° TN cell, the linearly polarized light rotated approximately by 90° by the cell can almost "completely" pass the rear polarizer; i.e., the analyzer. There is no doubt that the spectra of the E-mode as well as the O-mode in the CP scheme (configured by two linear polarizers whose transmission axes are orthogonal to each other) are nearly identical to their corresponding ones under the SP scheme. Figure 4 reveals a distinctive profile in the spectrum for the M-mode because the light passing through the cell becomes non-linearly polarized light. Particularly, an outstanding peak, whose intensity is the strongest among the vast of peaks within the PBG, is located near 700 nm as shown in Fig. 4. This unique feature enables to extend the use of the PC/TN cell in the application of a monochromatic selector. In accordance with the simulation results published in the literature, this remarkable defect-mode peak is attributable to the intrinsic transmission characteristic of a MTN cell (Lin et al., 2010).

1.3. Aim of this chapter

Recently, electro-optical devices in line with the idea of energy saving and/or low power consumption become a popular research topic in that the green concept is globally promoted due to the great concern for energy shortage nowadays. The most representative one for alternative energy is the solar cell which has the ability to transfer natural energy from the sun to electric power. In view of the recent development in 1D PC/LC cells, the demonstrated features such as wavelength tunability and transmission tunablity enable their use for the application in various electro-optical devices, as described in Section 1.2. However, aforementioned features in 1D PC/LC cells are realized by the continuous-varying of electric field due to the use of DMLCs as the defect layer so that applications in green products are very limited. Lately, a new design of 1D PC with MMLC as a defect layer that brings the notion of multistability in defect modes has been demonstrated (Hsiao et al., 2011a, 2011c; Wu et al., 2011). In the 1D PC/MMLC cell, the spectral properties of defect modes in a memory state persist without applied voltage; such a cell supports a pathway for designing photonic devices with green concept. To interpret how the 1D PC/MMLC operates, this chapter reviews some previous works primarily done by this group and builds up the following three sections in the main body of this article to explicitly clarify the characteristics of such 1D PC/MMLC cells.

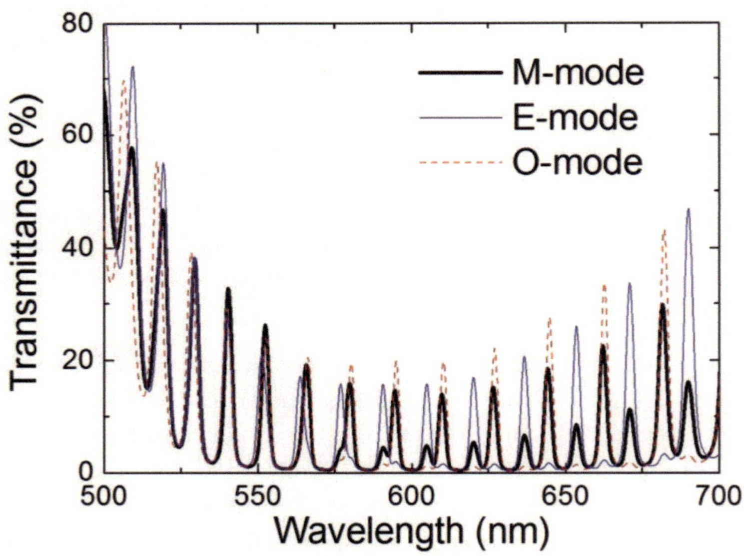

Figure 3. Transmission spectra of a PC/TN cell in the photonic bandgap in three different modes. The M-mode, E-mode and O-mode are represented by the thick solid line, thin solid line and dashed line, respectively (adapted from Lin et al., 2010).

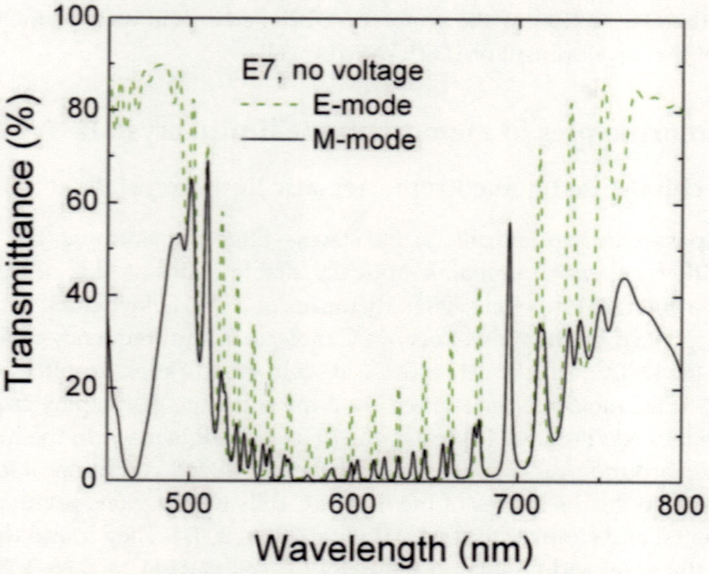

Figure 4. Transmission spectra of a 1D PC/TN cell at null voltage under crossed polarizers (adapted from Lin et al., 2010).

Among the recent development of MMLCs, the bistable or multistable cells using dual-frequency LCs (DFLCs) enable the switching between optically stable states by applying frequency-modulated voltage pulses. The DFLC is a kind of LC material whose sign of dielectric anisotropy ($\Delta\varepsilon$) can be varied by the frequency of an externally applied electric field (Xianyu et al., 2009). The DFLC has a certain crossover frequency (f_c) to discriminate the behavior of $\Delta\varepsilon$. While the frequency is lower than f_c, the $\Delta\varepsilon$ value is positive. Or it becomes negative if the frequency is higher than f_c. Based on this mechanism, various types of MMLC cells composed of DFLC are demonstrated in the literature (Hsiao et al., 2011b; Hsu et al., 2004; Jhun et al., 2006; Yao et al., 2009). Among them, the bistable chiral-homeotropic nematic (BHN) LC and dual-frequency cholesteric LC (DFCLC) have been used as defect layer in the 1D PC/MMLC cell (Hsiao et al., 2011a, 2011c; Wu et al., 2011). As such, Section 2 details the cell configuration and operation principles of these two MMLC modes so that one can grasp the switching mechanisms in the description in the next two sections. The configuration of a 1D PC/MMLC cell, including the design of multilayers is schematically depicted in Session 3. In addition, the optical properties and the tunability in the defect modes of the cell switching among the stable and voltage-sustained states are reported in Session 3 as well. To realize a low-power-consumption optical device, in Session 4, the characteristics of defect modes in various memory states of both PC/BHN and PC/DFCLC cells are further discussed. The features of defect-mode switching between two stable states are confirmed. Particularly, a new scheme of a tristable PC device based on DFCLC is demonstrated using a polymer-stabilized cholesteric texture (PSCT) as a defect layer (Hsiao et al., 2011). Finally, the key findings of the properties of the 1D PC/MMLC cells discussed in the preceding sections are summarized. In accordance with the concluding remarks, suitable device applications of the tunable 1D PC/MMLC hybrid structures are suggested in the last section of the chapter. A brief note will also be included on future perspectives of the development of 1D PC/MMLC cells.

2. Operation principles of memory-mode liquid crystals

2.1. Bistable chiral-tilted homeotropic nematic liquid crystals

The BHN LC possesses two optically stable states—tilted-homeotropic (tH) and tilted-twist (tT) states, either of which remains optically stable without the need of continuous application of a voltage (Hsu et al., 2004). By means of a DFLC, switching between these two states is accomplished by the flow effect of LC molecules and frequency-revertible dielectric anisotropy of the DFLC. To fabricate a BHN LC cell, the thickness-to-pitch ratio d/p and the pretilt angle θ of LC molecules, measured from the substrate plane, play crucial roles in the stability of the tH and tT states. Typically, the BHN LC cell is made in the high-pretilt-angle regime with d/p around unity. To understand the optimized condition of d/p and θ, Liang and Lin determined the d/p ranges of the BHN LC cells with various pretilt angles based on their experimental and simulation results (Liang & Lin, 2007). They found that the d/p range for obtaining stable tH and tT state in the BHN LC cell is 0.6–1.14, 0.68–1.21, and 0.82–1.24 when the pretilt angle is equal to 62°, 72°, and 80°, respectively. In additional to the conventional (0, 2π) BHN LC, the two stable states can also be found in the ($-\pi/2$, $-3\pi/2$) BHN LC under the condition of $\theta = 74°$ (Hsu, 2007).

According to the above-mentioned results, Fig. 5 illustrates the cell configuration of the BHN LC. The BHN cell is composed of a DFLC doped with a suitable concentration of a chiral agent sandwiched between two indium–tin-oxide (ITO)-coated glass substrates covered with alignment films. To achieve necessary tilted-homeotropic molecular orientation with a proper pretilt angle, one can simply coat a homeotropic polyimide or a mixed solution of homeotropic- and planar-alignment polyimide on the substrates as the aligning layers with the treatment of mechanical rubbing on the top and bottom substrates in anti-parallel direction. While the adaptable range of the pretilt angle is very limited in a BHN LC cell with homeotropic alignment, cells with mixed alignment enable the tunable pretilt angle in a wider range by adjusting its composition (Yeung et al., 2006).

On the basis of the switching mechanism of the BHN LC (Hsu et al., 2007; Liang et al., 2008), a brief illustration, shown in Fig. 6, is provided to summarize the operation between the two stable states in a BHN LC cell. Here, f_1 and f_2 are the frequencies satisfying the conditions of $f_1 < f_c$ and $f_2 > f_c$, and hence correspond to the positive and negative dielectric anisotropy ($\Delta\varepsilon$) of the DFLC, respectively. The solid and dash lines represent the texture transitions from tT to tH and tH to tT, respectively. Consider a tT state stabilized in the cell. The LC molecules will response to align perpendicular to the substrate when an electric field with frequency f_1 is applied vertically to the cell because the LC exhibits positive dielectric anisotropy ($\Delta\varepsilon > 0$). The molecules are then oriented homeotropically (i.e., vertically) in the cell at high voltage. This voltage-sustained state is called biased homeotropic (bH) state. As the electric field is switched off, the molecules relax to the tH state with very high pretilt angle. Accordingly, the switching from tT to tH state follows a transition process of tT–bH–tH. To switch the cell from tH to tT state, a connecting pulse with frequencies f_1 and f_2 is used. Note that the second pulse with frequency f_2 instantly follows the first one with frequency f_1. Once the connecting pulse is applied to the cell, the cell first induces the pulse with frequency f_1 which results in the switching of cell from tH to bH state. As the frequency of the connecting pulse changes from f_1 and f_2, where the dielectric anisotropy of the LC becomes negative ($\Delta\varepsilon < 0$), the dielectric coupling between the electric field and LC leads the molecules to align parallel to the substrate and induces flow effect. The cell is thus switched from bH state to another voltage-sustained state, called biased twisted (bT) state due to the backflow effect and applied electric field. The molecular in the bT state rotates about 2π in the azimuthal angle. The bT state subsequently relax to the stable tT state as the driving pulse ends. Both bT and tT state have 2π-twist molecular orientation but the tilted angle in bT state is higher than that in the tT state. As a result, a transition process of tH–bH–bT–tT is required within the switching from tH to tT state. It is worth reminding that the tH and tT states are of optical stability.

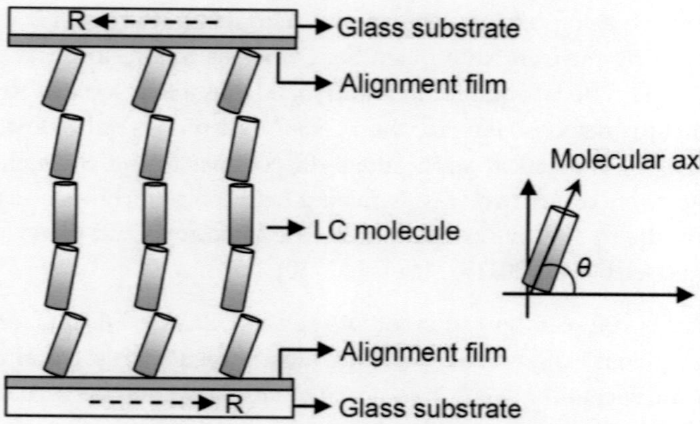

Figure 5. Schematic illustration of the configuration of a BHN LC cell. R, the rubbing direction.

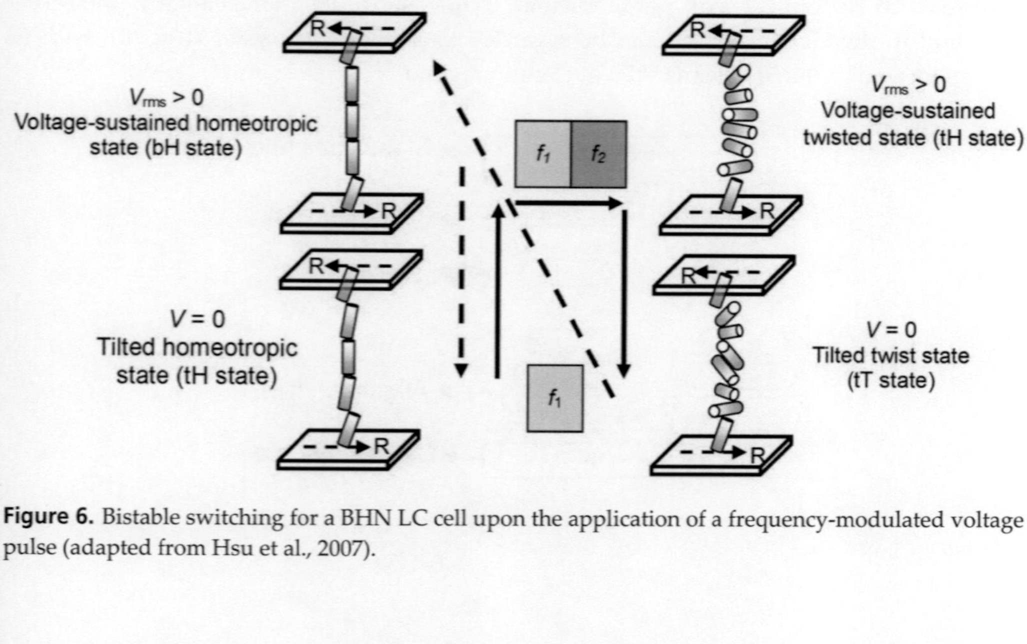

Figure 6. Bistable switching for a BHN LC cell upon the application of a frequency-modulated voltage pulse (adapted from Hsu et al., 2007).

2.2. Dual-frequency cholesteric liquid crystals

The optical bistability is one of the unique features in cholesteric LC (CLC) and its applications in electro-optical devices (Bao et al., 2009; Berreman & Heffner, 2006; Huang et al., 2003; Hsiao et al., 2011b; Xu & Yang, 1997). CLCs, also termed chiral nematic LCs because of their structural nature to be the chiral versions of the nematic molecules, have been widely investigated in the literature. In a typical bistable CLC cell, the two stable states are the transparent planar (P) state and light-scattering focal conic (FC) state. Switching between these two states can readily be achieved by adjusting the amplitude of applied electric field. For instance, when an electric field in square wave with a specific amplitude value is applied to the CLC cell, the texture will be changed from P to FC state. Further increasing the voltage over a critical value results in the transition of texture from FC to homeotropic (H) state. The texture in the H state can be switched back to the P state by turning off the field rapidly or to the light-scattering FC state by turning off the field slowly. As a result, the P-to-FC transition can be accomplished directly by the external applied field whereas the FC-to-P transition is indirect since an intermediate H state is required. In addition, it takes a few seconds to relax from H to P state, meaning that the response time for the FC-to-P transition is quite slow. This drawback limits the CLC device for practical applications. In contrast to the conventional CLCs, the DFCLC device empowers direct two-way switching between two cholesteric states due to its material property of the frequency-revertible dielectric anisotropy and thus yields a feature of fast response speed (Hsiao et al., 2011b; Ma et al., 2010).

Figure 7 depicts the cell configuration of a typical CLC or DFCLC cell prepared in homogeneous (i.e., planar) alignment. A (DF)LC host doped with a chiral agent in proper concentration is sandwiched between two conductive glass substrates coated with planar alignment films. The rubbing directions of the top and bottom substrates are set to be antiparallel. As a consequence, a planar texture is initially formed, exhibiting periodic helix structure with its optical axis perpendicular to the substrate plane. Such a molecular orientation in the CLC cell itself can be regarded as a photonic crystal structure with its bandgap located within a designated wavelength regime.

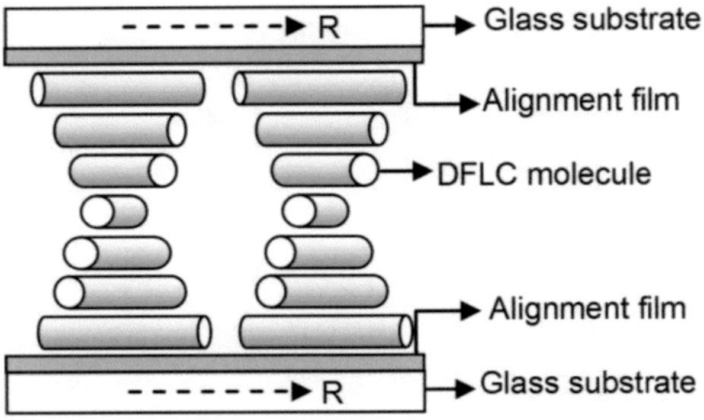

Figure 7. Schematic of the molecular configuration of a planar-alignment DFCLC cell. The helix axis is vertical in the initial state.

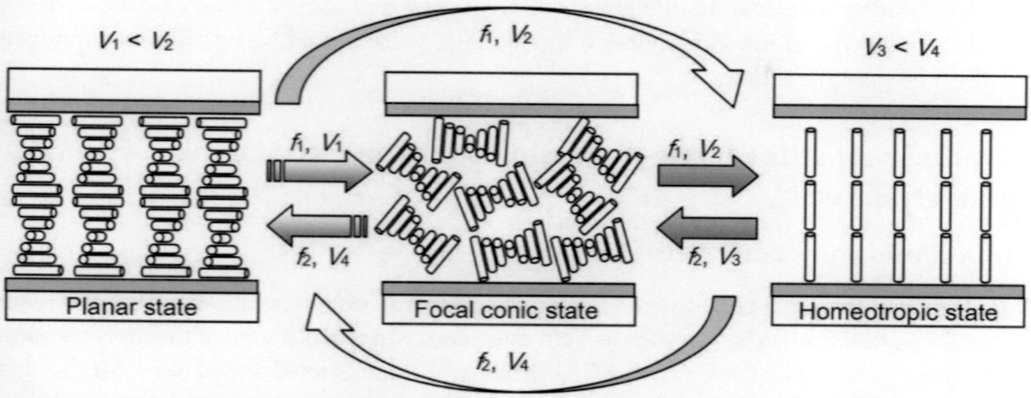

Figure 8. Operation principle and texture transitions of a DFCLC cell under the application of various voltage pulses (adapted from Hsiao et al., 2011a).

To clarity how the cholesteric textures are switched by frequency-modulated voltage pulses, Fig. 8 illustrates operations among three cholesteric states (i.e., P, FC, and H) in a DFCLC cell (Hsiao et al., 2011a). It should be noted here again that, when a voltage pulse is applied to the DFCLC cell, the dielectric anisotropy is positive at low frequencies (f_1) and negative at high frequencies (f_2) beyond the temperature-dependent crossover frequency (f_c). Once a voltage pulse with frequency f_1 is applied vertically to the DFCLC cell, the dielectric coupling leads the molecules to reorientation in parallel to the field direction and the helix tends to be unwound. Consequently, direct P-to-FC and P-to-H switchings can be induced at voltages V_1 and V_2, respectively. The switching mechanism of the DFCLC cell in the increasing-voltage process is the same as that of a conventional CLC one. It is worth mentioning here that while applying a pulse with frequency f_2, at which the molecules incline to align themselves homogeneously with respect to the substrate plane, the backward switching of the FC-to-P and H-to-P transitions can directly be accomplished at voltage V_3 and V_4, respectively. Noticeably, reversible direct-switching between FC and H state can also be achieved by individually applying a voltage pulse at voltage V_2 at f_1 and f_2. Note that V_2 is higher than V_1 ($V_2 > V_1$) and V_4 is higher than V_3 ($V_4 > V_3$) because switching between P and H states requires more energy to make the molecules reach balance. In the bistable DFCLC device, the P and FC textures are essentially the stable states whereas the H texture is a voltage-sustained state. Upon incorporating a certain content of photo-curable monomer (or prepolymer) into DFCLC, another type of CLC device called polymer-stabilized cholesteric texture (PSCT), can be created after adequate polymerization. In such a PSCT cell, the stabilization of polymer network throughout the bulk cell enables the possession of stable H state (Ma et al., 2010). Combining the switching mechanism of DFCLC with PSCT, a novel tristable PSCT in which the texture can permanently be stabilized optically in the P, FC, and H states, has been demonstrated very recently (Hsiao et al., 2011a). While thinking about the use of the DFCLC in practical applications there are some issues that should be taken into consideration. The common drawback of the DFCLC is its considerable sensitivity to temperature. Typically, high operation voltage results in dielectric heating, which shifts the crossover frequency to a higher value (Yin et al., 2006; Wen & Wu, 2005). On the other hand, a low-voltage pulse would be insufficient to induce the direct FC-to-P transition.

3. Optical properties of one-dimensional photonic-crystal/memory-mode-liquid-crystal cells

3.1. Definition of the cell configuration

A 1D PC structure incorporated with LC as a defect layer is well known to have the ability on tuning the optical characteristics such as light intensities and wavelengths of the defect modes. While most of investigations of 1D PC/LC cells are demonstrated based on DMLCs, this section is dedicated to a new classification of 1D PC/LC cells formed by infiltrating with a MMLC as a defect layer. The configuration of a 1D PC/MMLC hybrid structure is schematically depicted in Fig. 9 (Hsiao et al., 2011c). Referring to the conventional 1D periodical structure, the 1D PC here is constructed with two identical dielectric mirrors having multilayer films deposited on the electrically conductive glass substrates. The MMLC, for example, the DFCLC whose electro-optical characteristic exhibits bistable or multistable switching, is infiltrated into the 1D PC as a defect layer with a thickness d. The dielectric multilayer consists of a number of high- and low-index materials stacked alternately.

To provide high reflectivity, the OPL of these two materials is theoretically set to be equal to one-quarter wavelength of the incident light, i.e., $n_H d_H = n_L d_L = \lambda/4$,where n_H, n_L, and d_H, d_L are the refractive indices and thicknesses of the high- and low-index materials in a multilayer, respectively, and λ is the center wavelength in the PBG. Note that the reflectivity of the cell can further be enhanced by additionally coating a high-index film on the top of the multilayer. In other words, the total number of layers is odd (Hecht, 2002). The width of the PBG can be enlarged by increasing the refractive-index difference ($n_H - n_L$) and shifting the central wavelength to redder regime. Figure 10 compares the trans-

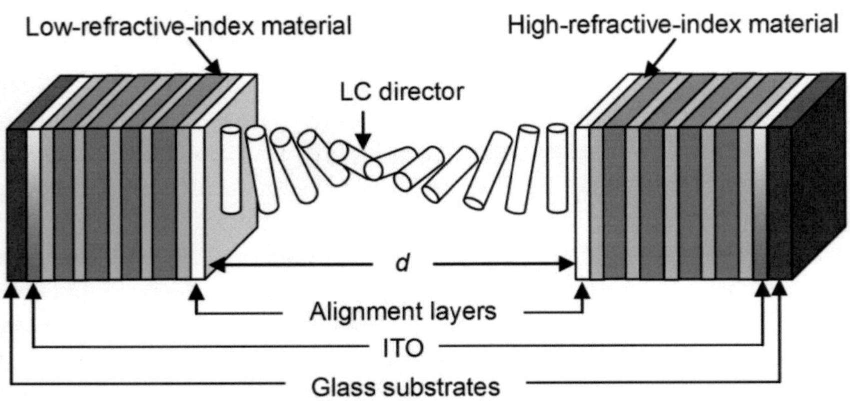

Figure 9. Cell configuration of a 1D PC/MMLC hybrid structure. The exemplary defect layer is a DFCLC (adapted from Hsiao et al., 2011c).

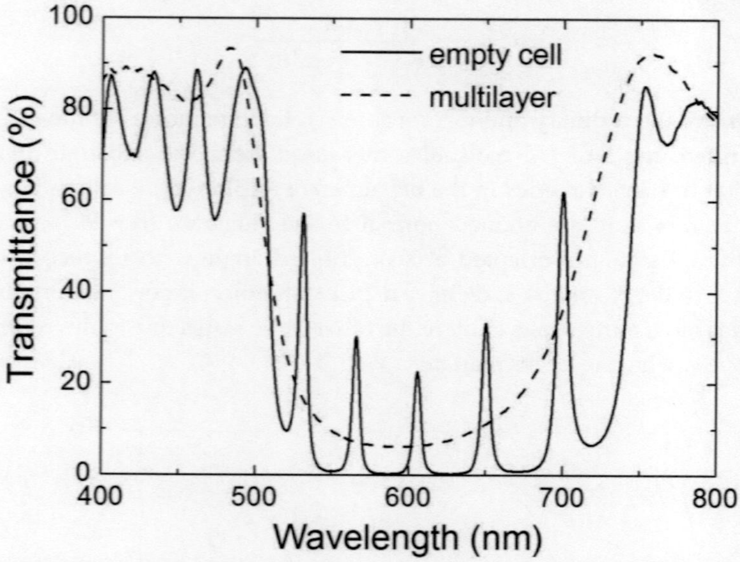

Figure 10. Transmittance of a dielectric mirror and an empty PC cell made of two identical dielectric mirrors.

mission spectra of a multilayer film on a glass substrate and a combination of two multilayer-coated substrates with a central air gap (empty cell). In this example, the multilayer film has nine layers, including five high-index layers and four low-index layers. Obviously, the width of PBG is around 250 nm with the central wavelength near 620 nm in the single multilayer. While sandwiching a defect layer between two multilayers, a number of defect modes are generated within the PBG. Note that the insertion of the defect layer does not essentially affect the width of the PBG. The profile of defect modes is influenced by some parameters. Ozaki and associates report that increasing the layer number in the multilayer decreases the line width and transmittance of defect-mode peaks because the Bragg-reflected light, dictated by the multilayer, is limited to reach the defect layer (Ozaki et al., 2004a). On the other hand, it is concluded by Zyryanov and coauthors that the number of defect modes is increased with increasing thickness of the defect layer or using the material with high refractive index as a defect layer (Zyryanov et al., 2008).

3.2. One-dimensional photonic-crystal/bistable-chiral-tilted-homeotropic-nematic cells

According to the operation of a BHN LC discussed in Section 2.1, applying an external electric field to the PC/BHN hybrid cell enables the switching among four specific states; i.e., tH, bH, tT, and bT states. Due to different molecular orientation of LC as well as the contributions of effective refractive index (n_{eff}), the transmission spectra of the 1D PC/BHN hybrid structure in four different states without employing any polarizer disclose dissimilar spectral profiles, as shown in Fig. 11. Recalling the mathematical expression of the effective refractive index of an optically uniaxial material (Yeh & Gu, 1999b)

$$n_{eff} = \frac{n_o n_e}{(n_o \cos^2\theta + n_e \sin^2\theta)^{1/2}}, \tag{2}$$

where n_o and n_e are the ordinary and extraordinary refractive indices of the LC, respectively and θ is the tilted angle of LC molecules measured from the substrate plane, one can comprehend that the defect modes in the bH state are attributed by sole n_o because most of LC molecules in this state are aligned normal to the substrate (θ = 90°). In contrast, the molecules in the tH state are oriented at a high tilted angle with respect to the substrate plane whereas both the tT and bT state have a 2π-twist molecular orientation. Consequently, the defect modes have more peaks in all of the other three states due to the contribution of n_e to the resulting n_{eff}, which is larger than n_o.

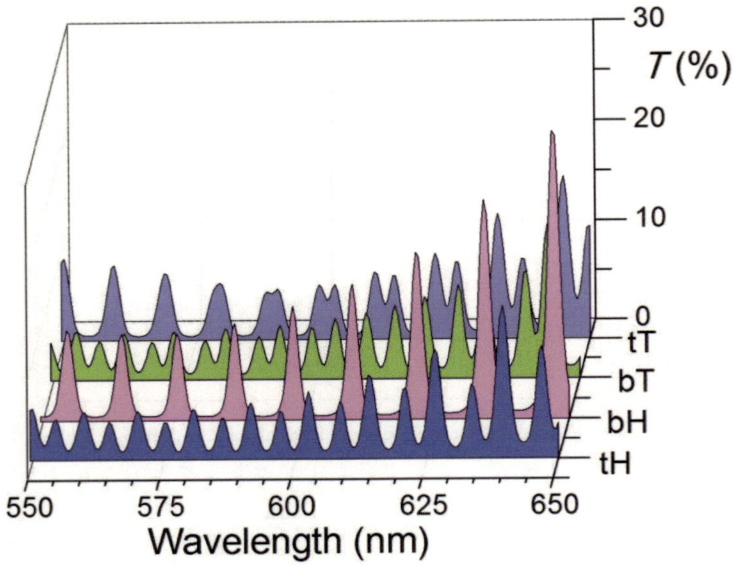

Figure 11. Transmission spectra of the four states of the PC/BHN LC cell within the photonic bandgap.

Similar to the tunable mechanism in the 1D PC/DMLC device, the tunability in defect modes of the PC/BHN cell can be realized by dynamic switching between two proper states. In the case of two homeotropic states, Fig. 12 shows defect modes of the tH and bH state in specific wavelength range without employing any polarizers. In the tH state, the peaks overlapped with those in the bH state characterizes the ordinary defect modes whereas other peaks are explained as the extraordinary defect modes. According to Eq. (2), the effective refractive index in the tH state is higher than that in the bH state, determined by their molecular orientation. In addition, it can also be conceived that operation between tH to bH state enables the control on the contribution of extraordinary index as well as the effective refractive index by the applied voltage. Accordingly, the transmission-intensity tunability in the extraordinary defect modes can be achieved by switching between these two states. For instance, the switching between tH and bH state is created by applying an electric field at low frequency f_1 (see Section 2.1). Considering a forward switching from tH to bH state, the intensity of peaks in the extraordinary defect modes is reduced with increasing voltage due to the decrease in the effective refractive index. The intensity then reaches minima as the tH state transfers completely to the bH state. Similarly, backward switching from bH to tH state reproduces the original intensity of peaks.

Furthermore, Fig. 13 reveals the wavelength tunability of the defect modes in the bT state of the PC/BHN cell. Switching to the bT state is accomplished from bH state by adjusting the frequency from low frequency f_1 to high frequency f_2 (see Section 2.1). In the bT state, the tilted angle of molecular director is higher than that in the stable tT state, as illustrated in Fig. 6. Thus, while increasing the high-frequency voltage, the tilted angle is reduced and increases the contribution of n_e component to the overall effective refractive index. As a result, redshift (Miroshnichenko et al., 2008a, 2008b; Zyryanov et al., 2008a) of the defect modes as a function of increasing voltage is observed, demonstrating the ability of defect mode tunability.

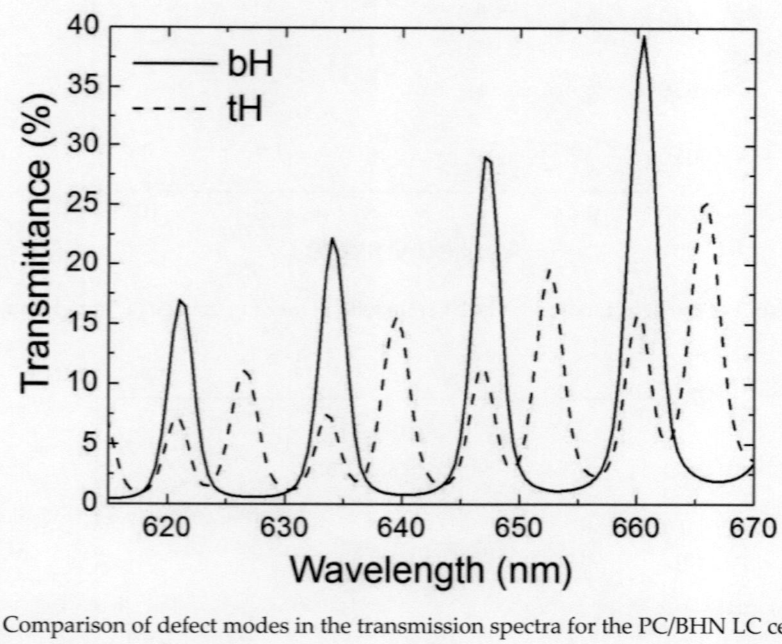

Figure 12. Comparison of defect modes in the transmission spectra for the PC/BHN LC cell in tH and bH state.

3.3. One-dimensional photonic-crystal/dual-frequency-cholesteric-liquid-crystal cells

The texture PC/DFCLC cell can be switched to three states, according to its operation mechanism illustrated in Fig. 7. These textures are P, FC, and H state. While the P and FC can be the stable states H state is the voltage-sustained state. Figure 14 shows the transmission spectra of the PC/DFCLC cell in two stable states. The cell in the transparent P state reveals a number of defect modes with transmission intensity around 25 to 60% in the

PBG. When the texture is switched from P to FC state by a voltage pulse of $V_1 = 20\ V_{rms}$ at f_1 = 1 kHz, the intensity of defect modes drops dramatically to about 1% due to the formation of randomly oriented poly-domain in this state. Accordingly, the bistble switching between P and FC state results in the tunable defect modes between turning on and off state that enable the PC/DFCLC to serve as an electrically tunable light filter with features of fast switching speed and low power consumption. When the cell is switched between P and H state the tunability on the wavelength of the defect modes can be realized.

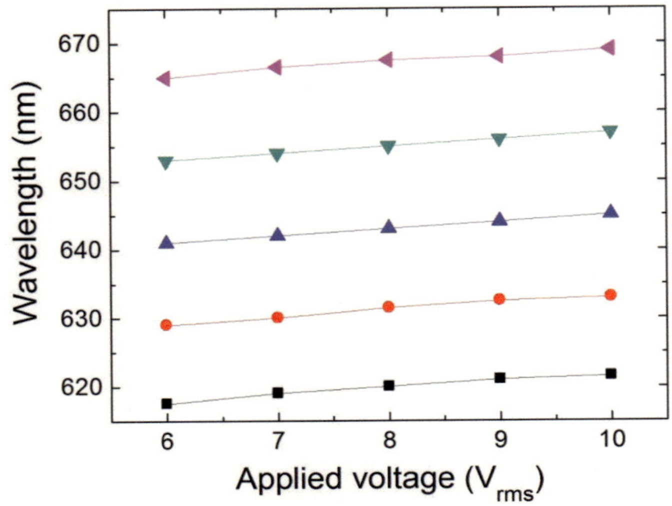

Figure 13. Redshift of the defect modes in the bT state with increasing voltage (adapted from Wu et al., 2011).

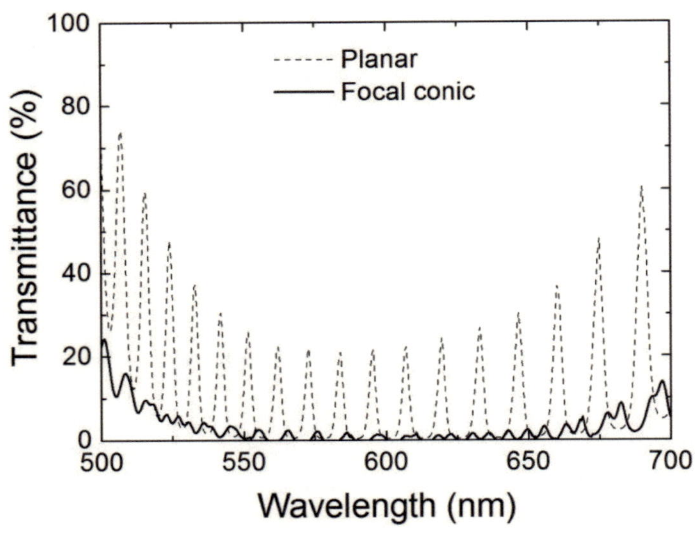

Figure 14. Transmission spectra of a PC/DFCLC in the photonic bandgap in two stable states (adapted from Hsiao et al., 2011c), demonstrating light-intensity tunability.

Figure 15 shows the transmission spectra of the PC/DFCLC cell in the P and H state. It can clearly be recognized that two separated sets of defect modes with comparable transmission strengths are obtained. In contrast to the defect modes in the P state, the defect modes in the H state shifts to the shorter wavelengths due to the decrease in the effective refractive index. Noticeably, the peaks of defect modes in one of the two states overlaps to the stop band in the other state indicating the complementary nature in wavelengths. Furthermore, Fig. 16 demonstrates a new approach for tuning the peak transmittance of the defect modes by varying the frequency of the applied voltage pulse. In this example, it is noted that the dielectric anisotropy of the DFLC is negative and it increases with increasing frequency in the frequency range from 20 to 100 kHz. Therefore, while a 20 kHz voltage pulse align the molecules in the FC state the cell subsequently transits to the planar state as the frequency increases due to the enhancement in the torque of dielectric coupling. As a result, the transmittance of peaks in the defect modes increases with increasing frequency due to the change of molecular orientation.

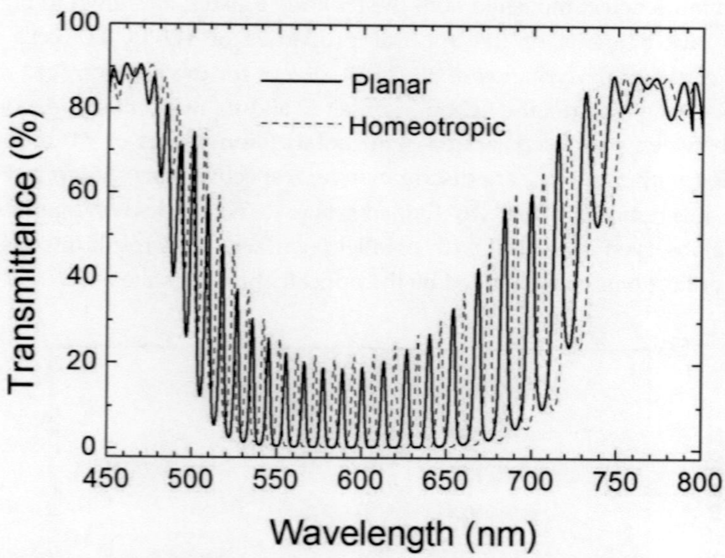

Figure 15. Transmittance of the PC/DFCLC in the photonic bandgap with two different sets of defect modes in the P and H states (adapted from Hsiao et al., 2011c).

4. Memorable multichannel devices based on one-dimensional photonic-crystal/memory-mode-liquid-crystal cells

4.1. Bistable one-dimensional photonic-crystal/chiral-tilted-homeotropic-nematic cells under parallel polarizers

It is clarified from Section 3.2 that the electrically tunable defect modes in a 1D PC/BHN hybrid cell can be realized even for the light propagating through the cell without employing any polarizers. For instance, the transmittance tunabiltiy in the extraordinary defect modes is performed in the cell while switching between the tH and bH states, suggesting a potential application in light filter. Moreover, operation in bT state as a function of increasing voltage results in the tunable wavelength, shifting to the longer wavelengths, in the defect modes. Although above mentioned mechanisms enable the PC/BHN to extent its use in various optical devices without employing any polarizers, all of them are performed by the dynamic switching between one stable state and voltage-sustained state. Here, a concept of wavelength-tunability in defect modes, characterized by the two stable states of the PC/BHN cell is proposed by setting the cell between two parallel polarizers (meaning linear polarizers with parallel transmission axes). Figure 17 depicts the transmission spectra of the PC/BHN cell in stable states with and without parallel polarizers. The angle between the rubbing direction of the cell and the transmission axis of either polarizer is denoted as β. While the incoming light passing through first polarizer at $\beta = 0°$ represents as E-ray, it becomes O-ray at $\beta = 90°$. It is known from Section 3.2 that the effective refractive index in the cell without any polarizers is contributed by both n_o and n_e; thus, forming minute defect modes in both the tH and tT states, as shown in Fig. 17(a). Along with pervious investigations of the spectral properties of 1D PC/LC cells with crossed polarizers (Lin et al., 2010; Zyryanov et al., 2010), except for the homeotropic state, the o-ray and e-ray propagating through the LC bulk senses n_o and n_e, respectively. Accordingly, when the cell is set between parallel polarizers with polarization angles of 90° and 0°, the defect modes associated with n_o and n_e are discriminative, respectively, as shown in Figs. 17(b) and (c). In addition, it is noticeable that two divided sets of defect modes corresponding to the two stable states are observed in the cell with parallel polarizers. This result implies a concept for multichannel applications, characterized by the optically bistable states.

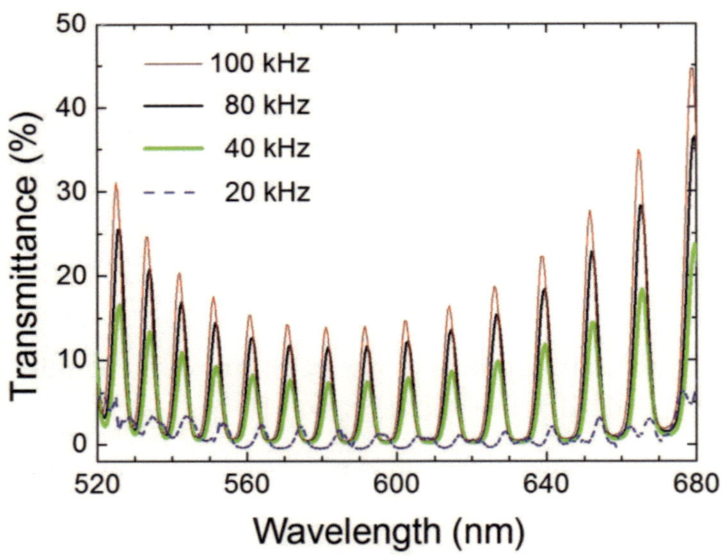

Figure 16. Transmittance of stable defect modes in the PBG induced by a 24.5-V voltage pulse at various frequencies.

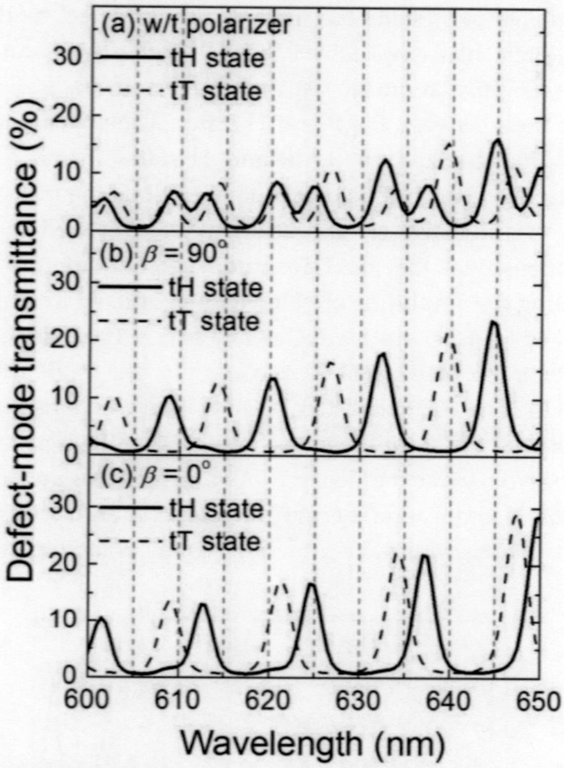

Figure 17. Transmittance of a PC/BHN cell (a) without polarizers and with parallel polarizers at (b) β = 90° and (c) β = 0°.

4.2. Tristable one-dimensional photonic-crystal cell with polymer-stabilized cholesteric textures

Based on the cell configuration of the DFCLC cell, a tristable PSCTs cell is created by incorporating photo-curable monomers into DFCLC with proper amounts due to the distribution of polymer networks throughout the cell. These three PSCTs are also referred to as the P, FC, and H states. According to the operation principle illustrated in Fig. 8, switching from one to another state in the PSCTs can be achieved by applying suitable frequency-modulated voltage pulses. In contrast to the electrical tunability mechanisms in the 1D PC/DFCLC cell, the 1D PC structure infiltrated with tristable PSCTs is expected to characterize the tunability on intensity and wavelength of defect modes by its tristable states. Figure 18 demonstrates the transmission spectra of the 1D PC/PSCTs cell in three stable states. Since the spectra profiles in these three states is revealed in Section 3.3, it can be understood from Fig. 18 that the mechanisms of tunable intensity and wavelength in the defect modes can also be performed in the PC/PSCT cell by operating between two stable states. In the case of intensity tunability, the FC state is certainly represented as the light off state because the transmission of defect modes in this state is the lowest. While switching the cell from FC to either P or H state, the transmission of defect modes becomes intense, denoting the light on state. It is emphasized that all of the three states can be stabilized permanently after voltage pulse removal. On the other hand, the position of defect modes can be shifted by switching between the P and H states. The appearance of two individual sets of defect modes, characterized by the P and H states in the cell, is applicable for enhancing the performance of the multichannel device. While the transmission of the peaks in defect modes can be adjusted by frequency-modulated voltage pulses with fixed amplitude, Fig. 19 demonstrates that the transmittance-modulation in the defect modes is performed by modulating the amplitude of voltage pulse at fixed frequency. In this case, the initial state of the cell is H state and the frequency is f_2 = 100 kHz, corresponding to the negative dielectric anisotropy of the DFLC. Accordingly, the PC/PSCTs cell is transferred from the initial P state to H-to-FC mixed state and FC state when the voltage increases from 0 to 40V. If the initial state is P state, it is comprehensible that the transmission tunability can be achieved by varying voltage at frequency f_1. As a result, this powerful photonic device has potential to extend the application, permitting its use as an electrically controllable and optically tristable multichannel filter without requiring any polarizers.

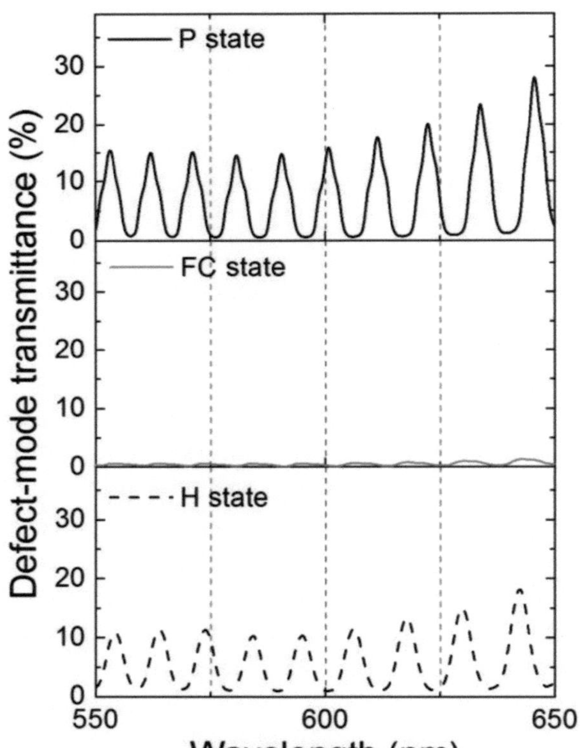

Figure 18. Transmission within the photonic bandgap of a PC/DFCLC cell in three stable states (adapted from Hsiao et al., 2011a).

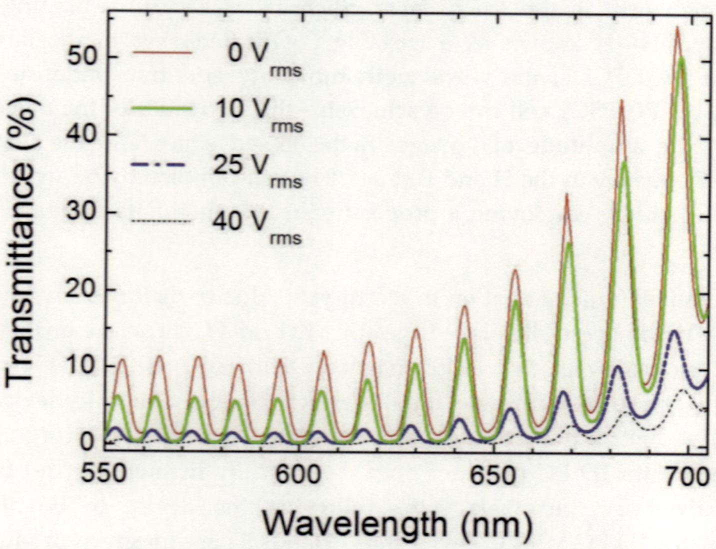

Figure 19. Transmittance of stable defect modes in the photonic bandgap induced by a voltage pulse at frequency f_l with various amplitudes, (adapted from Hsiao et al., 2011a). Note that the texture in the cell is transited from planar to focal conic state with increasing voltage.

5. Conclusions

In this chapter, two types of 1D PC/MMLC cells which exhibit both electrical tunability and optical bi- or tri-stability in the defect modes have been reviewed in accordance with our previously published papers. Several fascinating features have also been keynoted. In the case of the PC/BHN cell based on a chiral-agent-doped DFLC infiltrated as a defect layer, it can be switched in four different states by applying voltage pulses with designated waveforms. The dynamic switching in the voltage-sustained bT state results in the redshift of the defect modes with increasing voltage (at 100 kHz) due to the increase in the effective refractive index. Moreover, the tunable defect modes in the PC/BHN cell can be achieved by their two stable states (bH and bT) when the cell is set between parallel polarizers. Such bifunctional photonic devices pave a new pathway for the application in low-power-consumption multichannel optical switches and integrated photonic devices. On the other hand, switching among the P, FC, and H states in a PC/DFCLC cell can be regulated rapidly, directly and reversibly by using frequency-modulated voltage pulses. The ability of wavelength tunability of defect modes in the PC/DFCLC cell is achieved by the texture transition from the stable P state to the voltage-sustained H state. Owing to the frequency-dependent dielectric anisotropy of the DFLC, the transmittance of the defect modes can be tuned by both the voltage frequency and amplitude, providing a new way for intensity tunability through such a filter characterized by the defect modes. This device requires no polarizers and is of low power consumption. Noticeably, the voltage-sustained H state in the bistable DFCLC can further be a third optically stable state by incorporating an adequate amount of photo-curable monomer, oligomer, or prepolymer into the DFCLC material to create polymer networks in the defect layer. Such a DFCLC cell exhibiting three stable states—P, FC, and H—is known as a tristable PSCT. Referring to the aforementioned tunability in the PC/DFCLC, the wavelength tunability and transmittance tunability of defect modes in the PC/PSCT cell can be achieved—the intensities of the defect modes can be regulated by the amplitude of voltage in the mixed states and the wavelengths be switched by the frequency in the H and P states. The cell can directly be switched from one to another stable state by employing a proper frequency- modulated voltage pulse on the cell.

1D PC/LC has received much attention in recent years due to its tunability in defect modes within the PBG. It has been established that this hybrid PC structure enables the control over the defect modes by electric field, magnetic field, and the like as a stimulus. The resulting features make it applicable for designing tunable photonic devices such as a multichannel filter, light shutter, and optical modulator. Specifically, incorporating MMLC as a defect layer in the 1D PC provides both the optically memorable and tunable defect modes characterized by the stable states, allowing the device to be of low power consumption. Such a 1D PC/MMLC device thus extends its use for green products.

Author details

Po-Chang Wu
Department of Physics, Chung Yuan Christian University, Chung-Li, Taiwan, Republic of China

Wei Lee
Department of Physics, Chung Yuan Christian University, Chung-Li, Taiwan, Republic of China
College of Photonics, National Chiao Tung University, Guiren Dist., Tainan, Taiwan, Republic of China

Acknowledgement

The authors acknowledge the financial support from the National Science Council of the Republic of China (Taiwan) under grant Nos. NSC 98-2923-M-033-001-MY3 and NSC 98-2112-M-009-023-MY3, and are grateful to Yu-Cheng Hsiao, Yu-Ting Lin, Ivan Timofeev, Chong-Yin Wu, Yi-Hong Zou, and Victor Ya. Zyryanov for their assistance with the preparation of this manuscript.

6. References

Arkhipkin, V. G.; Gunyakov, V. A.; Myslivets, S. A.; Gerasimov, V. P.; Zyryanov, V. Ya.; Vetrov, S. Ya. & Shabanov, V. F. (2008). One-Dimensional Photonic Crystals with a Planar Oriented Nematic Layer: Temperature and Angular Dependence of the Spectra of Defect Modes. *Journal of Experimental and Theoretical Physics*, Vol. 106, No. 2, pp. 388–398

Arkhipkin, V. G.; Gunyakov, V. A.; Myslivets, S. A.; Zyryanov, V. Ya. & Shabanov, V. F. (2007). Angular Tuning of Defect Modes Spectrum in the One-Dimensional Photonic Crystal with Liquid-Crystal Layer. *European Physical Journal E,* Vol. 24, No. 3, pp. 297–302

Bao, R.; Liu, C.-M. & Yang, D.-K. (2009). Smart Bistable Polymer Stabilized Cholesteric Texture Light Shutter. *Applied Physics Express,* Vol. 2, pp. 112401

Berreman, D. W. & Heffner, W. R. (1980). New Bistable Cholesteric Liquid-Crystal Display. *Applied Physics Letters,* Vol. 37, No. 1, pp. 109–111

Blanco, A.; Chomski, E.; Grabtchak, S.; Ibisate, M.; John, S.; Leonard, S. W.; Lopez, C.; Meseguer, F.; Miguez, H.; Mondia, J. P.; Ozin, G. A.; Toader, O. & van Driel, H. M. (2000). Large-Scale Synthesis of a Silicon Photonic Crystal with a Complete Three-Dimensional Bandgap near 1.5 Micrometers. *Nature,* Vol. 405, No. 6785, pp. 437–440

Bobrovsky, A.Yu.; Boiko, N. I.; Shibaev, V. P. & Wendorff, J. H. (2003). Cholesteric Mixtures with Photochemically Tunable Circularly Polarized Fluorescence. *Advance Materials,* Vol. 15, No. 4, pp. 282–287

Choi, S. S.; Morris, Stephen M.; Huck, Wilhelm T. S. & Coles, H. J. (2009). Electrically Tuneable Liquid Crystal Photonic Bandgaps. *Advance Materials,* Vol. 21, No. 38, pp. 3915–3918

Chow, E.; Lin, S. Y.; Johnson, S. G.; Villeneuve, P. R.; Joannopoulos, J. D.; Wendt, J. R.; Vawter, G. A.; Zubrzycki, W.; Hou, H. & Alleman, A. (2000). Three-Dimensional Control of Light in a Two-Dimensional Photonic Crystal Slab. *Nature,* Vol. 407, No. 6807, pp. 983–986

Fleming, J. G. & Lin. S.-Y. (1999). Three-Dimensional Photonic Crystal with a Stop Band from 1.35 to 1.95 μm, *Optics Express,* Vol. 24 No. 1, pp. 49–51

Gooch, C. H. & Tarry, H. A. (1975). The Optical Properties of Twisted Nematic Liquid Crystal Structures With Twist Angles < 90 Degrees. *Journal Physics D Applied Physics,* Vol. 8, No. 13, pp. 1575–1584

Hecht., E. (2002). Applications of Single and Multilayer Films. In: *Optics,* Ch. 8, pp. 425-431, Addison Wesley, ISBN 0-321-18878-0, San Francisco

Hsiao, Y.-C.; Hou, C.-T.; Zyryanov, V. Ya. & Lee, W. (2011a). Multichannel Photonic Devices Based on Tristable Polymer-Stabilized Cholesteric Textures. *Optics Express,* Vol. 19, No. 8, pp. 7349–7355

Hsiao, Y.-C.; Tang, C.-Y. & Lee, W. (2011b). Fast-Switching Bistable Cholesteric Intensity Modulator. *Optics Express,* Vol. 19, No. 10, pp. 9744–9749

Hsiao, Y.-C.; Wu, C.-Y.; Chen, C.-H.; Zyryanov, V. Ya. & Lee, W. (2011c). Electro-Optical Device based on Photonic Structure with a Dual-Frequency Cholesteric Liquid Crystal. *Optics Letters,* Vol. 36, No. 14, pp. 2632–2634

Hsu, J.-S. (2007). Stability of Bistable Chiral-Tilted Homeotropic Nematic Liquid Crystal Displays. *Japanese Journal of Applied Physics,* Vol. 46, No. 11, pp. 7378–7381

Hsu, J.-S.; Liang, B.-J. & Chen, S.-H. (2004). Bistable Chiral Tilted-Homeotropic Nematic Liquid Crystal Cells. *Applied Physics Letters,* Vol. 85, No. 23, pp. 5511–5513.

Hsu, J.-S.; Liang, B.-J. & Chen, S.-H. (2007). Dynamic Behaviors of Dual Frequency Liquid Crystals in Bistable Chiral Tilted-Homeotropic Nematic Liquid Crystal Cell. *Applied Physics Letters,* Vol. 89, No. 5, pp. 091520

Huang, C. Y.; Fu, K.-Y.; Lo, K.-Y. & Tsai, M.-S. (2003). Bistable Transflective Cholesteric Light Shutters. *Optics Express*, Vol. 11, No. 6, pp. 560-565

Huang, Y.; Zhou, Y.; Doyle, C. & Wu, S.-T. (2006). Tuning the Photonic Band Gap in Cholesteric Liquid Crystals by Temperature-Dependent Dopant Solubility. *Optics Express*, Vol. 14, No. 3, pp. 1236–1244

Imada, M.; Noda, S.; Chutinan, A.; Tokuda, T.; Murata, M. & Sasaki, G. (1999). Coherent Two-Dimensional Lasing Action in Surface-Emitting Laser with Triangular-Lattice Photonic Crystal Structure. *Applied Physics Letters*, Vol. 75, No. 3, pp. 316–318

Jhun, C. G.; Chen, C. P.; Lee, U. J.; Lee, S. R.; Yoon, T. H. & Kim, J. C. (2006). Tristable Liquid Crystal Display with Memory and Dynamic Operating Modes. *Applied Physics Letters*, Vol. 89, No. 12, pp. 123507

John, S. (1987). Strong Localization of Photons in Certain Disordered Dielectric Superlattices. *Physical Review Letters*, Vol. 58, No. 23, pp. 2486–2489

Kasano M.; Ozaki, M.; Yoshino, K.; Ganzke, D. & Haase, W. (2003). Electrically Tunable Waveguide Laser based on Ferroelectric Liquid Crystal. *Applied Physics Letters*, Vol. 82, No. 23, pp. 4026–4028

Knight, J. C.; Broeng, J;. Birks, T. A. & Russell, P. St. J. (1998). Photonic Band Gap Guidance in Optical Fibers. *Science*, Vol. 282, No. 5393, pp. 1476–1478

Knight, Jonathan C. (2003). Photonic Crystal Fibres. *Nature*, Vol. 424, No. 6950, pp. 2486–2589.

Krauss, Thomas F.; De La Rue, Rchard M. & Brand, S. (1996). Two-Dimensional Photonic-Bandgap Structures Operating at Near-Infrared Wavelengths. *Nature*, Vol. 383, No. 6602, pp. 699–702

Liang B.-J. & Lin, C.-L. (2007). Crucial Influence on *d*/*p* Range in Bistable Chiral Tilted-Homeotropic Nematic Liquid Crystal Cells. *Journal of Applied Physics*, Vol. 102, No. 12, pp.124504

Liang, B.-J.; Hsu, J.-S.; Lin, C.-L. & Hsu, W.-C. (2008). Dynamic Switching Behavior of Bistable Chiral-Tilted Homeotropic Nematic Liquid Crystal Displays. *Journal of Applied Physics*, Vol. 104, No. 7, pp. 074509

Lin, T.-H.; Chen, Y.-J.; Wu, C.-H.; Fuh, Andy Y.-G.; Liu, J.-H. & Yang, P.-C. (2005). Cholesteric liquid crystal laser with wide tuning capability. Applied Physics Letters. Vol. 86 pp. 161120

Lin, T.-H.; Jau, H.-C.; Chen, C.-H.; Chen, Y.-J.; Wei, T.-H.; Chen, C.-W. & Fuh, Andy Y.-G. (2006). Electrically Controllable Laser based on Cholesteric Liquid Crystal with Negative Dielectric Anisotropy. *Applied Physics Letters*, Vol. 88, No. 6, pp. 061122

Lin, Y.-T.; Chang, W.-Y.; Wu, C.-Y.; Zyryanov, V. Ya. & Lee, W. (2010). Optical Properties of One-Dimensional Photonic Crystal with a Twisted-Nematic Defect Layer. *Optics Express*, Vol. 18, No. 26, pp. 26959–26964

Ma, J.; Shi, L. & Yang, D.-K. (2010). Bistable polymer stabilized cholesteric texture light shutter. *Apply Physics Express*, Vol. 3, No. 2, pp. 021702

Matsuhisa, Y.; Ozaki, R.; Takao, Y. & Ozaki, M. (2007). Linearly Polarized Lasing in One-Dimensional Hybrid Photonic Crystal containing Cholesteric Liquid Crystal. *Journal of Applied Physics*, Vol. 101, No. 3, pp. 033120

Matsuhisa, Y.; Ozaki, R.; Yoshino, K. & Ozaki, M. (2006). High Q Defect Mode and Laser Action in One-Dimensional Hybrid Photonic Crystal containing Cholesteric Liquid Crystal. *Applied Physics Letters*, Vol. 89, No. 10, pp. 101109

Miroshnichenko A. E.; Brasselet, E. & Kivshar, Y. S. (2008a). All-Optical Switching and Multistability in Photonic Structures with Liquid Crystal Defects. *Applied Physics Letters*, Vol. 92, No. 25, pp. 253306.

Miroshnichenko A. E.; Brasselet, E. & Kivshar, Y. S. (2008b). Light-Induced Orientational Effects in Periodic Photonic Structures with Pure and Dye-Doped Nematic Liquid Crystal Defects. *Physical Review A*, Vol. 78, No. 5, pp. 053823

Morris, S. M.; Ford, A. D.; Pivnenko, M. N. & Coles, H. J. (2005). Enhanced Emission from Liquid-Crystal Lasers. *Journal of Applied Physics*, Vol. 97, No. 2, pp. 023103

Nelson, B. E.; Gerken, M.; Miller, David A. B.; Piestun, R.; Lin, C.-C. & Harris, James S. (2000) Use of a Dielectric Stack as a One-Dimensional Photonic Crystal for Wavelength Demultiplexing by Beam Shifting. *Optics Express*, Vol. 25, No. 20, pp.1502–1504

Ozaki, R.; Matsuhisa, Y.; Ozaki, M. & Yoshino, K. (2005). Low Driving Voltage Tunable Laser based on One-Dimensional Photonic Crystal Containing Liquid Crystal Defect Layer. *Molecular Crystal and Liquid Crystals*, Vol.441, No. 1, pp. 441, 87–95

Ozaki, R.; Matsui, T.; Ozaki, M. & Yoshino, K. (2002). Electro-Tunable Defect Mode in One-Dimensional Periodic Structure containing Nematic Liquid Crystal as a Defect Layer. *Japanese Journal of Applied Physics*, Vol. 41, No. 12B, pp. L1482–L1484

Ozaki, R.; Matsui, T.; Ozaki, M. & Yoshino, K. (2003a). Electrically Color-Tunable Defect Mode Lasing in One-Dimensional Photonic-Band-gap System containing Liquid Crystal. *Applied Physics Letters*, Vol. 82, No. 21, pp. 3593–3595

Ozaki, R.; Matsui, T.; Ozaki, M. & Yoshino, K. (2004a). Optical Property of Electro-Tunable Defect Mode in 1D Periodic Structure with Light Crystal Defect Layer. *Electronics and Communications in Japan*, Vol. 87, No. 5, pp. 24–31

Ozaki, R.; Moritake, H.; Yoshino, K. & Ozaki, M. (2007). Analysis of Defect Mode Switching Response in One-Dimensional Photonic Crystal with a Nematic Liquid Crystal Defect Layer. *Journal of Applied Physics*, Vol. 101, No. 3, pp. 033503

Ozaki, R.; Ozaki, M. & Yoshino, K. (2003b). Defect Mode Switching in One-Dimensional Photonic Crystal with Nematic Liquid Crystal as Defect Layer. *Japanese Journal of Applied Physics*, Vol. 42, No. 6B, pp. L669–L671

Ozaki, R.; Ozaki, M. & Yoshino, K. (2004b). Defect Mode in One-Dimensional Photonic Crystal with In-Plane Switchable Nematic Liquid Crystal Defect Layer. *Japanese Journal of Applied Physics*, Vol. 43, No. 11B, pp. L1477–L1479

Painter, O.; Lee, R. K.; Scherer, A. Yariv; A. O'Brien, J. D.; Dapkus, P. D. & Kim, I. (1999). Two-Dimensional Photonic Band-gap Defect Mode Laser. *Science*, Vol. 284, No. 5421, pp. 1819–1821

Park, B.; Kim, M.; Kim, S. W. & Kim, I. T. (2009). Circularly Polarized Unidirectional Lasing from a Cholesteric Liquid Crystal Layer on a 1-D Photonic Crystal Substrate. *Optics Express*, Vol. 17, No. 15, pp. 12323–12331

Park, S. H.; Xia, B. & Gates, Y. (1999). A Three-Dimensional Photonic Crystal Operating in the Visible Region. *Advance Materials*, Vol. 11, No. 6, 462–466

Timofeev, I. V.; Lin, Y.-T.; Gunyakov, V.A.; Myslivets, S. A.; Arkhipkin, V. G.; Vetrov, S. Ya.; Lee, W. & Zyryanov, V. Ya. (2012). Voltage-Induced Defect Mode Coupling in a One-Dimensional Photonic Crystal with a Twisted-Nematic Defect Layer. *Physical Review E*, Vol. 85, No. 1, pp. 011705

Wen, C.-H. & Wu, S.-T. (2005). Dielectric Heating Effects of Dual-Frequency Liquid Crystals. *Applied Physics Letters*, Vol. 86, No. 23, pp. 231104

Wu, S.-T. & Wu, C.-S. (1996). Mixed-Mode Twisted Nematic Liquid Crystal cells for Reflective Displays. *Applied Physics Letters*, Vol. 68, No. 11, pp. 1455–1457

Wu, C.-Y.; Zou, Y.-H.; Timofeev, I.; Lin, Y.-T.; Zyryanov, V. Ya.; Hsu, J.-S. & Lee, W. (2011). Tunable Bi-functional Photonic Device based on One-Dimensional Photonic Crystal infiltrated with a Bistable Liquid-Crystal Layer. *Optics Express*, Vol. 19, No. 8, pp. 7349–7355

Xianyua, Haiqing; Wu S.-T. & Lin, C.-L. (2009). Dual Frequency Liquid Crystals: a Review. "*Liquid Crystals*, Vol. 36, No. 6-7, pp. 717-726

Xu, M. & Yang, D.-K. (1997). Dual Frequency Cholesteric Light Shutters. *Applied Physics Letters*, Vol. 70, No. 6, pp. 720–722

Yablonovitch, E. (1987). Inhibited Spontaneous Emission in Solid-State Physics and Electronics. *Physical Review Letters*, Vol. 58, No. 20, pp. 2059–2062

Yao, I.-A.; Yang, C.-L.; Chen, C.-J.; Pang, J.-P.; Liao, S.-F.; Li, J.-H. & Wu, J.-J. (2009). Bistability of Splay and π-Twist States in a Chiral-Doped Dual Frequency Liquid Crystal Cell. *Applied Physics Letters*, Vol. 94, No. 7, pp. 071104

Yeh, P. H. & Gu, C. (1999a). Normal Modes of Propagation in a General TN-LC. In: *Optics of Liquid Crystal Display*, pp. 130-136, John Wiley & Sons, ISBN 0-471-18201-X, Canada

Yeh, P. H. & Gu, C. (1999b). Light Propagation in Uniaxial Media. In: *Optics of Liquid Crystal Display*, pp. 63-68, John Wiley & Sons, ISBN 0-471-18201-X, Canada

Yeung, Fion S.-Y.; Xie, F.-C.; Wan, Jones T.-K.; Lee, F. K.; Tsui, Ophelia K. C.; Sheng, P. & Kwok, H.-S. (2006). Liquid Crystal Pretilt Angle Control using Nanotextured Surfaces," *Journal of Applied Physics*, Vol. 99, No. 12, pp. 124506

Yin, Y.; Shiyanovskii, S. V. & Lavrentovich, O. D. (2006). Electric Heating Effects in Nematic Liquid Crystals. *Journal of Applied Physics*, Vol. 100, No. 2, pp. 024906

Yokoyama, S.; Mashiko, S.; Kikuchi, H.; Uchida, K. & Nagamura, T. (2006). Laser Emission from a Polymer-Stabilized Liquid-Crystalline Blue Phase. *Advance Materials*, Vol. 18, No. 1, pp. 48–51

Yu, H.; Tang, B.; Li, J. & Li, L. (2005). Electrically Tunable Lasers made from Electro-Optically Active Photonics Band gap Materials. *Optics Express*, Vol. 13, No. 18, pp. 7243–7249

Zyryanov, V. Ya.; Gunyakov, V. A.; Myslivets, S. A.; Arkhipkin, V. G. & Shabano, V. F. (2008). Electrooptical Switching in a One-Dimensional Photonic Crystal. *Molecular Crystal and Liquid Crystals*, Vol. 488, No. 1, pp. 118–126

Zyryanov, V. Ya.; Myslivets, S. A.; Gunyakov, V. A.; Parshin, A. M.; Arkhipkin, V. G.; Shabanov,V. F. & Lee, W. (2010). Magnetic-Field Tunable Defect Modes in a Photonic-Crystal/Liquid-Crystal Cell. *Optics Express*, Vol. 18, No. 2, pp. 1283–1288

Chapter 5

Self-Organized Three-Dimensional Optical Circuits and Molecular Layer Deposition for Optical Interconnects, Solar Cells, and Cancer Therapy

Tetsuzo Yoshimura

Additional information is available at the end of the chapter

http://dx.doi.org/10.5772/48221

1. Introduction

Photonics is being coupled with the molecular nanotechnology to be a main technology in the information processing/communication, solar energy conversion, and bio/medical systems. We have developed two original core technologies for optical interconnects so far. One is Self-Organized Lightwave Network (SOLNET) [1-3] and the other is Molecular Layer Deposition (MLD) [4-6]. SOLNET is self-aligned optical waveguides formed in photo-induced refractive index increase (PRI) materials. MLD is a monomolecular-step growth process of tailored organic materials. Recently, we expanded our scope for the SOLNET/MLD applications toward the solar energy conversion and the bio/medical fields.

In the present chapter, after SOLNET and MLD are reviewed, their applications to optical interconnects within computers [1,7-9], solar cells [4,10-12], and cancer therapy [4,10,13] are presented with some proof-of-concepts demonstrated by the finite difference time domain (FDTD) method and preliminary experiments.

2. Core technologies

2.1. SOLNET

2.1.1. Concept of SOLNET

The concept of SOLNET is shown in Figure 1 [1-3]. In one-beam-writing SOLNET, a write beam is introducqed into a PRI material from an optical device such as an optical waveguide, an optical fiber, a laser diode (LD), etc. In the PRI material, refractive index increases by write beam exposure. When the write beam propagates in the PRI material, the write beam propagation is affected by the refractive index distribution, which is generated by the write beam itself. Then, the write beam is concentrated along the propagation axis, inducing the self-focusing to construct an optical waveguide from the optical device. This is the SOLNET. SOLNET is also constructed by free-space light beams.

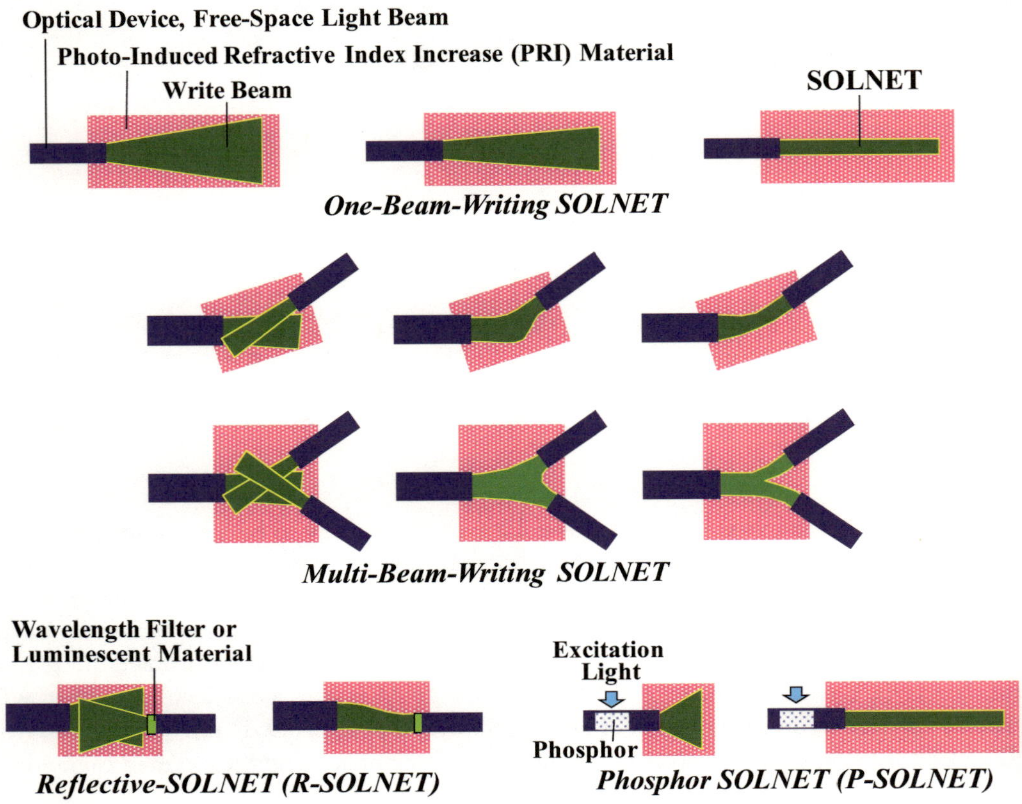

Figure 1. Concept of SOLNET

In multi-beam-writing SOLNET, a plurality of optical devices are put into a PRI material. Write beams are introduced into the PRI material from them. The write beams are attracted to each other and merge by the self-focusing to construct self-aligned coupling waveguides between the optical devices automatically. The coupling waveguides can be formed even when the optical devices are misaligned and have different core sizes.

In reflective SOLNET (R-SOLNET), some of the write beams in the multi-beam-writing SOLNET are replaced with reflected write beams from reflective elements on core edges of optical devices. In the example shown in Figure 1, a write beam from an optical waveguide and a reflected write beam from a wavelength filter on an edge of another optical waveguide overlap. In the overlap region, the refractive index of the PRI material increases, pulling the write beam to the wavelength filter location more and more. We call this effect the "pulling water" effect. Finally, by the self-focusing, a self-aligned coupling waveguide is formed between the optical waveguides.

Here, the wavelength filter can be replaced with a luminescent material. When a write beam from an optical waveguide is introduced onto the luminescent material through the PRI material, luminescence is generated from the material. The luminescence acts like the reflected write beam to induce the "pulling water" effect.

In phosphor SOLNET (P-SOLNET), phosphor is doped in a part of an optical waveguide. By exposing the doped phosphor to an excitation light, a write beam generated from the phosphor propagates in the optical waveguide to be emitted into the PRI material. P-SOLNET is effective when write beams cannot be introduced from outside; for example, when SOLNET is formed in inner parts of three-dimensional (3-D) optoelectronic (OE) platforms.

2.1.2. Photo-induced refractive index increase (PRI) materials

The PRI materials are, for example, photopolymers, photo-definable materials, photo-refractive crystals, etc. Figure 2 shows the mechanism of the refractive-index increase in a photopolymer. High-refractive-index monomers and low-refractive-index monomers are mixed. The high-refractive-index ones have higher photo-reactivity to write beams than the low-refractive-index ones. When the photopolymer is exposed to a write beam, high-refractive-index monomers are combined by photo-chemical reactions to produce new molecules. Then, high-refractive-index monomers diffuse into the exposed region from the surrounding area to compensate for the reduction of the high-refractive-index monomer concentration. Repeating this process increases the refractive index of the exposed region. The wavelength of the write beams typically ranges from ~350 to ~900 nm. The spectral response can be adjusted by sensitizers.

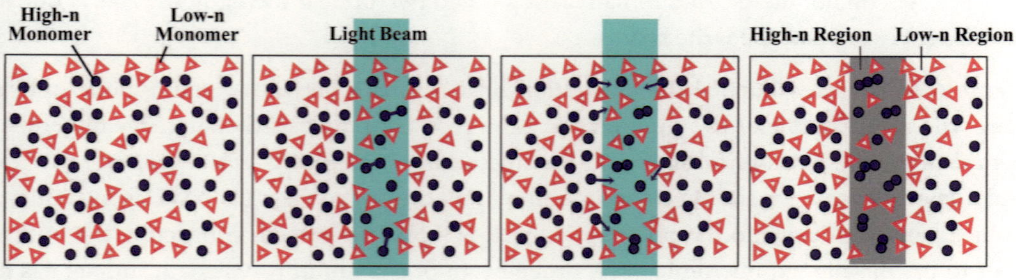

Figure 2. Example of PRI materials

Some molecules are known to exhibit two-photon absorption [14]. As shown in Figure 3, in the four-level model, a photon with a wavelength of λ_1 excites an electron from S_0 state to S_n state in a molecule. The excited electron in the S_n state transfers to T_0 state. Then, a photon with another wavelength of λ_2 further excites the electron to T_n state to induce chemical reactions. The chemical reactions occur only in places, where both λ_1 photons and λ_2

photons exist. By using the two-photon absorption in the multi-beam-writing SOLNET, smooth self-focusing is expected in the overlap regions. R-SOLNET might be formed by two-wavelength write beams using luminescent materials. Namely, R-SOLNET grows by a write beam of λ_1 from an optical waveguide and luminescence of λ_2 from the luminescent material.

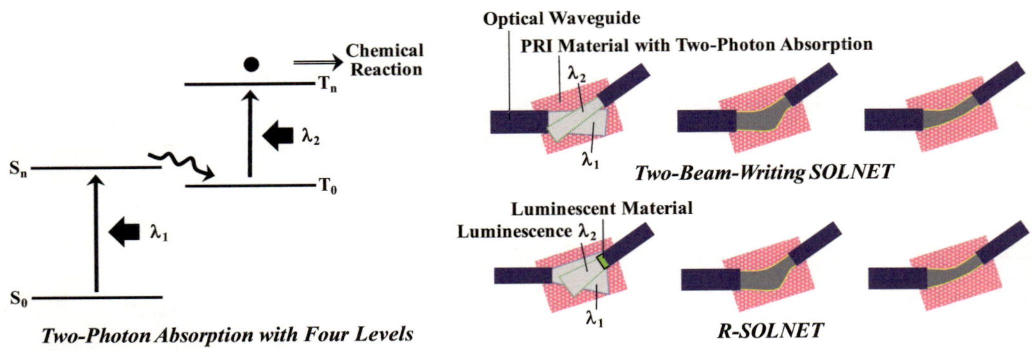

Figure 3. Concept of R-SOLNET using two-wavelength write beams

2.1.3. Experimental demonstrations of SOLNET

Figure 4 shows SOLNET formed in a PRI sol-gel thin film with a 405-nm write beam at 200°C. SOLNET grows toward the right-side edge. Figure 5 shows two-beam-writing SOLNET connecting two optical fibers with a core diameter of 9.5 μm in a photopolymer. For lateral misalignments of 1~9 μm, self-aligned coupling waveguides of SOLNET are constructed. For a lateral misalignment of 20 μm, optical waveguides grown from the two optical fibers do not merge, but, remain as separated two optical waveguides. This is due to insufficient overlap of the write beams.

Figure 6 shows experimental demonstration of R-SOLNET between a multi-mode optical fiber with a core diameter of 50 μm and an Al micromirror deposited on an edge of another optical fiber [15]. The optical fiber and the micromirror are placed with a gap of ~800 μm and a lateral misalignment of 60 μm in a photopolymer. When a write beam is introduced from the optical fiber into the photopolymer, R-SOLNET is formed, connecting the optical fiber to the misaligned micromirror. A probe beam of 650 nm in wavelength propagates in the S-shaped self-aligned waveguide of R-SOLNET.

Figure 4. SOLNET formed in PRI sol-gel thin film

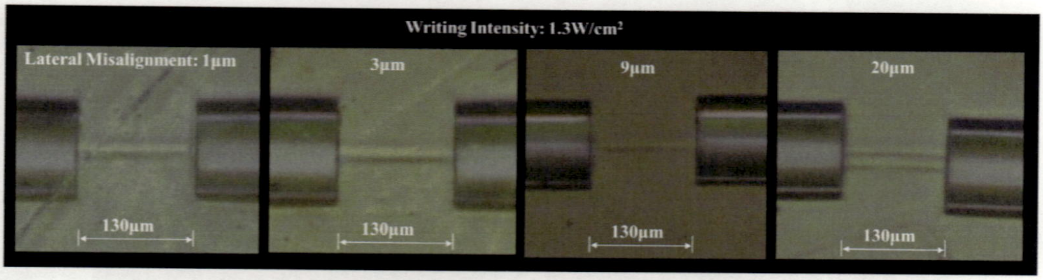

Figure 5. Two-beam-writing SOLNET connecting two optical fibers in photopolymer

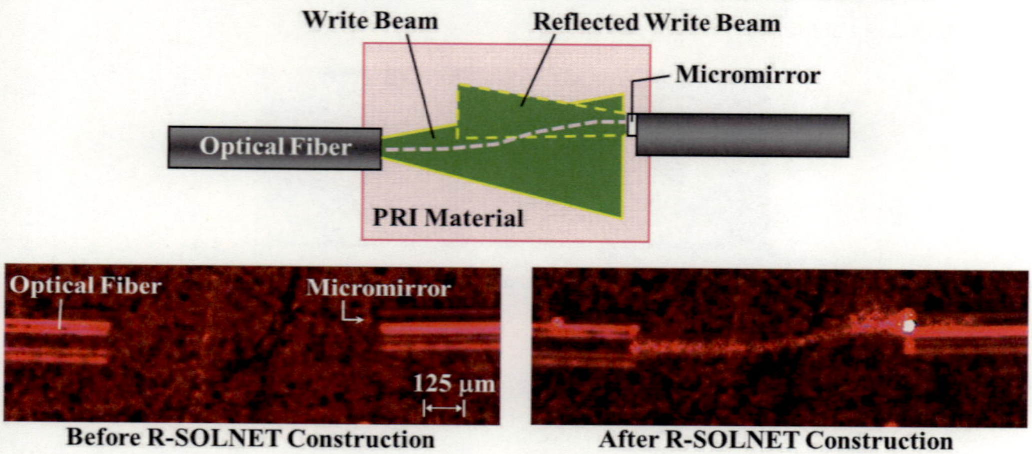

Figure 6. R-SOLNET formed between an optical fiber and a micromirror

2.2. MLD

MLD is a method to grow organic materials with designated molecular sequences as shown in Figure 7 [4-6]. The source molecules are designed so that the same molecules cannot be combined while different molecules can be combined by utilizing selective chemical reactions or the electrostatic force between them. Molecule A is provided onto a substrate surface to form a monomolecular layer of A. Once the surface is covered with A, the deposition of the molecules is automatically terminated by the self-limiting effect similarly to the atomic layer deposition (ALD) [16]. By switching source molecules sequentially as A, B, C, D,..., materials with molecular sequences like A/B/C/D/... are obtained.

MLD can grow organic tailored materials such as the molecular wire and the polymer multiple quantum dot (MQD), which means a polymer wire containing MQD. MLD can also grow ultra-thin/conformal organic layers on 3-D surfaces such as deep trenches, porous layers, particles etc. as schematically depicted in Figure 8.

Figure 9 shows an example of MLD using pyromellitic dianhydride (PMDA) and 4,4'-diaminodiphenyl ether (DDE) [6]. Monomolecular-step growth to synthesize polyamic acid is observed.

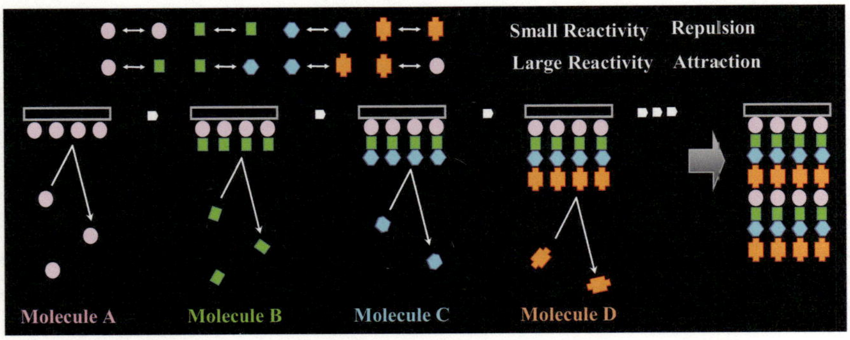

Figure 7. Concept of MLD

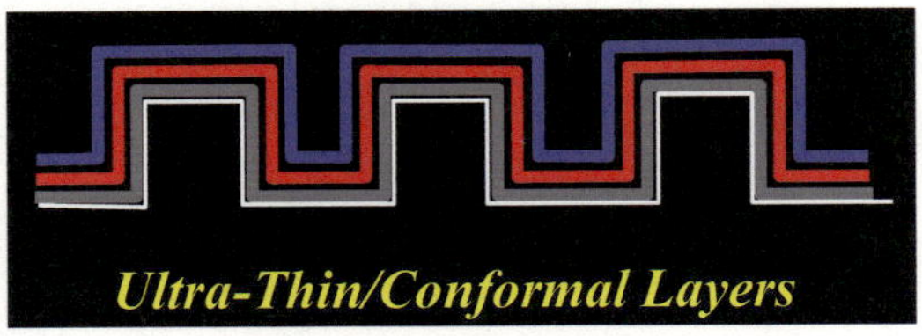

Figure 8. Deposition of ultra-thin/conformal layers on 3-D surfaces by MLD

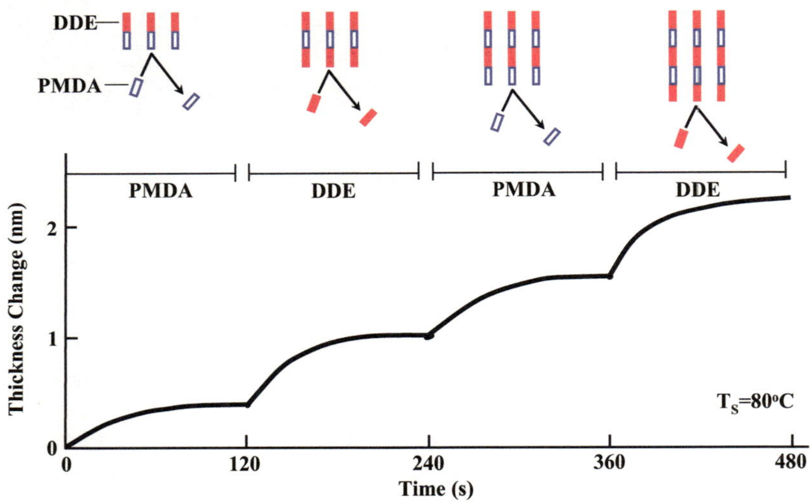

Figure 9. Experimental demonstration of MLD.

3. Applications to optical interconnects within computers

3.1. Integrated 3-D optical interconnects based on S-FOLM

3.1.1. Concept and advantages of S-FOLM

A future image of high-performance computers is shown in Figure 10. Optical signals are generated at outputs of large-scale integrated circuits (LSIs) and transmitted through optical waveguides of OE multi-chip module (MCM), OE printed circuit board (PCB) and OE backplane (BP) to reach inputs of LSIs.

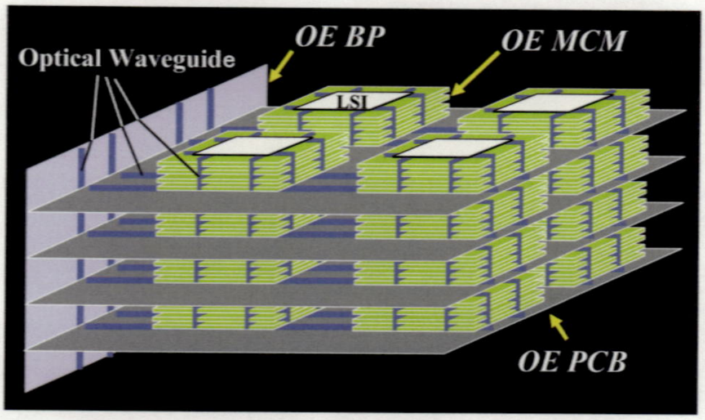

Figure 10. Future image of high-performance computers

The 3-D stacked OE MCM shown in Figure 11 is the core unit for the integrated 3-D optical interconnects [1, 7-9]. OE films, in which thin-film flakes of light modulators and photodetectors (PDs) are embedded, are stacked together with films containing thinned LSIs. Optical signals generated in the light modulators driven by electrical signals from LSI outputs propagate to PDs, being converted into electrical signals for LSI inputs. In some cases, the optical signals are coupled to optical waveguides of the OE board by the back-side connection to be transmitted to other OE MCMs. The 3-D structure also provides the 3-D optical switching systems [1,17], where thin-film flakes of optical switches are embedded, for the on-board reconfigurable network.

Since heat generation is a serious problem in the 3-D stacked OE MCM, light sources of cw or pulse trains are placed outside of the OE MCM. Implementation of the OE amplifier/driver-less substrate (OE-ADLES), where E-O and O-E conversions are respectively carried out by light modulators directly driven by LSI output signals and by PDs directly generating LSI input signals, is effective to realize "low power dissipation" and "high data rate" capability [1,7,8,18,19]. By implimenting light sources with a plurality of wavelengths and wavelength filters, wavelength division multiplexing (WDM) interconnects can be performed [1,2,7].

The integrated 3-D optical interconnects are built from the scalable film optical link module (S-FOLM) shown in Figure 12 [1,7-9]. A set of OE films, in which thin-film device flakes are embedded, are prepared. The thin-film devices include optical waveguides, light modulators, optical switches, wavelength filters, vertical cavity surface emitting lasers (VCSELs), PDs, photovoltaic devices, interface ICs, LSIs, and so on. By combining the films in stacked configurations, various kinds of 3-D OE platforms are constructed. By stacking a VCSEL/PD-embedded film on an LSI, a smart pixel will be made. When a VCSEL/PD-embedded optical waveguide film and an interface-IC-embedded film are stacked, an optical interconnect board will be made. If LSI-embedded films are added in the stack, 3-D stacked OE MCM will be made. Such scalability will contribute to system cost reductions.

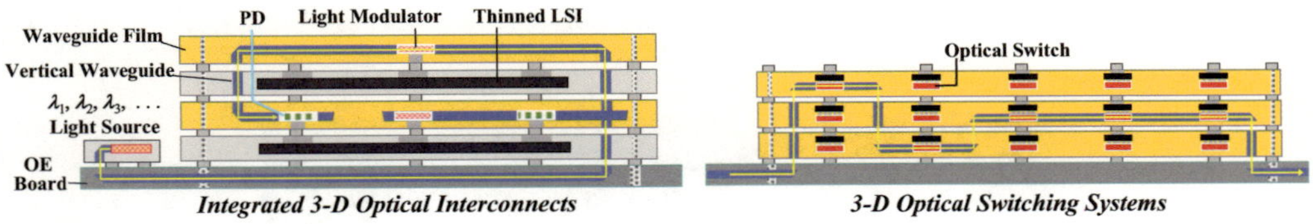

Figure 11. 3-D stacked OE MCM

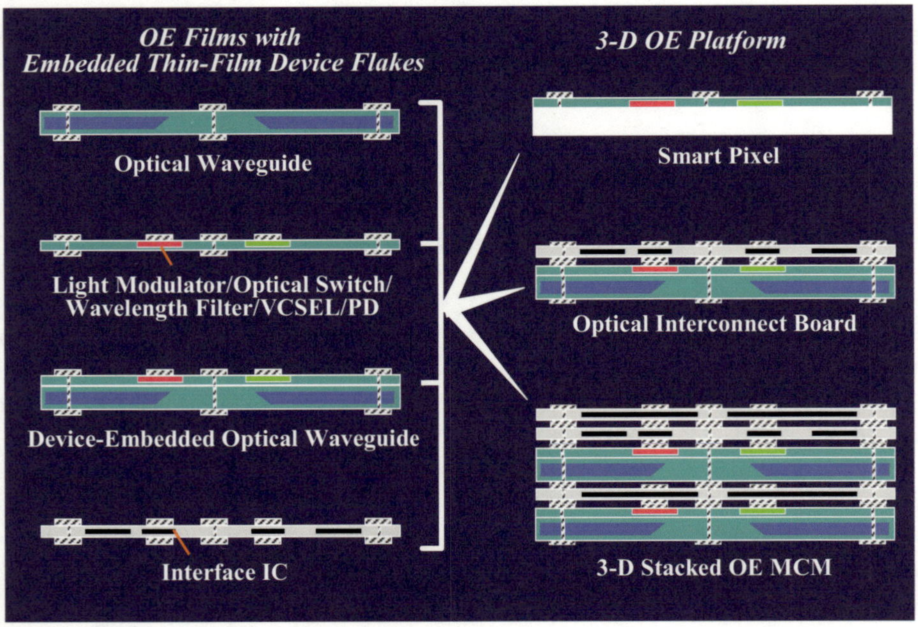

Figure 12. Concept of S-FOLM

S-FOLM gives the following advantages over the conventional flip-chip-bonding-based packaging.

- Excess space is not necessary on the surface.
- Long metal lines are eliminated, resulting in noise/delay reductions.
- Assembling cost is reduced by embedding the device flakes using the all-photolithographic process like PL-Pack with SORT [1,17].
- Semiconductor material cost is reduced because the material exists only at places where it is necessary.
- WDM module cost is reduced because different-wavelength devices can easily be put together into OE films by SORT [1,17].
- Standardized interface is available by inserting interface-IC-embedded films, enabling computer designers to treat optics as transparent agents for interconnect-spec improvement without knowledge of optics.
- Any-place/any-shape placement capability is available by attaching the OE films selectively on substrates.

3.1.2. *Self-organized 3-D integrated optical circuits with SOLNET*

In future optical interconnects, a large number of optical couplings and optical Z-connections will be involved. In such cases, following two issues should be considered.

1. Enormous alignment efforts with micron or submicron accuracy
2. Vertical waveguide formation for the 3-D optical circuits

As a solution for them, the self-organized 3-D integrated optical circuit shown in Figure 13 is proposed [1,20]. After a precursor with optical devices distributed three-dimensionally is prepared, optical waveguides are formed between the devices in a self-aligned manner to construct 3-D optical circuits. This will be achieved by SOLNET. Optical solder and Optical wiring in free spaces of SOLNET are respectively applicable for issue 1) and 2).

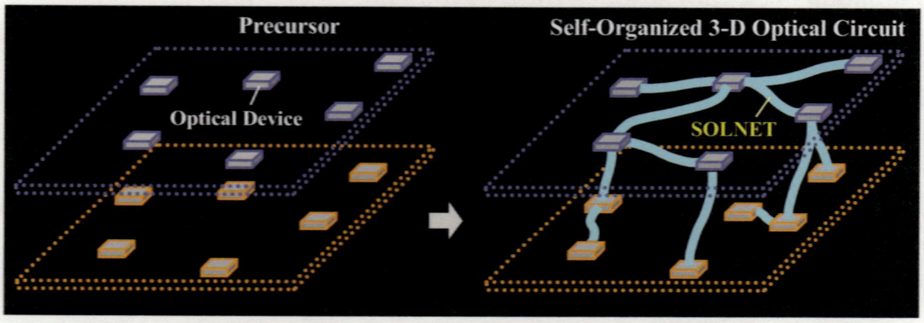

Figure 13. Concept of self-organized 3-D integrated optical circuits utilizing SOLNET

Optical devices can be connected with self-aligned optical couplings by putting the optical solder of SOLNET betweent them. Figure 14 shows examples of the optical solder of R-SOLNET for couplings of devices that cannot emit write beams such as PDs, and light modulators. Wavelength filters or luminescent materials are placed on the devices to pull the SOLNET to the target locations.

Figure 15 shows an example of the free space optical wiring of R-SOLNET for optical Z-connections in 3-D optical circuits. A luminescent target is placed at a vertical mirror location of a waveguide film. By introducing a write beam from a waveguide in an OE board into a free space filled with a PRI material, the luminescent target generates luminescence to construct a self-aligned vertical waveguide of R-SOLNET between the waveguide film and the OE board even when a misalignment and a core size mismatching exist.

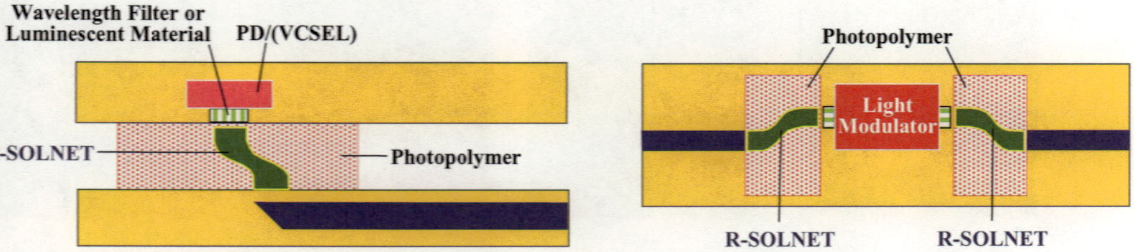

Figure 14. Optical solder of SOLNET

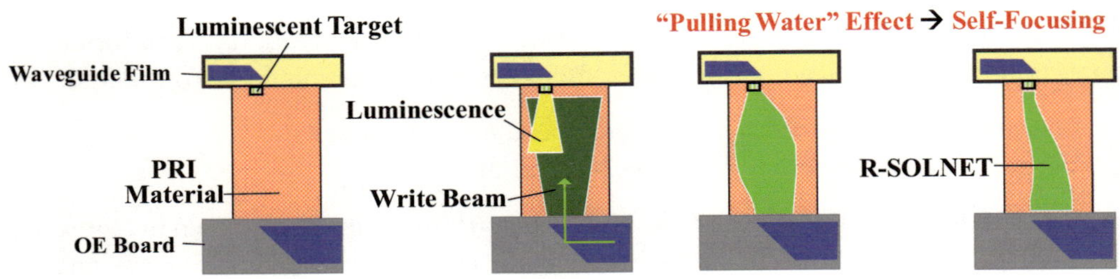

Figure 15. Self-organized vertical waveguides of R-SOLNET

3.1.3. Simulation of self-organized vertical waveguides of R-SOLNET

SOLNET simulator is based on the FDTD method as described in detail elsewhere [1,21]. Figure 16 shows an example of a simulation. A 650-nm write beam emitted from an input waveguide with core width of 500 nm is initially reflected upward with diffraction by a 45° wavelength filter. Then, the self-focusing gradually appears in the exposed part. Finally, the write beam is focused by SOLNET growth along the center axis of the write beam propagation.

Figure 17 shows a simulation of an optical Z-connection construction with a vertical waveguide of R-SOLNET [21]. A 2 μm-thick core with a total internal reflection (TIR) 45° mirror is on a 0.5 μm-thick under cladding layer to form an optical waveguide film. Two optical waveguide films are stacked with a gap filled with a PRI material. A wavelength filter is deposited at the TIR 45° mirror aperture in the upper optical waveguide film. Refractive index of the PRI material varies from 1.5 to 1.7 with write beam exposure. Wavelengths of write beams and probe beams are 650 nm and 850 nm, respectively. Polarization is *E*//*z*.

In Figure 17, negative and positive lateral misalignments respectively represent left-side and right-side dislocations of the upper optical waveguide film. In a lateral misalignment range from –0.12 to 0.75 μm, R-SOLNET is led to the wavelength filter location by the "pulling water" effect.

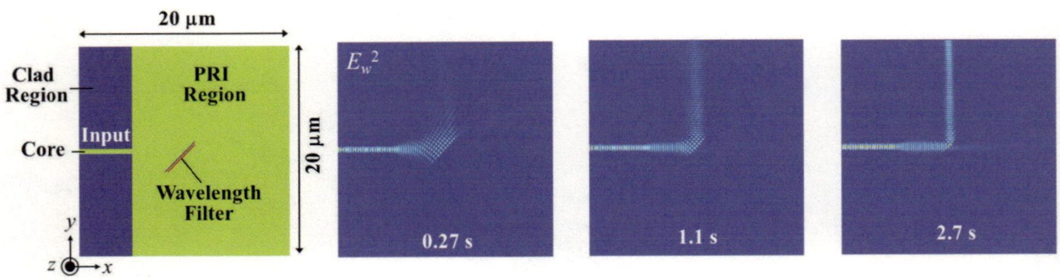

Figure 16. Simulation of the self-focusing of a write beam

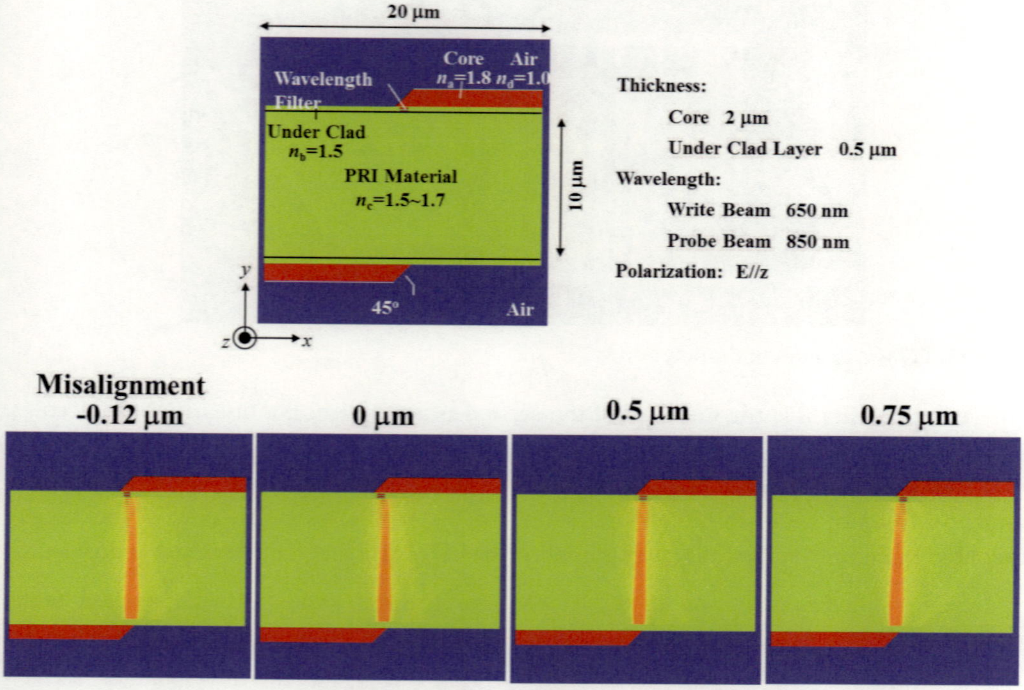

Figure 17. Simulation of optical Z-connections with vertical waveguides of R-SOLNET

3.2. Enhancement of pockels effect by controlling wavefunction shapes

In the integrated 3-D optical interconnects, high-speed/small-size optical modulators and optical switches are the key components, for which high-performance electro-optic (EO) materials are required. So far, EO materials such as $LiNbO_3$ (LN) and quantum dots of III-V compound semiconductors [22] have been developed. As the next generation EO material, organic materials with π-conjugated systems attract interest because they have both "large optical nonlinearity" and "low dielectric constant" characteristics. For example, the styrylpyridinium cyanine dye (SPCD) thin film was found to exhibit a large EO coefficient r of 430 pm/V [23], about 14 times that of LN.

Organic EO materials are classified into poled polymers, molecular crystals, and conjugated polymers. Among these, the polymer MQD with conjugated polymer backbones seems most promising because it enables a wavefunction control between the one dimensional and the zero-dimensional to enhance the EO effect. In order to apply the polymer MQD to EO waveguides, locations and orientations of the polymer wires, as well as molecular arrangements, should be controlled by using MLD as shown in Figure 18.

In the below part of this subsection, the enhancement of the Pockels effect, which is an EO effect inducing refractive index changes proportional to applied electric fields, in the polymer MQD by controlling wavefunction shapes is theoretically predicted [4,24] using the molecular orbital calculations.

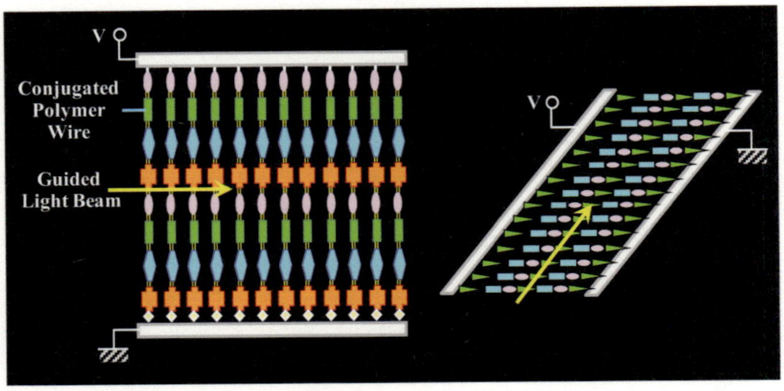

Figure 18. EO waveguides of the polymer MQD

Hyperpolarizability β is the measure of the second-order optical nonlinearity of molecules. The EO coefficient for the Pockels effect is proportional to β. In the two level model, $\beta \propto r_{gn}{}^2 \Delta r$. Here, r_{gn} and Δr are respectively the transition dipole moment and the dipole moment difference between the ground state and the excited state. Then, as schematically illustrated in Figure 19, the following guideline for EO coefficient enhancement is derived.

- Promote wavefunction overlap between the ground state and the excited state, increasing r_{gn}.
- Promote wavefunction separation between the ground state and the excited state, increasing Δr.

For three wavefunction shapes shown in Figure 19, from left to right, the wavefunction separation increases while the wavefunction overlap decreases, i.e., Δr increases and r_{gn} decreases. Therefore, β is expected to have its maximum in a wavefunction shape of intermediate wavefunction separation with an appropriate conjugated system dimensionality existing somewhere between zero and one, i.e., between a quantum dot (QD) condition and a quantum wire condition.

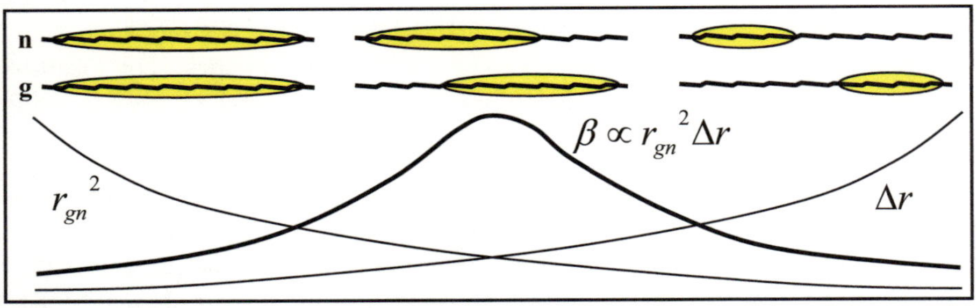

Figure 19. Guidelines for EO coefficient enhancement

Such optimized wavefunction conditions can be obtained by adjusting the QD lengths and donor/acceptor substitution sites in the QDs using the push/pull effects of the donors/acceptors. Three types of models for conjugated polymer wires with poly-diacetylene (PDA) backbones are considered as shown in Figure 20. $-NH_2$ is the donor (D) and $-NO_2$ the acceptor (A). DA, DAAD, and DDAA represent the types of donor/acceptor substitution, and the numbers following them indicate the number of carbon sites N_C, which corresponds to molecular lengths. The calculation, for which the details are described in articles published elsewhere [4,24], revealed that wavefunction separations increase in the order of DA34, DAAD34, and DDAA30.

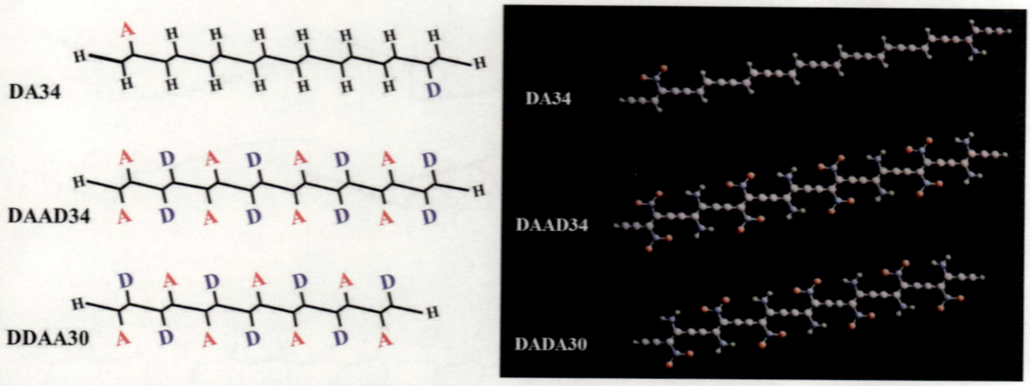

Figure 20. Models of OE conjugated polymer wires

Figure 21(a) shows Δr, r_{gn}, and ρ_β of these model molecules. Here, ρ_β is β per 1 nm in molecular lengths. Δr increases and r_{gn} decreases in the order of DA34, DAAD34, and DDAA30, which makes ρ_β of DAAD34 in between maximum. This parallels to the tendency of the qualitative guideline shown in Figure 19. It is found from Figure 21(b) that the molecular length affects ρ_β and that adjusting the length improves the nonlinear optical effect. In DA and DAAD molecules, ρ_β reaches the maximum values near $N_C = 20$, corresponding to ~2 nm in molecular length. Assuming PDA wire density of 1.3×10^{14} 1/cm^2, the expected maximum EO coefficient of the DA and DAAD is about 3000 pm/V, which is 100 times larger than the EO coefficient r_{33} of LN. It is therefore concluded that a large EO effect will be obtained by controlling the wavefunction dimensionality and separation.

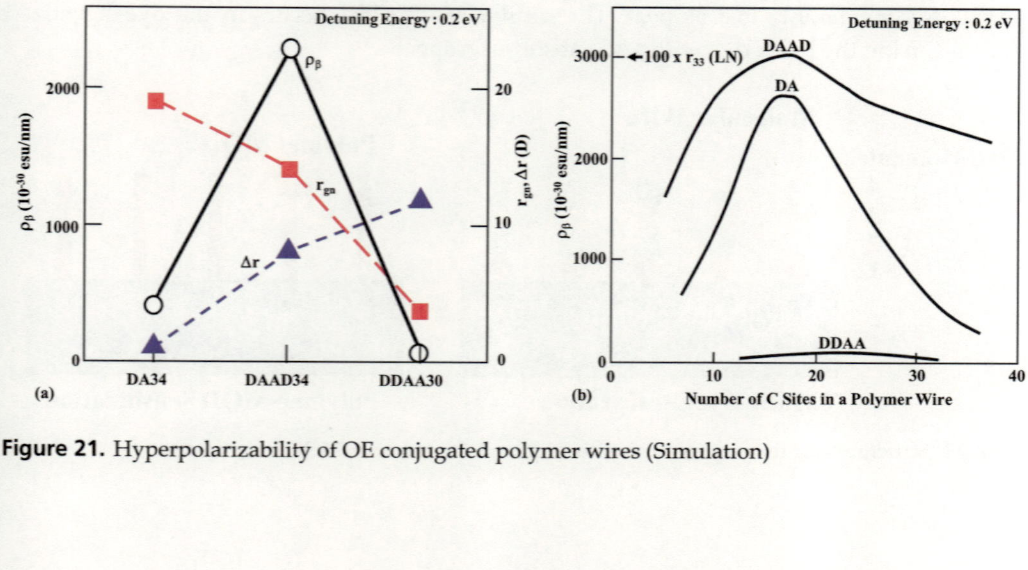

Figure 21. Hyperpolarizability of OE conjugated polymer wires (Simulation)

The molecular orbital calculations revealed that energy gaps of DAAD molecules are smaller than those of PDA with no donor/acceptor substitution. Using the phenomena, it is possible to insert many DAAD molecules into a PDA backbone to construct a DAAD-type polymer MQD shown in Figure 22 [4].

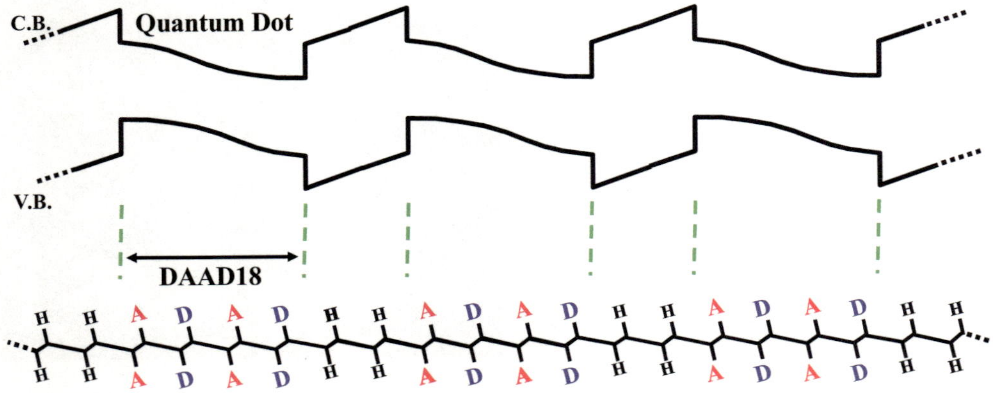

Figure 22. Polymer MQD containing DAAD18-type QDs

4. Applications to solar cells

4.1. Sensitized solar cells

4.1.1. Concept of multi-dye sensitization and polymer-MQD sensitization

Structures for the multi-dye sensitization and the polymer-MQD sensitization are shown in Figure 23 [4,10,11]. In the former [4,10], molecular wires consisting of dye molecules, and in the latter [4,11], polymer wires having MQDs with different dot lengths are grown on semiconductor surfaces by MLD.

In Si solar cells, as shown in Figure 24(a), since the band gap is narrow, excess energy of photons is considerably lost as heat. The similar energy loss occurs in the dye-sensitized solar cells using the black dye with wide absorption spectra.

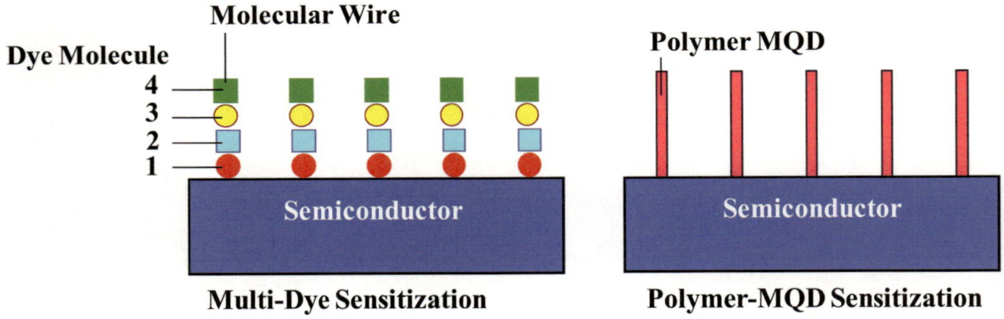

Figure 23. Structures for the multi-dye sensitization and the polymer-MQD sensitization

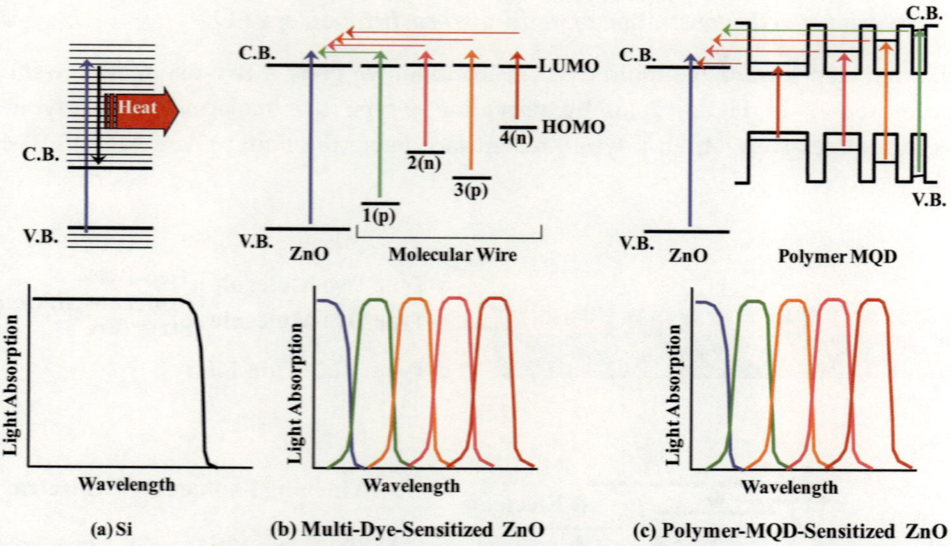

Figure 24. Schematic illustrations of energy levels and absorption spectra in Si, multi-dye-sensitized ZnO, and polymer-MQD-sensitized ZnO

In the multi-dye sensitization of ZnO, where molecular wires consisting of p-type and n-type dyes are grown on ZnO, the energy diagrams and absorption spectra are schematically drawn as Figure 24(b). The absorption wavelength region can be divided into narrow absorption bands of individual dyes, suppressing the energy loss arising from the excess photon energy. In the polymer-MQD sensitization, the similar effect is expected as drawn in Figure 24(c).

In Figure 25, the z-scheme-like sensitization is shown [4]. An electron excited by a photon with wavelength of λ_1 in molecule 1 is injected into ZnO. An electron excited by a photon with λ_2 in molecule 2 is transferred to HOMO of molecule 1. The hole left in molecule 2 is compensated by a redox system. This sensitization mechanism suppresses the energy loss arising from the excess photon energy, and at the same time, it increases the difference in energy between the Fermi level of ZnO and the standard electrode potential of the redox system to increase the generated voltage in the photo-voltaic device. The similar z-scheme-like sensitization might arise from molecules with the four-level two-photon absorption characteristics [4].

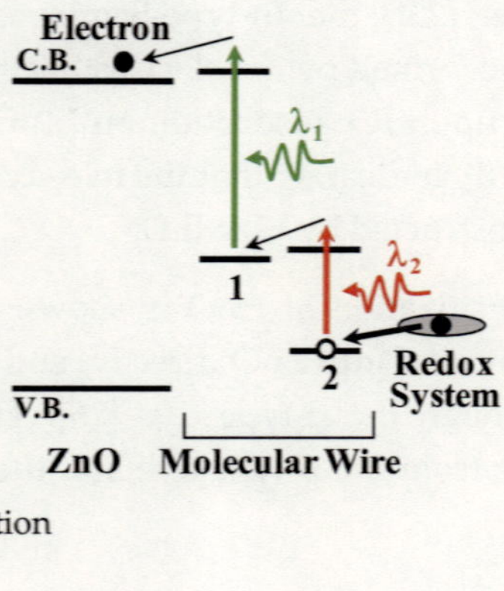

Figure 25. Z-scheme-like sensitization

4.1.2. Experimental demonstration of multi-dye sensitization of ZnO

As the first step toward the multi-dye sensitization, we grew a two-dye-molecule-stacked structure shown in Figure 26(a) by providing p-type dye molecules and n-type dye molecules successively on an n-type ZnO surface using the liquid-phase MLD (LP-MLD) [4,10].

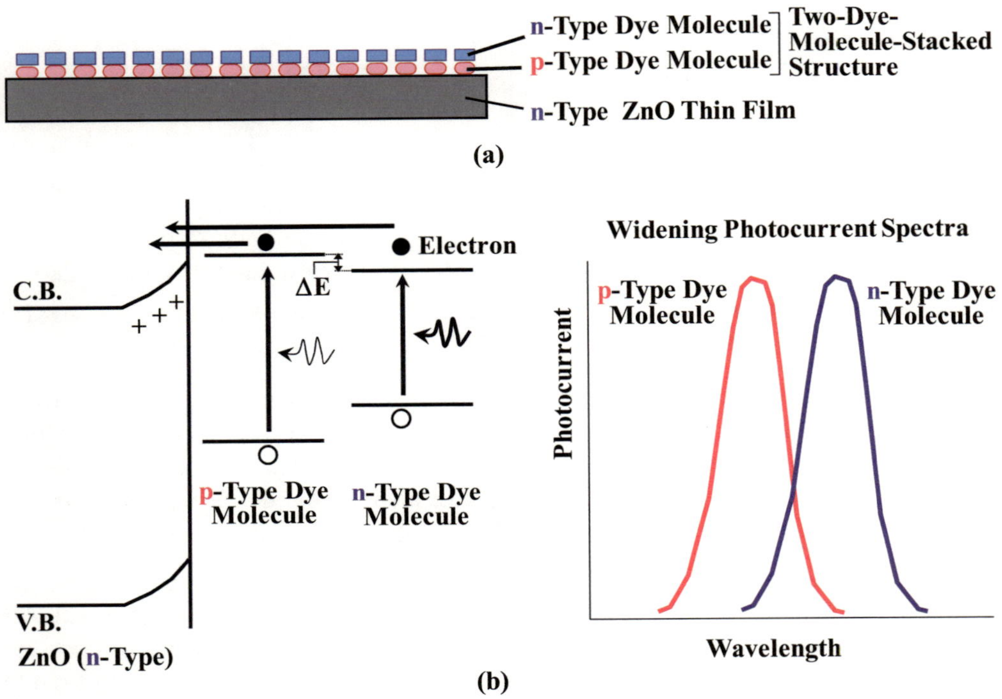

Figure 26. (a)Structure and (b) mechanism of two-dye sensitization

In the present experiments, we used rose bengal (RB) for the p-type dye and crystal violet (CV) for the n-type dye. Because the p-type dyes tend to accept electrons and the n-type dyes tend to donate electrons [25], the [n-type-Semiconductor/p-type-Dye/n-type-Dye] structure of [ZnO/RB/CV] can be formed by the electrostatic force. The surface potential was found to shift to negative side upon RB adsorption on [ZnO], and to shift to positive side upon CV adsorption on [ZnO/RB], indicating that the two-dye-molecule-stacked structure of [ZnO/RB/CV] was definitely constructed by LP-MLD.

Mechanism of the two-dye sensitization of ZnO is shown in Figure 26(b). The electrons excited in the p-type dye are injected into ZnO directly, and electrons excited in the n-type dye are injected into ZnO through the p-type dye [26]. Thus, by using the p/n-stacked structure, photocurrents arising from the p-type dye and the n-type dye are superposed to widen the spectra.

As Figure 27(a) shows, absorption spectrum of CV is located in a longer wavelength region comparing with that of RB. Consequently, while the photocurrent is not observed at 633 nm in [ZnO/RB], the photocurrent spectrum extends to 633 nm in [ZnO/RB/CV], exhibiting the spectral widennig arising from the superpositon of RB and CV.

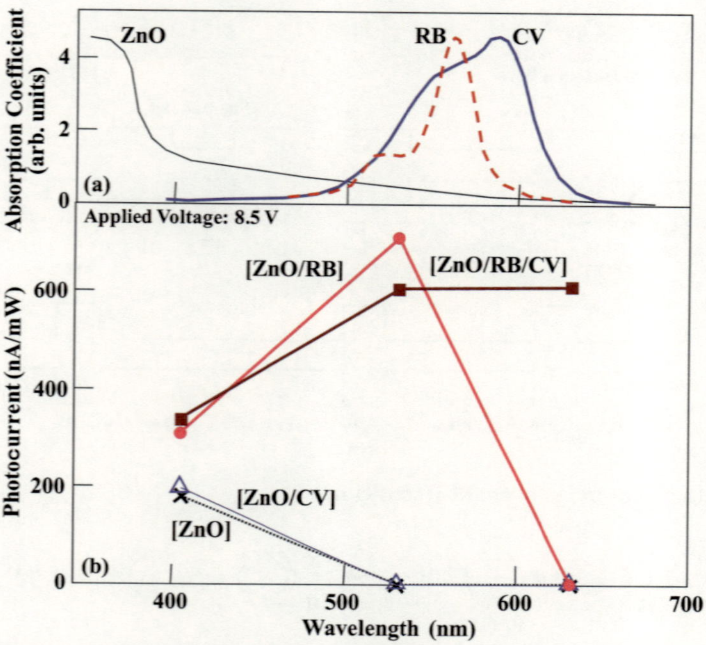

Figure 27. (a)Absorption spectra of RB and CV, and (b) photocurrent spectra of sensitized ZnO

4.1.3. Polymer Multiple Quantum Dots (MQDs)

In Figure 28, polymer MQD structures of OTPTPT, OTPT, OT, and 3QD [4,11], which contains the OT, OTPT and OTPTPT structures in a wire, are shown, as well as a structure of poly-azomethine (poly-AM) quantum wire. These are formed by using terephthalaldehyde (TPA), *p*-phenylenediamine (PPDA) and oxalic dihydrazide (ODH) as source molecules with chemical reactions shown in Figure 29. In poly-AM, TPA and PPDA are alternately connected, so the wavefunction of π-electrons is delocalized across the polymer wire. For OTPTPT, molecules are connected in the sequence of -ODH-TPA-PPDA-TPA-PPDA-TPA-ODH---. The regions between ODHs are regarded as QDs of ~3 nm long. For OTPT, the molecular sequence of -ODH-TPA-PPDA-TPA-ODH--- yields QDs of ~2 nm long. For OT, ODH and TPA are alternately connected, resulting in very short QDs of ~0.8 nm long.

Figure 30 shows the absorption spectra of polymer MQDs. The absorption peak shifts to shorter wavelengths in the trend: OTPTPT, OTPT, and OT. This trend follows that of decreasing QD length, being attributed to the quantum confinement of π-electrons in the QDs. In 3QD, a broad absorption band extending from ~480 to ~300 nm appears. The measured spectrum is fairly coincident with the predicted spectrum that is the superposition of absorption bands of OT, OTPT and OTPTPT.

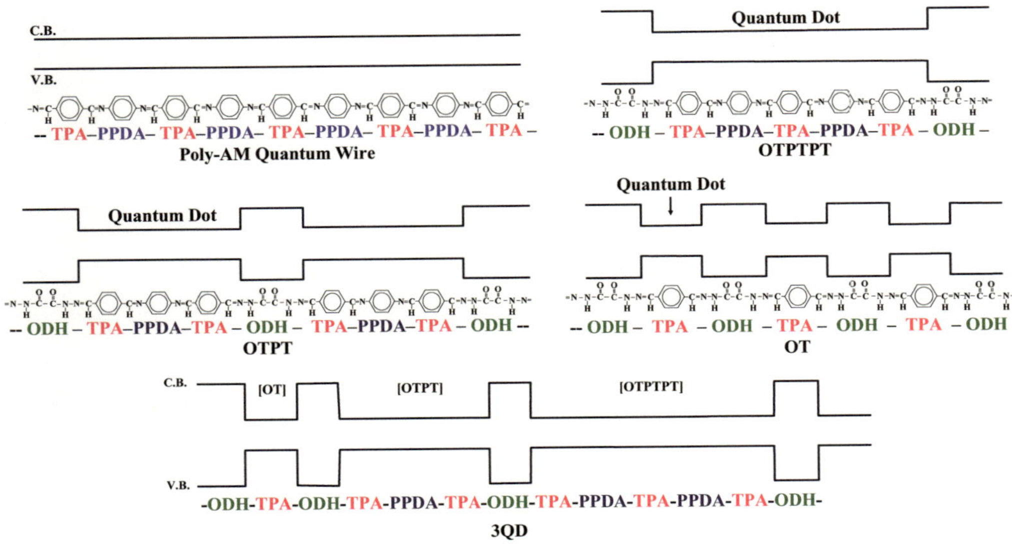

Figure 28. Quantum wire and polymer MQD structures

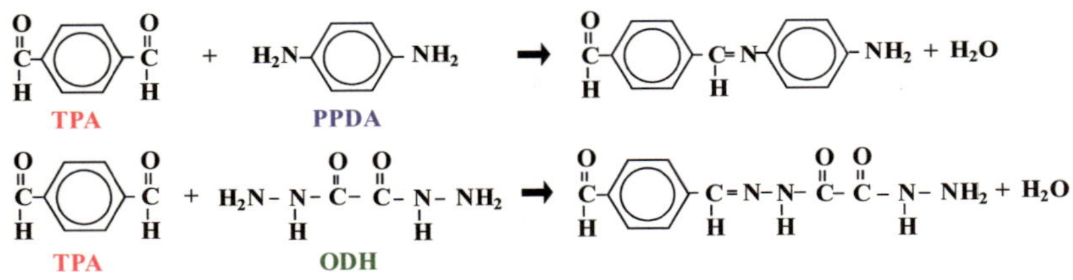

Figure 29. Source molecules and chemical reactions

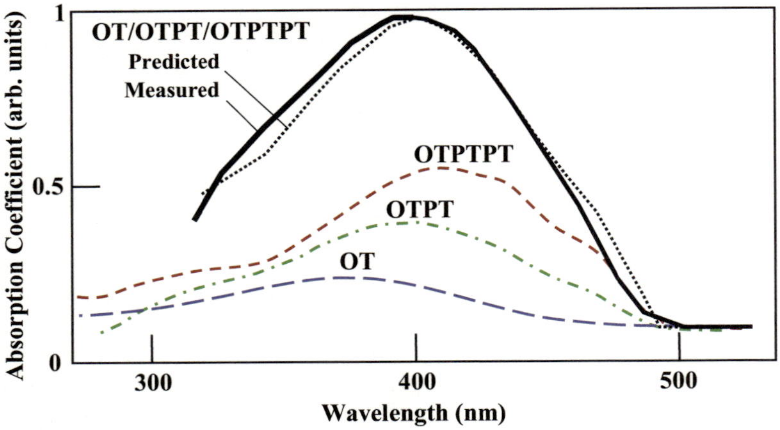

Figure 30. Absorption spectra of polymer MQD structures

4.2. Waveguide-type sensitized solar cells

In the conventional dye-sensitized solar cells, in order to increase the number of adsorbed dye molecules on semiconductor surfaces, porous semiconductors are used. In this case, the crystallinity of the semiconductors is degraded and the porous structure narrows the electron transporting channels, causing an increase in the internal resistivity.

In the waveguide-type sensitized solar cell shown in Figure 31 [4,10], a thin-film semiconductor with a flat surface and high crystallinity is used. So, the internal resistivity decreases. However, if the "normally-incident light" configuration is used for light absorption, the light passes through only monomolecular layer of dye, resulting in very small light absorption. In the "guided light" configuration the light passes through a lot of dye molecules to enhance light absorption. Thus, high-performance sensitized solar cells are expected.

We estimated the effect of the "guided light" configuration on the photocurrent enhancement using setups shown in Figure 32 [4,10]. Photocurrents were measured by a slit-type Al electrode. For the "normally-incident light" configuration, the light was introduced onto the ZnO surface from an optical fiber. For the "guided light" configuration, the light was introduced into the ZnO thin film from the edge. As shown in Figure 33, photocurrent enhancement by a factor of 5~15 is observed in the "guided light" configuration.

In the waveguide-type sensitized solar cells, the optical coupling to the thin-film semiconductor is a concern. SOLNET might be one of the solutions, enabling light beams to be coupled efficiently into the film [4].

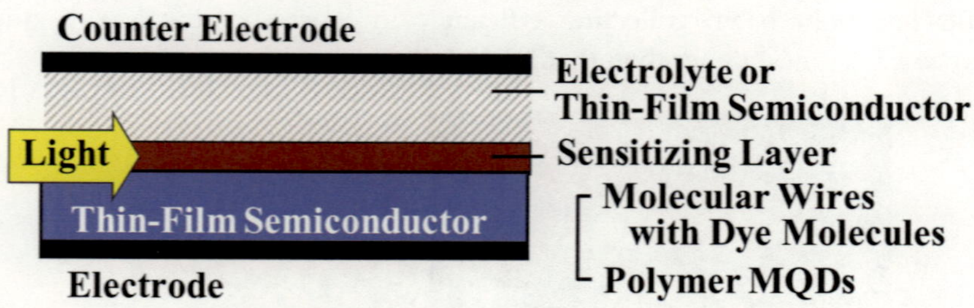

Figure 31. Waveguide-type sensitized solar cell

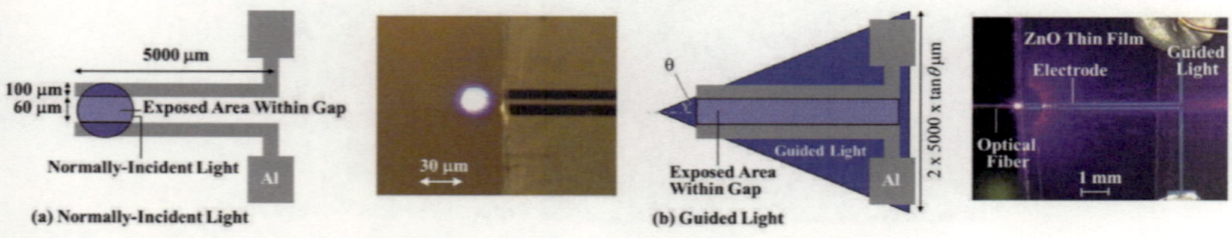

Figure 32. Setups of photocurrent measurements

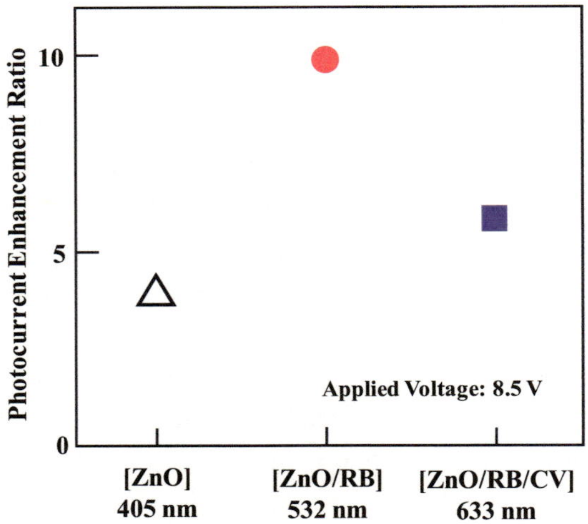

Figure 33. Photocurrent enhancement induced by "guided light" configuration

4.3. Film-based integrated solar cells

Conventional solar cells are material-consuming since semiconductors are placed all over the modules. To reduce the semiconductor material consumption and to provide wide-angle light beam collecting capability to the systems, and at the same time, to make systems flexible and compact, we proposed the film-based integrated solar cell with optical waveguides [1,4]. Figure 34 shows the schematic illustration of the proposed solar cell, in which semiconductor flakes are placed partially in a light beam collecting film by the heterogeneous integration process such as PL-Pack with SORT [17]. Figure 34 also shows a photograph of an array of tapered vertical waveguides fabricated by the built-in mask method [20]. The light beam collecting efficiency in tapered vertical waveguides was estimated to be 1.5-4 times higher than that in straight vertical waveguides.

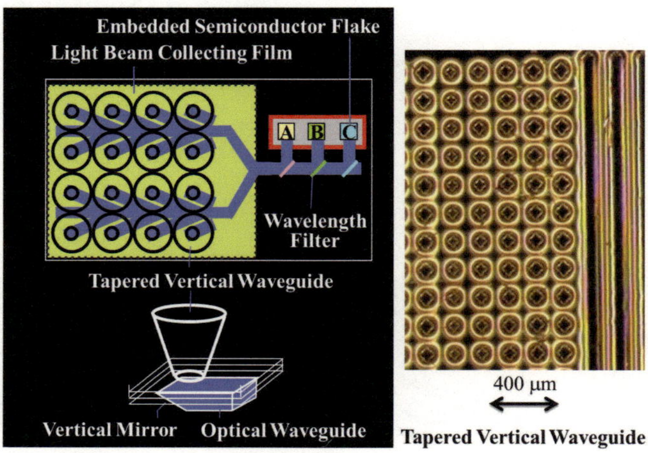

Figure 34. Film-based integrated solar cells utilizing optical waveguides

5. Cancer therapy

Figure 35 depicts how LP-MLD can be applied to cancer therapy [4,11,13]. In step 1, Molecule A is injected into a human body. Molecule A is adsorbed in cancer cells selectively, and extra Molecule A is excreted. In step 2, Molecule B is injected to be connected to Molecule A. Similarly, by injecting Molecule C and D successively, tailored materials having a structure of A/B/C/D can be constructed at the cancer sites.

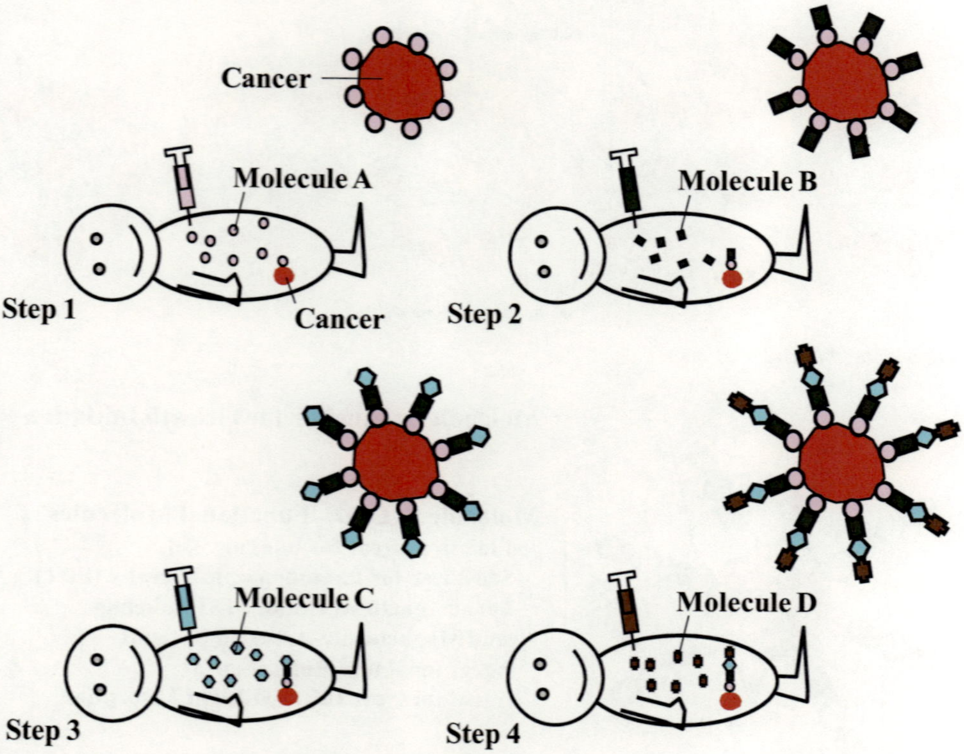

Figure 35. Application of LP-MLD to cancer therapy

5.1. *In-situ* selective drug synthesis

LP-MLD can be regarded as a kind of *in-situ* synthesis within human bodies. When a molecule of a drug is too large, as schematically illustrated in Figure 36(a), it might be difficult for the drug to reach deep inside the cancer through narrow channels. The deep drug delivery might become possible by building up the large drug from small component molecules at cancer sites *in situ* by LP-MLD as shown in Figure 36(b). For toxic drugs, they might be delivered without attacking normal cells by building them up from non-toxic components *in situ* at cancer sites.

Furthermore, using LP-MLD, multi-functional tailored structures are expected to grow selectively at cancer sites as Figure 37 shows.

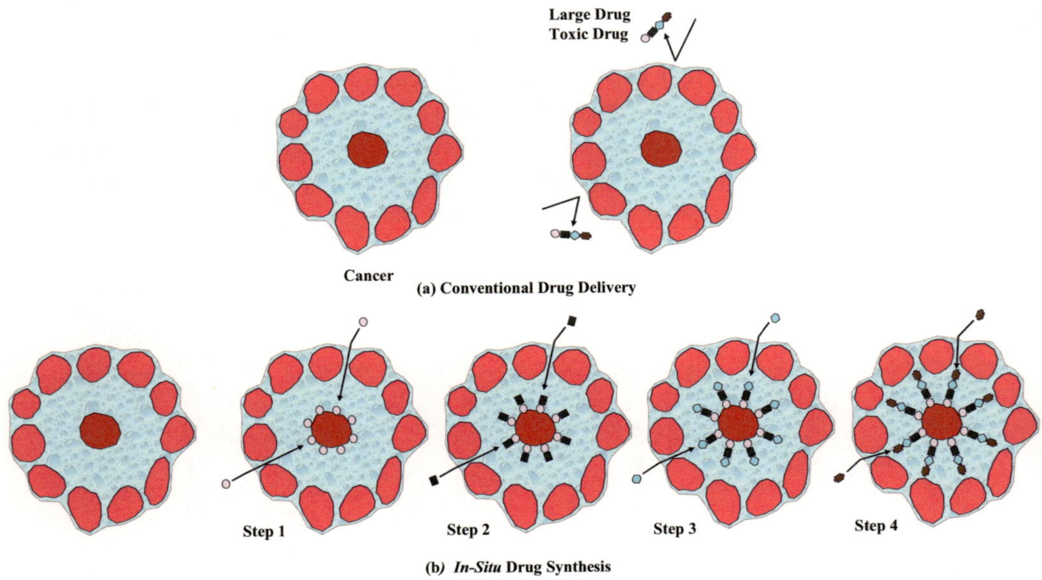

Figure 36. *In-situ* drug synthesis

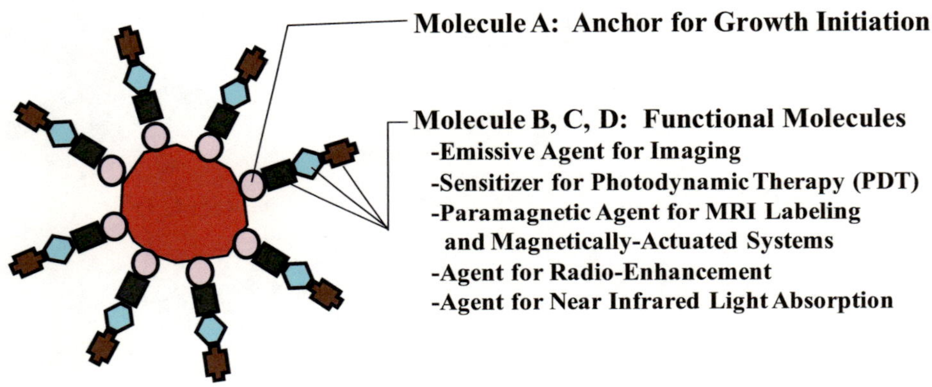

Figure 37. Selective delivery of multi-functional materials to cancer cells by LP-MLD

5.2. SOLNET-assisted laser surgery

5.2.1. Concept of SOLNET-assisted laser surgery

Figure 38 shows the concept of the SOLNET-assisted laser surgery. Luminescent molecules are attached to cancer cells by LP-MLD. After inserting an optical fiber and a PRI material around the cancer sites, a write beam is introduced from the optical fiber to construct R-SOLNET between the optical fiber and the cancer sites. By introducing surgery beams into the R-SOLNET via the optical fiber, cancer cells are destroyed selectively. By detecting the luminescence emitted from the luminescent molecules, *in-situ* monitoring might be possible.

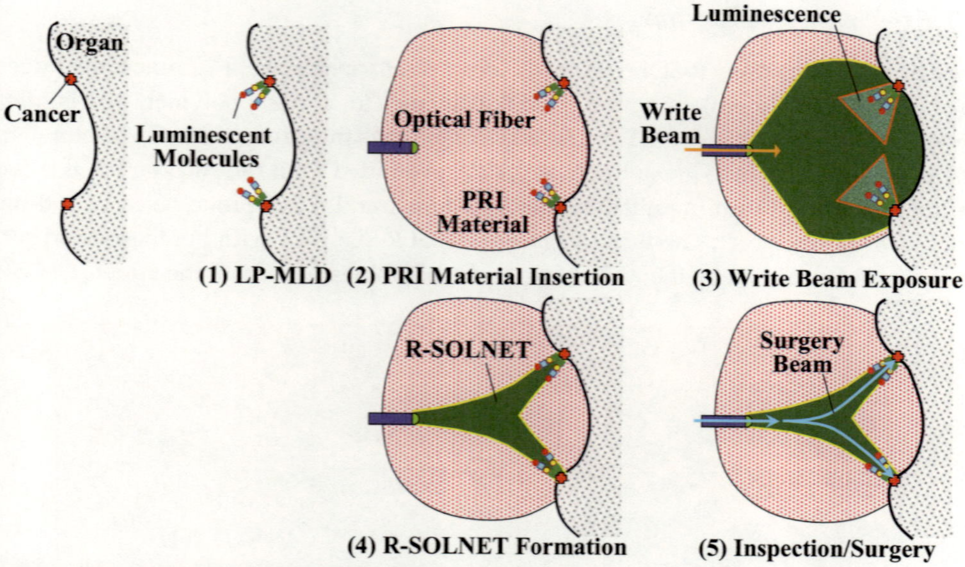

Figure 38. Concept of SOLNET-assisted laser surgery

5.2.2. Simulation by FDTD method

In the model shown in Figure 39, a 600-nm wide luminescent target is placed in a PRI material with a lateral offset of 600 nm from the axis of the optical waveguide with core width of 1.2 μm. The wavelengths of the write beam, the luminescence, and the surgery beam are 650, 700, and 650 nm, respectively. Luminescence efficiency is 0.36. It is found that, with writing time, R-SOLNET is gradually constructed between the optical waveguide and the target. As a result, the surgery beam is guided to the target site [1,13].

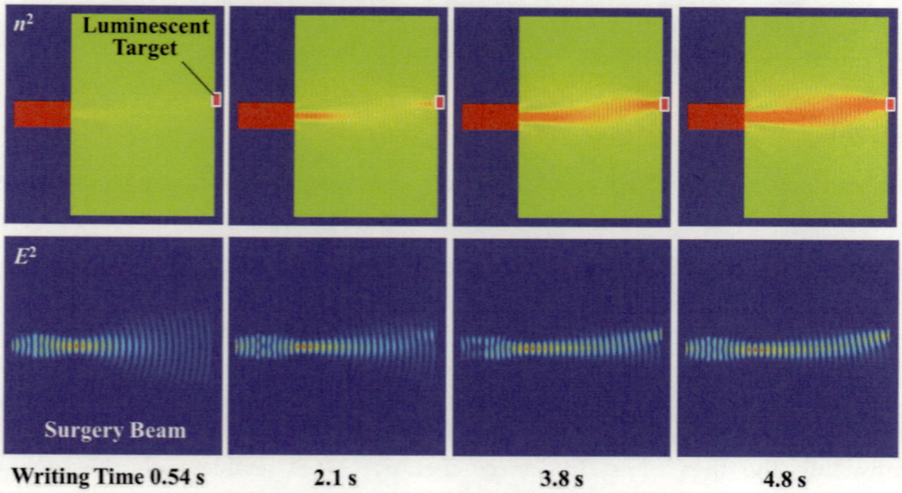

Figure 39. Simulation of R-SOLNET using a luminescent target

5.2.3. Experimental demonstration

In order to demonstrate R-SOLNET with luminescent materials, a luminescent target of tris(8-hydroxyquinolinato)aluminum (Alq3) was put in a photopolymer. As Figure 40 shows, when a write beam of 405 nm in wavelength was introduced into the photopolymer from an optical fiber, blue-green luminescence was emitted from the Alq3 target. It is found that the blue write beam is pulled to the Alq3 target and a red probe beam of 650 nm is guided toward the target. These results indicate that R-SOLNET with luminescent materials was successfully formed by the write beam and the luminescence from the target [1,13,27].

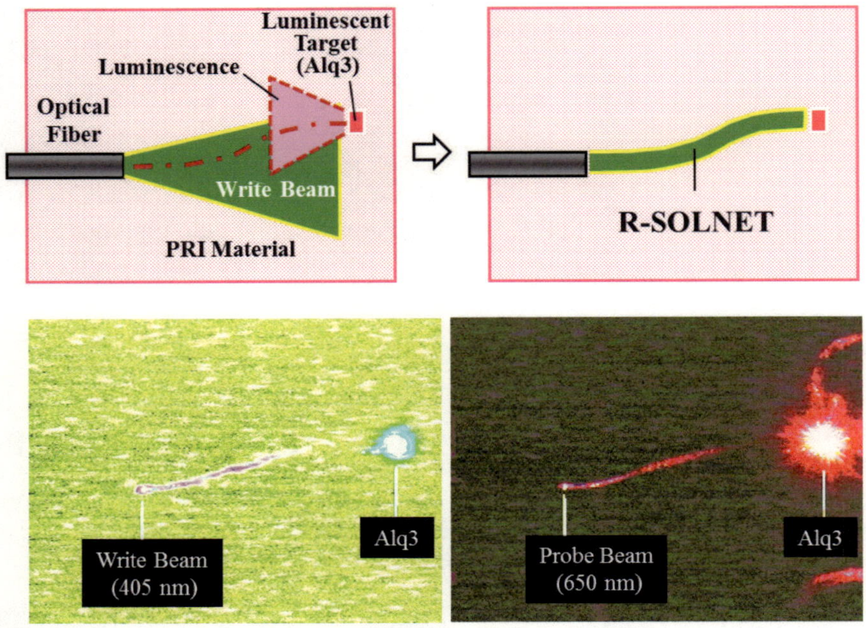

Figure 40. R-SOLNET formed between an optical fiber and a luminescent target

6. Summary

Our core technologies, SOLNET and MLD, were reviewed, and their applications to optical interconnects within computers, solar cells, and cancer therapy were presented.

In the integrated optical interconnects within computers based on the self-organized 3-D optical circuits, SOLNET is used for vertical waveguides in optical Z-connections and the optical solder for self-aligned optical couplings. MLD is expected to contribute to improving the light modulator/optical switch performance. In the sensitized solar cells, MLD is used for growth of the multi-dye molecular wires and the polymer MQDs that are sensitizing layers on thin films of oxide semiconductors. SOLNET is expected to couple light beams efficiently into the thin films. In cancer therapy, MLD enables selective delivery of large/toxic drugs and multi-functional materials to cancer cells, as well as the SOLNET-assisted laser surgery for the self-aligned laser beam guiding to cancer sites.

Author details

Tetsuzo Yoshimura

Tokyo University of Technology, School of Computer Science, Hachioji, Tokyo, Japan

7. References

[1] Yoshimura T (2012) Optical Electronics: Self-Organized Integration and Applications. Pan Stanford Publishing Pte. Ltd., Singapore.

[2] Yoshimura T, Sotoyama W, Motoyoshi K, Ishitsuka T, Tsukamoto K, Tatsuura S, Soda H, Yamamoto T (2000) Method of producing optical waveguide system, optical device and optical coupler employing the same, optical network and optical circuit board. U.S. Patent 6,081,632.

[3] Yoshimura T, Roman J, Takahashi Y, Wang W.V, Inao M, Ishitsuka T, Tsukamoto K, Motoyoshi K, Sotoyama W (2000) Self-Organizing Waveguide Coupling Method 'SOLNET' and Its Application to Film Optical Circuit Substrates. Proc. 50th Electron. Components Technol. Conf. (ECTC): 962–969.

[4] Yoshimura T (2011) Thin-Film Organic Photonics: Molecular Layer Deposition and Applications. CRC/Taylor & Francis, Boca Raton, Florida.

[5] Yoshimura T, Yano E, Tatsuura S, Sotoyama W (1995) Organic functional optical thin film, fabrication and use thereof. US Patent 5,444,811.

[6] Yoshimura T, Tatsuura S, Sotoyama W (1991) Polymer films formed with monolayer growth steps by molecular layer deposition. Appl. Phys. Lett. 59: 482-484.

[7] Yoshimura T, Takahashi Y, Inao M, Lee M, Chou W, Beilin S, Wang W.V, Roman J, Massingill T (2002) Systems Based on Opto-Electronic Substrates with Electrical and Optical Interconnections and Methods for Making. U.S. Patent 6,343,171 B1.

[8] Yoshimura T, Roman J, Takahashi Y, Lee M, Chou B, Beilin S.I, Wang W.V, Inao M (1999) Proposal of Optoelectronic Substrate with Film/Z-Connection Based on OE-Film. Proc. 3rd IEMT/IMC Symposium: 140–145.

[9] Yoshimura T, Roman J, Takahashi Y, Lee M, Chou B, Beilin S.I, Wang W.V, Inao M (2000) Optoelectronic Scalable Substrates Based on Film/Z-connection and Its Application to Film Optical Link Module (FOLM). Proc. SPIE. 3952: 202–213.

[10] Yoshimura T, Watanabe H, Yoshino C (2011) Liquid-Phase Molecular Layer Deposition (LP-MLD): Potential Applications to Multi-Dye Sensitization and Cancer Therapy. J. Electrochem. Soc. 158: 51-55.

[11] Yoshimura T, Ebihara R, Oshima A (2011) Polymer Wires with Quantum Dots Grown by Molecular Layer Deposition of Three Source Molecules for Sensitized Photovoltaics. J. Vac. Sci. Technol. A. 29: 051510-1-6.

[12] Shioya R, Yoshimura T (2009) Design of Solar Beam Collectors Consisting of Multi-Layer Optical Waveguide Films for Integrated Solar Energy Conversion Systems. J. Renewable Sustainable Energy. 1: 033106 1-15.

[13] Yoshimura T, Yoshino C, Sasaki K, Sato T, Seki M (2012) Cancer Therapy Utilizing Molecular Layer Deposition (MLD) and Self-Organized Lightwave Network (SOLNET)

-Proposal and Theoretical Prediction-. IEEE J. Select. Topics in Quantum Electron. 18. Biophotonics 1. May/June [to be published].

[14] Brauchle C, Wild U.P, Burland D.M, Bjorkund G.C, Alvares D.C (1982) Two-Photon Holographic Recording with Continuous-Wave Lasers in the 750–1100 nm Range. Opt. Lett. 7: 177–179.

[15] Yoshimura T, Kaburagi H (2008) Self-Organization of Optical Waveguides between Misaligned Devices Induced by Write-Beam Reflection. Appl. Phys. Express. 1: 06200.

[16] Pessa M, Makela R, Suntola T (1981) Characterization of Surface Exchange Reactions Used to Grow Compound Films. Appl. Phys. Lett. 31: 131-133.

[17] Yoshimura T, Ojima M, Arai Y, Asama K (2003) Three-Dimensional Self-Organized Micro Optoelectronic Systems for Board-Level Reconfigurable Optical Interconnects — Performance Modeling and Simulation. IEEE J. Select. Top. Quantum Electron. 9: 492–511.

[18] Yoshimura T, Roman J, Takahashi Y, Beilin S.I, Wang W.V, Inao M (1999) Optoelectronic Amplifier/Driver-Less Substrate <OE-ADLES> for Polymer-Waveguide-Based Board-Level Interconnection — Calculation of Delay and Power Dissipation. Nonlinear Opt. 22: 453–456:.

[19] Yoshimura T, Suzuki Y, Shimoda N, Kofudo T, Okada K, Arai Y, Asama K (2006) Three-Dimensional Chip-Scale Optical Interconnects and Switches with Self-Organized Wiring Based on Device-Embedded Waveguide Films and Molecular Nanotechnologies. Proc. SPIE. 6126: 612609-1-15.

[20] Yoshimura T, Inoguchi T, Yamamoto T, Moriya S, Teramoto Y, Arai Y, Namiki T, Asama K (2004) Self-Organized Lightwave Network Based on Waveguide Films for Three-Dimensional Optical Wiring Within Boxes. J. Lightwave. Technol. 22:2091-2100.

[21] Yoshimura T, Wakabayashi K, Ono S (2011) Analysis of Reflective Self-Organized Lightwave Network (R-SOLNET) for Z-Connections in Three-Dimensional Optical Circuits by the Finite Difference Time Domain Method. IEEE J. Select. Topics in Quantum Electron. 17: 566-570.

[22] Yoshimura T, Futatsugi T (2001) Non-linear optical device using quantum dots. U.S. Patent 6,294,794B1.

[23] Yoshimura T (1987) Characterization of the EO effect in styrylpyridinium cyanine dye thin-film crystals by an ac modulation method. J. Appl. Phys. 62: 2028-2032.

[24] Yoshimura T (1989) Enhancing Second-Order Nonlinear Optical Properties by Controlling the Wave Function in One-Dimensional Conjugated Molecules. Phys. Rev. B40: 6292–6298.

[25] Yoshimura T, Kiyota K, Ueda H, Tanaka M (1979) Contact Potential Difference of ZnO Layer Adsorbing p-Type Dye and n-Type Dye. Jpn. J. Appl. Phys. 18: 2315-2316.

[26] Yoshimura T, Kiyota K, Ueda H, Tanaka M (1981) Mechanism of Spectral Sensitization of ZnO Coadsorbing p-Type and n-Type Dyes. Jpn. J. Appl. Phys. 20: 1671-1674.

[27] Seki M, Yoshimura T (2012) Proposal and FDTD Simulation of Reflective Self-Organizing Lightwave Network (R-SOLNET) Using Phosphor. Proc. SPIE. 8267 82670V-1 - 9.

Chapter 6

Bio-Inspired Photonic Structures: Prototypes, Fabrications and Devices

Feng Liu, Biqin Dong and Xiaohan Liu

Additional information is available at the end of the chapter

http://dx.doi.org/10.5772/50199

1. Introduction

Like the ability of electron regulation of electronic semiconductors, the photonic analogs usually considered as photonic structure materials are regarded as essential for light manipulation [1, 2]. With particular designs of photonic structures, they are expected to achieve different far-field and near-field optical features and thus lead to a perspective in all-optical circuit [3]. Though humankind has entered the nano-scale realm several decades ago, it is still a hard task for engineers to explore novel optical functional devices due to the limited experiences and originalities on artificial photonic structures design and the desired optical features. Additionally, it is also great challenges to fabricate photonic structures, owing to their sub-optical-wavelength to sub-micron featured sizes, especially in a high dimensional way by nano-fabrication technologies today.

By contrast, nature are found to develop photonic structures millions of years before our initial attempts. Diversified photonic structures, most of which are sophisticated and hierarchic, are revealed in beetles, butterflies, sea animals and even plants in recent surveys [4–10]. The exhibited optical features are regarded to have particular biological functions such as signal communications, conspecific recognition, and camouflage, which are optimized under selection pressure. Naturally, the occurring photonic structures provide us ideal 'blueprints' on design and stimulate similar optical functional devices. Various fabrication methods of bio-inspired photonic structures are explored [11–16]. By chemical methods (e.g. Sol-Gel, colloidal crystallization, chemical systhesis), nanoimprint lithography and nanocasting, physical layer deposition (PLD), atomic layer deposition (ALD), and etc., bio-inspired photonic structures, their reverse counterparts, and applications are achieving greater success than ever before.

This Chapter will review the typical bio-inspired photonic structures and focus on the biomimetic fabrications, the corresponding optical functions and the prototypes of optical devices. It is organized as follows: four subsections are introduced in Section 2, in which each describes one catagory of bio-inspired optical functional devices. In every subsection, the nature prototype is introduced first, then followed by biomimetic fabrication methods and

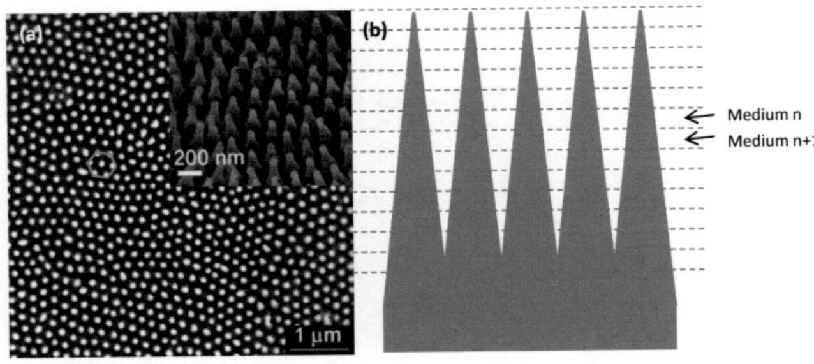

Figure 1. Antireflection structures. (a) Hexagonal arranged tapered pillars root in the surface of cicada wings in top view (tilted view of the pillar array is showed in the inset) [19]; (b) The tapered pillars lead to gradual refraction index variation in view of effective theory and then minimize the reflection according to Fresnel relations.

optical features of artificial analogs, finally closed on the bio-inspired optical devices. A brief perspective is given in Section 3.

2. Bio-inspired optical functional devices

2.1. Anti-reflection devices

2.1.1. Prototypes

In arthropodal animals such as butterflies, nipple arrays with typical spacing of optical wavelength or subwavelength are commonly found on surfaces of their compound eyes, which are believed helpful to the light-harvest efficiency of the biological visual system [17]. With optical impedance matching to the ambience, the light transmission are enhanced. Another analogous examples are the transparent wings of some lepidoptera insects like hawkmoths [18] and cicada [19] (Fig.1 (a)). The tapered pillars lead to gradual changes of refraction index in view of effective medium theory (Fig.1 (b)) and therefore play a key role in minimizing the reflection over broadband and large viewangles. In order to physically explain the anti-reflection origin, the Fresnel equations are given as follows.

$$r_s = |\frac{n_1 cos i_1 - n_2 cos i_2}{n_1 cos i_1 + n_2 cos i_2}|, \tag{1}$$

$$r_p = |\frac{n_1 cos i_2 - n_2 cos i_1)}{(n_1 cos i_2 + n_2 cos i_1}|, \tag{2}$$

where r_s and r_p refer to reflection coefficients of s polarised light (the electric field of the light perpendicular to the incident plane) and p polarised light (the electric field in the incident light plane), n_1 and n_2 are refraction indices of neighboring mediums, respectively. The relationship between incident angle i_1 and refraction angle i_2 is given by Snell's Law $n_1 \sin i_1 = n_2 \sin i_2$. From the equations, e.g., for normal incidence, r_s and r_p are suppressed to near zero if n_1 and n_2 have very close values [20, 21].

2.1.2. Bio-inspired fabrications

Inspired by nature, many efforts are made in exploring techniques for fabricating the anti-reflection nanostructures and a variety of methods, e.g., conformal-

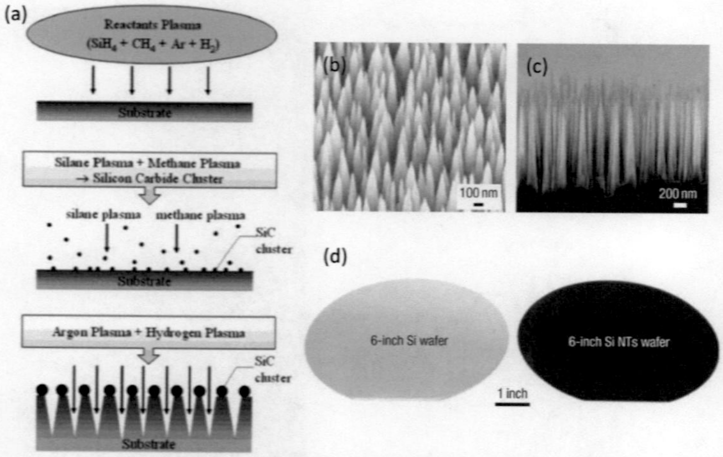

Figure 2. Improved antireflection of biomimetic nanotips by ECR plasma etching technique [25, 27]. (a) Schematic diagram for the silicon nanotip formation on silicon wafer; (b) and (c) show a tilted top view and a cross-sectional view of silicon nanotips, respectively; (d) Compared with polished silicon wafer (left), 6-inch silicon wafer coated with silicon nanotips (right) show greatly improved broadband and quasi omnidirectional anti-reflection.

evaporated-film-by-rotation (CEFR), colloidal lithography, self-masked dry etching, nanoimprint lithography (NIL), ALD and other approaches, are realized [16].

Oblique angle deposition (OAD) is usually employed to fabricate anisotropic film which is originated from the oblique growth of contained nanorods with a tilted angle to the substrate surface normal. The so-called CEFR method, which rotates the substrates at a high speed under OAD, leads to the straight growth of a dense array columns to substrate surface rather than helical structures which are formed under low rotation rate. That is, CEFR method is suitable for conformal replication of the photonic structure with a curved surface even under thick film deposition. With compound eyes of the fruit fly as bio-templates, it is reported artificial replica is successfully fabricated and hence similar optical features are inherited, respectively [22].

Many lithography techniques are applied for fabricating antireflection structures, among which colloidal lithography is a much simpler approach [23]. With colloidal crystals as masks, the silicon substrate is etched by reaction ion etching (RIE). During the fabrication duration, the colloidal spheres are etched by RIE gradually firstly, leading to a reduced transverse cross section of the spheres and thus an increasing exposure of the substrate. Attributed to the features of RIE, the etching rates of the apex and the junction parts of the spheres are not uniform, resulting in the etching morphology modification from frustum to cone arrays on the substrate finally [24].

Electron cyclotron resonance (ECR) plasma etching technique is employed by researchers to fabricate antireflection structures with much higher aspect ratio surfaces [25–27]. With the selected gas-mixture consisting of SiH_4, CH_4, Ar, and H_2, one step and self-masked dry etching are realized for fabricating high density nanotip arrays on a 6-inch silicon wafer. The fabrication progress is illustrated in the schematic diagram of Fig. 2(a). In brief, SiC clusters, which size and density can be tuned via process temperature, gas pressure, and composition, are formed on surfaces of the silicon substrate due to the reaction of SiH_4 and CH_4 plasma.

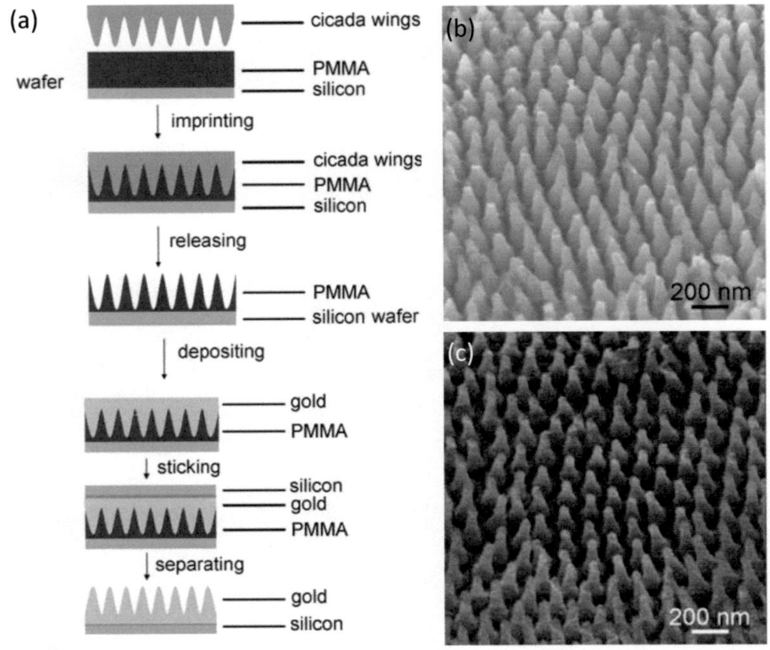

Figure 3. Cicada wing structure fabrication by NIL [19]. (a) Schematic diagram of NIL using cicada wings as bio-templates; (b) The fabricated structures and (c) the natural photonic structures show similar morphology from the SEM images.

Ar and H_2 are responsible for the dry etching process. The SiC clusters then act as nanomasks or nanocaps to protect the underlying substrate from etching, thus forming an aperiodic array of silicon nanotips with their lengths varying from $\sim$1000 nm to $\sim$ 16 μm finally (Fig. 2(b) and (c)). Even superior to natural prototypes, the biomimetic antireflection structures exhibit striking omnidirectional low reflection shown in Fig. 2(d) over a broad range of wavelengths from ultraviolet to terahertz region, irrespective of polarization.

Avoiding time-consuming and complicated mask fabrication, scientists also attempt to directly use cicada wings or insect eyes as bio-templates [19, 28, 29]. For example, the nipple arrays on wing surfaces are stamped under certain pressure on glass-phase PMMA, which is at higher degree than its glass-transition temperature, supported by silicon wafer. A release process makes the polymer reverse nanostructures of the bio-templates. With the patterned PMMA as a mask or a mold, inverse or similar structures of the cicada wing are achieved by RIE or thermodeposition. A schematic diagram of the NIL, the fabricated structures, and the natural templates for comparison are illustrated in Fig. 3(a), Fig. 3(b), and Fig. 3(c), respectively. It is also worthy to note that an extra advantage of using bio-templates in NIL process is the notable low-surface-tension, which is vital for the release process, due to the wax layer commonly found on surfaces of plants and insects.

Taking advantages of accurate thickness control and three-dimensional (3D) fabrication of ALD, conformal replica is accomplished after ALD growth and sintering the hybrid structures with fly eyes as bio-templates, achieving similar anti-reflective features in the artificial analog finally [29].

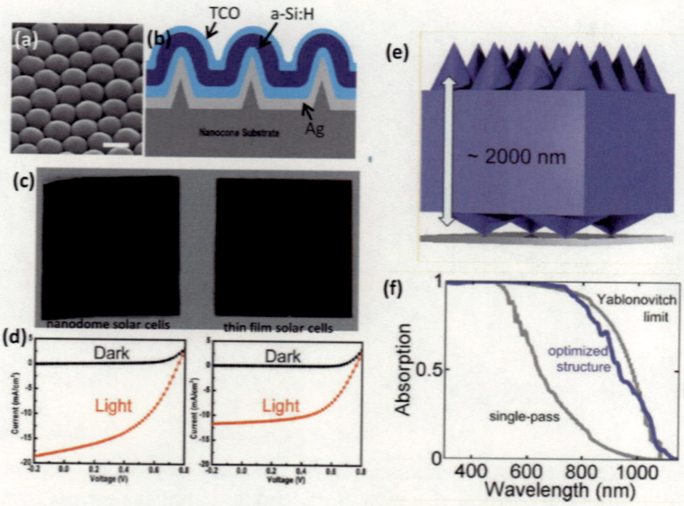

Figure 4. (Color online) Solar cells coating antireflection structures [32, 34]. (a) Nanodome solar cells in top view. Scale bar 500 nm; (b) Schematic diagram of the cross-sectional structures of the solar cells; (c) The photographs of nanodome solar cells (left) and flat film solar cells (right); (d) Dark and light I-V curve corresponding to (c); (e) and (f) The optimized double-sided nanostructure yields a photocurrent close to the Yablonovitch limit at an equivalent thickness of 2 μm.

2.1.3. Potential applications

Due to the high reflectance of silicon solar cells (more than 30%) induced by the high index contrast of silicon and air according to the Fresnel equations, scientists are already aware of the vital roles of high-quality antireflection coatings at early ages of solar cell fabrications [30]. Inspired by antireflection structures found in moth eyes, nanodomes or similar architectures are reproduced on surfaces of solar cells, leading to a dramatic light absorption increase and therefore a superior efficiency improvements than that of quarter-wavelength antireflection coating [31–33], as shown in Fig. 4(a)-(d). The recent theoretical investigations even report a high light trapping close to the Yablonovitch limit in the silicon solar cell by optimizing a double-sided antireflection structure design (Fig. 4(e) and (f))[34].

The biomimetic antireflection structures can also play a key role in light extraction of light-emitting devices (LEDs) [35–37]. Because of the total internal reflection and the waveguiding modes in the glass substrate, only about 20% amount of the generated light can irradiate from the LEDs. By fabricating silica cone arrays on the surfaces of the ITO glass substrate to modulate the above two bottleneck factors, the light luminance efficiency of white LEDs is significantly improved by a factor of 1.4 in the normal direction and even larger enhancement for large viewing angles.

Another fascinating application of the inspired antireflection structures is the use in the micro Sun sensor for Mars rovers [38]. On the basis of the recorded image by an active pixel sensor, the location coordinates of the rover can be calculated. However, the ghost image originating from the multiple internal reflection of the optical system leads to severe limitation of the accuracy. By fabricating dense nanotip arrays on the surfaces of the sensor, the internal reflection is minimized to be nearly 3 orders of magnitude lower than that of no treatments, resulting in a more reliable three-axis attitude information.

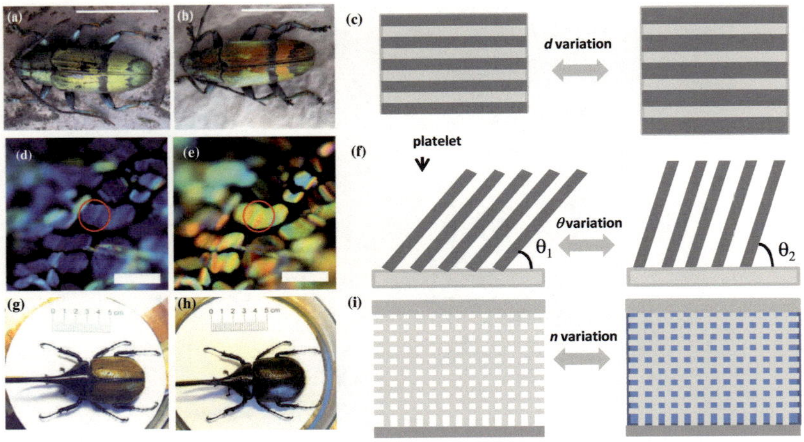

Figure 5. (Color online) Color tuning mechanisms found in nature [50, 53, 56]. The coloration of (a) longhorn beetles, (d) iridophore of tropic fish neon tetra, and (g) hercules beetles can reversibly change their coloration to (b), (e), and (h), which are intrinsically induced by (c) period d, (f) tilting angle θ, (i) refraction index n variation, respectively. Scale bars: (a) and (b) 10 mm, (d) and (e) 20 μm.

2.2. Color-tunable devices

2.2.1. Prototypes

Besides the well known coloration change strategy via migrations and volumes change of pigment granules such as chameleons, nature develops a second approach which is known as structural coloration change (SCC). By varying photonic structure characterizations, incident light angle, or the refraction index contrast of the color-produced optical system via the environmental stimuli, reversible coloration changes, which are basically passive, are revealed in fishes, beetles, and birds [39–46]. Most structural basis of SCC are attributed to the one-dimensional (1D) reflectors. For example, the damselfish *Chrysiptera cyanea* can change its color from blue to ultraviolet rapidly under stressful conditions, which is triggered by the simultaneous change in the spacing of adjoining reflecting plates made of guanine in the iridophore cell [47–49]. In insect world, the coloration change of longhorn beetles *Tmesisternus isabellae* from golden to red is revealed to originate from the swollen multilayer after water absorption [50] (Fig. 5(a), (b) and (c)). Another origin of structural coloration change is the tilted angle variability of the nanoplates with respect to incident light, which is found in tropic fish neon tetra *Paracheirodon innesi* [51–53] (Fig. 5(d), (e), and (f)). Physically, the underlying mechanism of the coloration change in the mentioned 1D biological photonic structures can be understood according to the given formula

$$\lambda_{max} = 2(n_1 d_1 cos\theta_1 + n_2 d_2 cos\theta_2), \tag{3}$$

where d is the layer thickness, n is refraction index, and θ is angle of refraction. The subscripts represent the layer index. The angle of refraction at different layers θ_1 and θ_2 can be obtained from Snell's law $n_1 sin\theta_1 = n_1 sin\theta_1 = sin\theta_0$, where θ_0 is the incident angle from air. From the equation, it is easy to elucidate that either n, d, or θ_0 is related to the optical path in the multilayer and thus leads to shifting interference peaks at different wavelengths.

In biological system, high-dimensional photonic structures responsible for the coloration change are also discovered, though they are rare. An intriguing example is hercules beetles

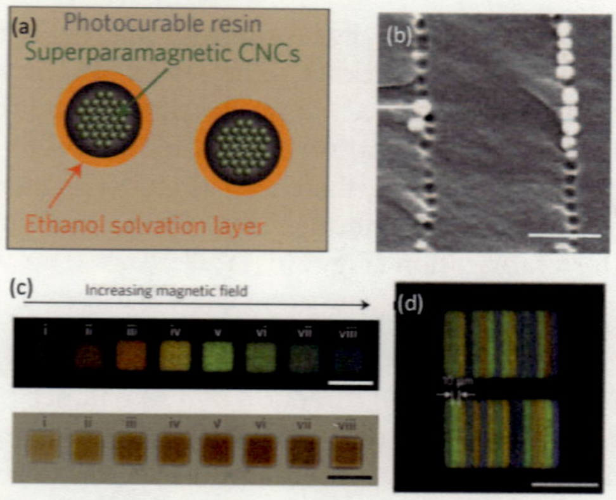

Figure 6. (Color online) Structural color printing using 'M-Ink' [57]. (a) Three-phase material system of 'M-Ink'; (b) 'M-Ink' particles align in a chain under external magnetic field, which acts as basic coloration units of color pattern after fixing by UV light; (c) Reflection images of multicolored structural colors (upper) and transmission photographs of the same sample (below) by gradually increasing magnetic fields; (d) High-resolution multiple structural color patterns. Scale bars: (b) 1μm, (c) and (d) 100 μm.

Dynastes hercules which can alter their appearance from khaki-green to black providing the ambience changes from dry to a high humidity level [54–56] (Fig. 5(g) and (h)). The 3D photonic crystal structures (nanoporous structures) are filled with water instead of air voids in dry status, rendering different refraction index contrast and thus the variation of Bragg scattering (Fig. 5(i), here only 2D cross section is illustrated). However, the underlying physics of coloration change induced by high-dimensional photonic structures is no other than that of their 1D counterparts.

2.2.2. *Bio-inspired fabrications and applications*

Due to the obvious appearance changes which are easy to be picked up with the naked eye, color-tunable devices are explored to identify the status changes by temperature, vapor, solvent, humidity in ambience, the applied mechanical force, electric field, magnetic field and etc. Besides the sensors, some novel writing system ('paper and ink') are developed. The key idea is to modify period (d), refraction index (n), viewangle (θ) or their combinations of photonic structures, just like nature shows us.

2.2.2.1. d variation

Because of relatively simple control by the environmental stimuli and large spectral variation which can be recognized by the naked eye, approaches on the modulation of photonic structure period are always of scientist interests in obtaining tunable color applications.

'M-Ink' is a mixture of colloidal nanocrystal clusters (CNCs), solvent and photocurable resin (Fig. 6(a)). With the superparamagnetic Fe_3O_4 nanocrystals encapsulated by silica shell, 'M-Ink' can response to external magnetic fields. The role of the resin is to provide repulsive force which balances the attractive force of the CNCs. Without external magnetic fields, CNCs are randomly dispersed (infinite period) in liquid resin. The exhibited coloration is consistent with the magnetite, to be brown. After applying magnetic fields, the CNCs are assembled to form chain-like structures along the magnetic field lines (Fig. 6(b)). The additional magnetic force, the intrinsic force among the CNCs, and the repulsive force by resin establish dynamic balance with variation of the external magnetic fields, tuning the distance between the neighboring CNC (finite period). The switchable period then determines the color of the light diffracted from the CNC chain, leading to a full color show (Fig. 6(c)). The final step is to fix the desired coloration. After exposure to ultraviolet (UV) light at different exerted magnetic fields locations, the chain-like CNCs can be frozen in the solidified resin instantaneously, remaining the periods of the chains undistorted and accomplishing high-resolution color pattern fabrication (Fig. 6(d)) [57].

With similar principle, the electric field-driven tunable color sensor is also realized by highly charged polystyrene (PS) colloids which form non-close-packed face-centered cubic (fcc) lattice [58]. Tuning the period along [111] direction by the balance of the exerted electrostatic force and the repulsive force, the exhibited coloration changes as a result of the applied electric field. The so-called 'P-Ink' is an electroactive material which consists of inverse opal inside polyferrocenylsilane (PFS) derivatives matrix. Such ink fabrication includes 3 primary steps: An opal film made of silica spheres is deposited onto glass substrate first by self-assembly; UV light is exposed to the sample in order to solidify the matrix and then form a stable PFS/silica composite; With diluted HF, inverse opal structure are realized in the elastomeric polymer matrix. By applying tunable voltage, macroscopic swelling and shrinking of the polymer matrix and microscopic Bravais lattice change responsible for the reverse coloration occur [59]. Stimulated by electrical forces, quite a few switchable coloration devices or sensors based on other materials or circumstances can be found elsewhere [60–62].

Many pressure-based photonic and even laser devices are reported [63–66]. Using monodispersed PS spheres to form cubic close packing (ccp) structures which are embedded in polydimethylsiloxane (PDMS) matrix, reverse colors are observed simply by stretching and releasing the rubber sheet. Upon mechanical stress, the lattice is elongated along the applied force direction, while the interplanar spacing in the perpendicular direction (i.e. distance between the (111) planes) decreases because of the nearly invariance volume of the rubber. The compressed distance leads to a blue-shift, e.g., from red to green [63]. Such opal rubber is believed to have practical applications such as a color indicator, tension meter or elongation strain sensor. The inverse opal structures (filled by air voids) in elastomer network are also synthesized [64, 67] (Fig. 7(a)). The porous elastomeric photonic crystals (EPCs) show highly reversible optical response to compressive force, e.g., 60 nm spectral blue-shift under a compressive pressure of $\sim$ 15 kPa in the structures having 350-nm void size. Although the coloration change can be attributed to the Bravais lattice deformability, like the mechanism mentioned before. However, it is especially noteworthy that porous EPCs remain nearly undeformed in orthogonal directions when an external pressure is exerted in one direction, which can be ascribed to the high filling factor of air voids. The air voids enable the distortion of the cross-sectional void spaces from roughly circular to elliptical shape and a reduction of the air volume fraction under pressure. The elastic deformation feature of such structures helps to reduce the redistribution of stress along lateral directions when compressed by a patterned surface, leading to novel biometric applications such as the fingerprint recognition devices, as shown in Fig. 7(b). Additionally, air voids of porous EPCs provide a platform to incorporate with other functional materials for us to explore new applications. For instance, filling PbS quantum dots in the air voids, the photoluminescence (PL) emission, which leads to many potential applications in the near-infrared region, can be modified by overlapping with the forbidden bandgap of the inverse opal structures (Fig. 7(c)).

The chemical solvents are also used to be as stimuli. An interesting example is invention of new type 'photonic paper/ink' system [68, 69]. With novel soft materials consisting of

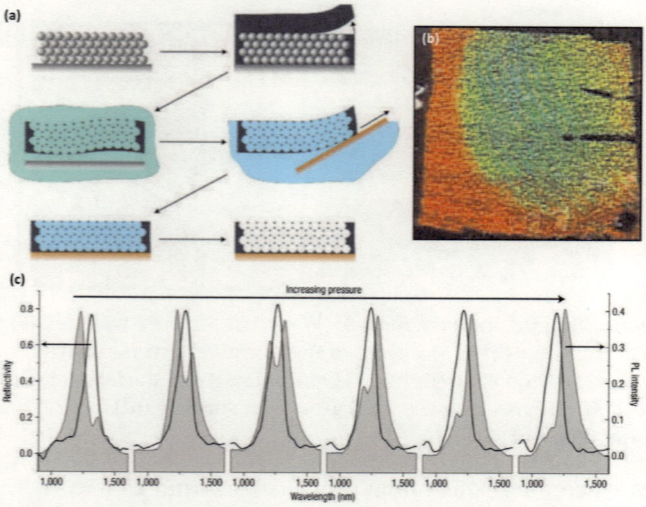

Figure 7. (Color online) From color fingerprinting to the control of photoluminescence in EPC films [67]. (a) Schematic diagram of the elastic inverse opal structure fabrication; (b) A captured still image of the EPC film under compression by an index finger; (c) NIR-emitting PL emission of colloidal PbS quantum dots which are incorporated into voids of the EPC can be tuned by overlapping the forbidden gap (solid black line) with the PL peak (grey filled curve).

closely packed PS and polymer elastomer (colloids and PDMS), the exhibited coloration can be altered reversely by immersing the materials into silicone liquid ('writing process') and an evaporation process ('erasing process'). The spacing between the (111) planes is adjusted by the strength of interaction between PDMS matrix and the contained silicone oligomers with different molecule weight in the liquid. With different solvents ('ink'), the swelling and shrinkage of the matrix show a featured reversible shift of Bragg diffraction peak. A multilayer based on alternating Teflon-like layer and Au nanoparticle/Teflon-like layer composite layer operates not in visible but optical telecommunication wavelength range [70]. When the structure is exposed to different organic solvent vapor (e.g. acetone, ethanol, methanol, water, chloroform, and etc.), the molecules enter the metal/polymer composite inside the holes and microvoids in its structure, resulting in the swelling up, e.g., for acetone vapors at a molar fraction of 0.25, with an relative increase of 12.5% in thickness to an equilibrium state while leaving the Teflon-like layer unchanged due to its inert chemical features. The measured reflectance show a large variation of 0.2 μm, which is advantageous to the detection of the organic/inorganic vapors.

Other external stimuli such as UV light, heat, or chemical reaction are applied to trigger and fabricate coloration sensitive materials as well by reversely controlling the spacing of the responsible photonic structures. Detailed information can be found in some references and recent reviews [14, 15, 61, 71–74].

2.2.2.2. n variation

Besides d, the refraction index n is another attribute of photonic structures. Different approaches to alter refraction index, in which infiltration is most regular, are revealed in order to achieve novel visual applications. Inspired by beetles *D. hercules*, 3D nanoporous structures are fabricated by so-called dip-coating deposition, achieving humidity sensing [75]. The 3D architecture, which is inverse opal structure, is reported to have its reflective peak

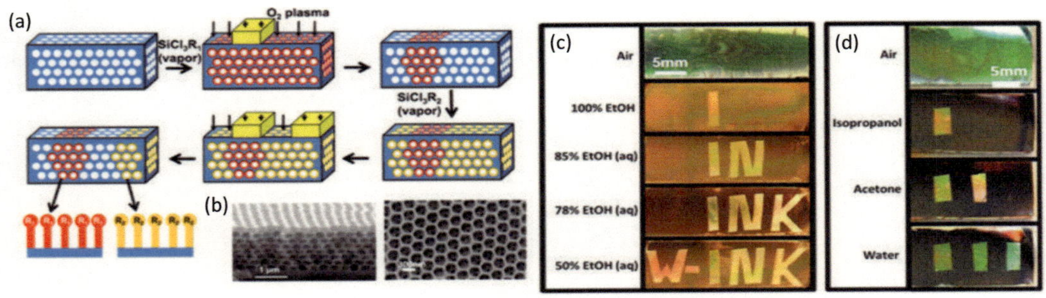

Figure 8. (Color online) Silica inverse opal films of 'Watermark-Ink' system [76]. (a) Schematic procedure of chemical encoding; (b) SEM images of the fabricated inverse opal structures in cross-sectional view (left) and top view (right); (c) Optical images of the fabricated film in which the word 'W-INK' is encoded via functional chemical groups on surfaces. (d) Optical images of different encoded patterns under different solvents.

red-shifting 14 nm when the relative humidity changes from 25% to 98%. It is worth noting that after treatments by O_2 plasma, the 3D photonic structures exhibit large bandgap shift of 137 nm and dramatic coloration change from bluish green in dry state to red in fully wet state. The modified hydrophilic feature play an important role in water collection in air voids at different level of humidity, leading to larger variation of refraction index contrast. The other clew for the remarkable coloration change can be ascribed to the high filling factor of air voids of inverse opal structures. The high value of $\sim 76\%$ results in a wider modulation of refraction index contrast by the amount of solvent absorption. Tuning hydrophilic features by chemical groups, the inverse opal film can even be used as 'Watermark-Ink' system [76]. In the studies, the inside porous structures are functionalized with chemical group ($R1$, $R2$, $R3$, or $R4$) through vaporizing an alkylchlorosilane. The chemical functionalities of the structures then are erased and the surface is reactivated by O_2 plasma exposure, leaving a patterned functionalized region where is masked by a PDMS polymer. With iterations of such kind of functionalization and reactivation, the opal structures are locally patterned by different chemical groups, leading to differentiate hydrophilic feature (Fig. 8(a) and (b)). When immersing the film into specific fluids, regulated infiltration by the wettability occurs spatially in the film and thus surveys different optical responses (i.e. different patterns, Fig. 8(c) and (d)). The fabricated structures are expected to have applications in encryption as well as colorimeter.

Besides infiltration, phase-transition materials can also induce refraction index variation and switchable coloration, providing the transition conditions are satisfied. The best-known phase-transition material maybe is liquid crystals (LCs). Above the phase-transition temperature of 34°, LCs change their nematic phase which is anisotropic to isotropic phase, leading to a significant change of refraction index and coloration, e.g., in inverse opal structures. By mixing the active material into LCs, sensitive reflection and polarization triggered by UV light are reported by a series of subsequent researches [77–79]. Besides LCs, various sensitive materials, including Ag_2Se, WO_3 and ferroelectric ceramics, are also found functionally in incorporating with diversified photonic structures to obtain tunable structural coloration materials and thus fabricate thermo- or electro-sensors [80–83].

2.2.2.3. θ variation

Iridescence is a characteristic of structural coloration, which is determined by band dispersions of photonic structures. With various incident angle or observation direction, the

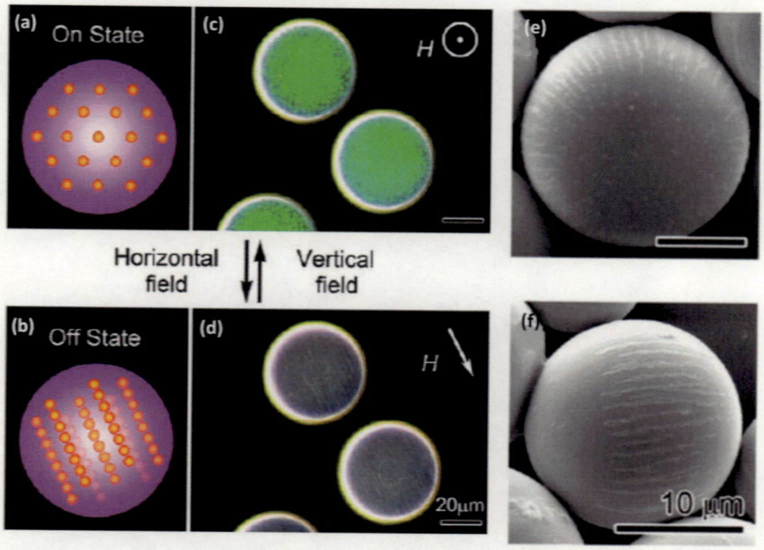

Figure 9. (Color online) Magnetochromatic microspheres switched between 'on' and 'off' states by rotating external fields [84, 85]. Schematic illustrations, optical images, and the corresponding SEM photographs of 'on' and 'off' states are (a), (c), (e), and (b), (d), (f), respectively.

perceived coloration is changing. Hence, it is feasible to explore novel tunable color devices by tuning orientations of the photonic structures [84–86]. Just like the fish neon tetra *P. innesi*, magnetochromatic microspheres alter their appearance by tilting the inner photonic structures with respect to the incidence (Fig. 9). As mentioned before, with superparamagnetic Fe_3O_4 particles coated by silica in PEGDA emulsions, the applied magnetic field guides to form photonic chains, in which the interpaticle distances depend on the balance of the attractive and repulsive force. Polymerized by UV light, microspheres are solidified and the inner periodic photonic chains, which give rise to structural coloration, are fixed permanently. Dispersing the microspheres into liquids, various colors are observed in top view due to the random orientations of the photonic chains in the microspheres. By applying external magnetic fields, the microspheres tend to rotate in order to keep the magnetic chains inside in the direction of the magnetic vector, leading to a homogenous coloration of green exhibited. The feature of the switchable coloration 'on' and 'off' status by the external stimuli is believed to have applications in color display, signage, bio- and chemical detection, and magnetic field sensing.

2.3. Structural color mixing and applications

2.3.1. Prototypes

Structural coloration results from the interaction of light and photonic structures with featured size of visible wavelengths. It is even more widespread than pigmentary coloration in animal world. Some literatures have well reviewed the field comprehensively. In the Chapter, we do not plan to pay attention to the overall structural coloration but only focus on a specific subject 'structural color mixing' [87–92]. In tiger beetles *Cicindela oregona*(Fig. 10(a)-(c)) [87, 88], the honeycomb-like pits are found on surfaces of the elytra. Under microscope, the brown morph actually includes blue-green patches in the red background, while the black one consists of magenta patterns surrounded by dull green. The microscopic colors are the results of

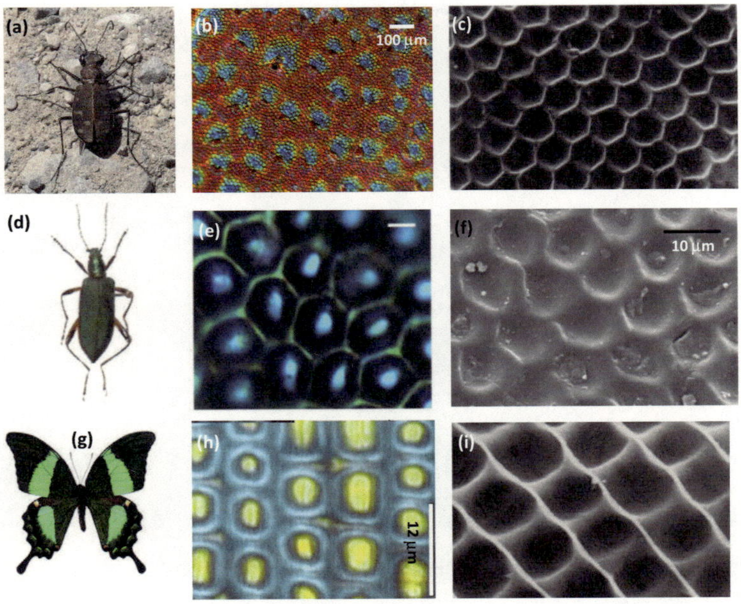

Figure 10. (Color online) Structural Color Mixing in Nature [87, 90, 91]. Photographs, optical microscopy and SEM images of tiger beetles *C. oregona* ((a), (b), (c)), long-jointed beetles *C. obscuripennis* ((d), (e), (f)), and swallowtail butterflies *P. palinurus* ((g), (h), (i)), respectively.

light interference by the multilayer structures in the cuticle. However, due to the colored patches (40-80 μm across) are too small to be resolved by the unaided eye (the resolution d is determined by the Rayleigh criterion $d{=}l \times 0.61\lambda / D$, where l is distance of distinct vision, λ the wavelength and D the average pupil diameter of humankind), the perceived coloration is a mixture of the discrete colors, leading to a totally different exhibition from that equipped with microscope. From the investigations of photonic structures of beetles *Chlorophila obscuripennis* (Fig. 10(d)-(f)), we revealed the color is a juxtaposition in a smaller region ($\sim 10{\times}10\ \mu\text{m}^2$) by green on the ridges and cyan in centers of the pits. Furthermore, the pits on the elytra surface give rise to diffused light reflected over a wide large of angles, leading to inconspicuous coloration shown [91]. In butterflies *Papilio palinurus* (Fig. 10(g)-(i)), the scanning electronic micrographs show surfaces in the scales comprise a similar two-dimensional (2D) pits (4-6 μm in diameter and 3 μm at the greatest depth). The flat regions between and in pits appear yellow, and the inclined region contributes to blue color. Because of the limitation of human eyes' resolution, the butterfly displays a mixture color of green, i.e. yellow plus blue turns to be green. Due to sufficient depth of the pits, light which normally incidents on the inclined side can experience dual reflection and is back-reflected. The retro-reflection is found to play a key role in the polarization conversion of incident light, which is crucial in some novel optical applications. Thanks to symmetric feature of the pits, no macroscopic polarization effects can be observed [90]. On the cover scale surfaces of butterflies *Suneve coronata*, however, the natural occurring triangular grooves array not symmetric but in 1D way with a period of $\sim$ 2 μm, giving rise to polarization conversion of the normal incident light in a specific direction. Intriguingly, it is revealed the coloration macroscopically remains unconscious change under different polarized illumination. The reason is the curly scales, which induce polarization conversion between the neighboring rows of scales but in the

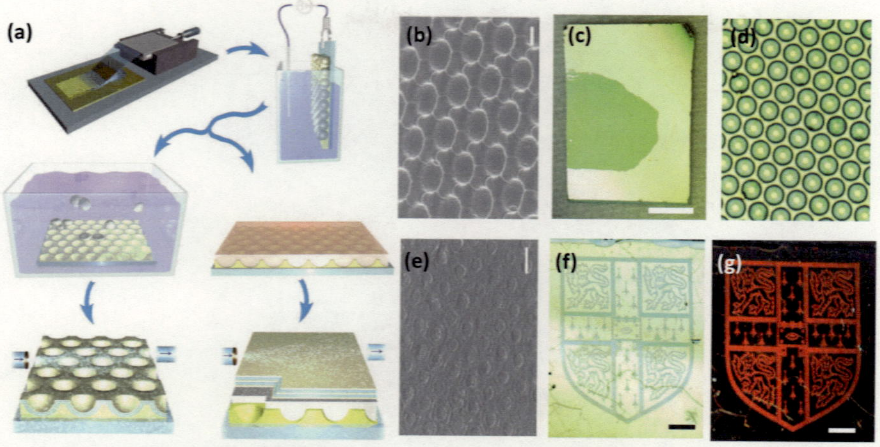

Figure 11. (Color online) Mixed structural coloration and applications inspired by nature [95]. (a) Schematic diagram of artificial samples mimicking the scale structures of *P. blumei*; (b) SEM photographs in top view show the concavities on surface of the replica which is (c) green macroscopically but (d) resolved yellow and green microscopically; (e) Modifications of the concavities morphology by melting colloidal spheres embedded in the concavities lead to a striking change in color of the sample from blue to red viewed (f) in direct specular reflection and (g) in retro-reflection. Scales: (a) 2 μm, (c)(f)(g) 5 mm, and (e) 5μm.

orthogonal direction. The nanostructures having different magnitude in size therefore cancel out the macroscopic polarization effects (Fig. 12(a)-(d)) [89].

2.3.2. *Bio-inspired fabrications and applications*

Structural coloration may be especially crucial for future color and related industry because of the non-fading feature (if the photonic structures are undeformed) [93, 94] and environmental friendliness. Naturally, structural coloration mixing inherits the advantages. Mimicking the nature, mixed structural color and its application can be obtained. For example, PS colloids with a diameter of 5 μm are assembled on a gold-coated silicon substrate. A 2.5-μm-thick layer of platinum or gold is then deposited to fill the interspaces of the colloids by electrochemical approach, creating a negative replica. After removal of the PS colloids by ultrasonic waves and a sputtering thin carbon film, a multilayer of quarter-wave titania and alumina films is grown by ALD (Fig. 11(a)), inheriting the morphology of hexagonally arranged pits of the negative replica (Fig. 11(b)). The sample exhibits similar color mixing with that of butterflies *P. blumei* (Fig. 11(c) and (d)). Moreover, by modifying the surface morphology of the imitation (Fig. 11(e)), even visual information, e.g., a picture, can be encoded into the photonic structures which display a striking appearance change from pale blue in the specular direction to red in retro-reflection (Fig. 11(f) and (g)). The bio-inspired work is expected to find applications in security labelling field or color industry such as painting and coating [95].

Inspired by color mixing researches, some novel applications based on polarization conversion are designed [89]. For instance, by etching periodic triangular-like grooves on surfaces of a flat multilayer, the film displays a coloration of green, which is actually a mixed color from yellow in the flat region and blue in the grooves in normal incidence, as shown in Fig. 12(e). Because of the broken symmetry of the 1D structures, the blue color can be suppressed by using a polarizer on light path, leading to the exhibited coloration change to

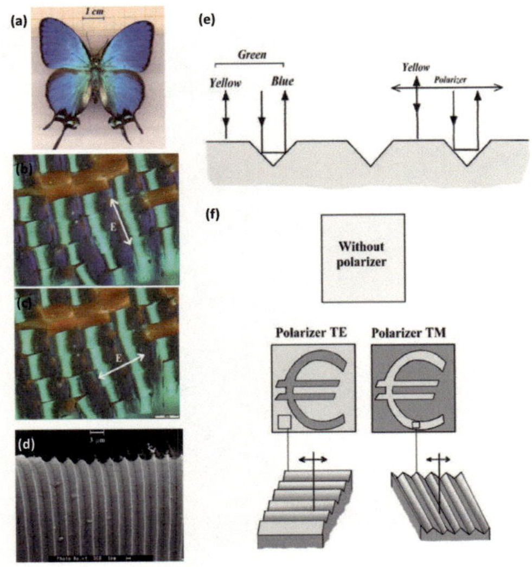

Figure 12. (Color online) (a) Optical image of a male *S. coronata* butterfly in dorsal view; The appearances under (b) TM and (c) TE polarization light show distinguished difference of the blue component, which can be attributed to (d) the broken symmetry of the triangular grooves in the surface plane; Two kinds of anti-fake or encryption designs, which apply (e) colored effect or (f) grey level change linked to the polarization and the broken surface symmetry, are inspired [89].

yellow (i.e. the hue in flat regions). That is, the grooved areas turn the light polarization $\pi/2$ and thus are shadowed by the inserted polarizer. Following the ideas, an encoded pattern which comprises grooved areas can be distinguished with the background consisting of perpendicular grooves via the luminosity level under different polarized light incidence (Fig. 12(f)). The two imaged examples inspired by the grooved structures of the butterfly show us great prospective in fields such as anti-fake and encryption.

2.4. Other bio-inspired fabrications and applications

Besides the main categories, some other bio-inspired photonic applications are also reported [96–100]. For examples, the natural scales of *Morpho sulkowskyi* are extremely sensitive to the different environmental vapors, which lead to dramatically improved responses as units of potential sensor applications compared with that of current devices[101]. Using black scales of the butterfly *Papilio paris* and *Thaumantis diores* as templates, hierarchically periodic microstructure titania replica was systhesized by chemical procedures. The high surface area inherited from the natural templates is great advantages to the light harvesting efficiency and dye sorption when the replica is used as photoanode in dye-sensitized solar cell [100]. With *Morpho* as bio-templates, alumina replicas show the potential applications in waveguides and beam-splitters under thin film coating by ALD and sintering [97]. Under thick film coating, the hybrid structure is found not to inherit the 'Christmas-tree' structures but develop a 'Pyramid-like' structures which have potential applications in light trapping [99]. With the complex photonic structures (multilayer plus 2D amorphous structures) of beetles *Trigonophorus rothschildi varians* as 'blueprints', the artificial counterpart which is fabricated by FIB etching holes through a 5×SiO/SiGe multilayer structure show similar optical features like non-specularity and only slightly angle-dependent reflectance [98]. Further bio-inspired work is undergoing which is stimulated by a recent revealed 3D architecture [102], aiming at the ultra-negative angular dispersion of diffraction and potential novel dispersive optical elements.

3. Perspectives

In the Chapter, several important kinds of bio-inspired photonic applications are reviewed, including antireflection devices, color-tunable sensors, structural color mixing applications and etc. The nature nourishes scientists the functional optical applications either the blueprints of photonic architecture or directly the bio-templates. Due to the higher index of inorganic materials used, the mimicking photonic structures even show better optical performances as well as enhanced mechanical properties of high temperature tolerance, stability and infrangibility. The biomimetic applications are anticipated to help our life better in the near future. However, complicated photonic structures (e.g. those of high-dimensional, hierarchic, amorphous features in nature) still remains hardly reproduced or, if they are fabricated successfully, the efforts involved are so great using the traditional fabrication ways that optical devices can not commercially explored. Thorough physical mechanism understanding as well as better fabrication approach explorations may help to simply the structure fabrications, achieve similar optical functions and realize commercial applications. In addition, adding substances such as functional chemical groups, fluorescence particles, metal, or other active materials, the mimicking photonic structures allow the properties of interest to be augmented, which may open a new window of novel optical device exploration. Although the photonic biomimicry is in its infancy, we believe that the bio-inspired optical device would surely have profound impacts on our modern society.

Author details

Feng Liu
Laboratory of Opto-electrical Material and Device, Department of Physics, Shanghai Normal University, Shanghai, China

Biqin Dong
Department of Mechanical Engineering, Northwestern University, Evanston, USA

Xiaohan Liu
Department of Physics, Fudan University, Shanghai, China

4. References

[1] Yablonovitch, E. (1987). Inhibited Spontaneous Emission in Solid-State Physics and Electronics. *Phys. Rev. Lett.*, Vol.58: 2059–2062.
[2] John, S. (1987). Strong localization of photons in certain disordered dielectric superlattices. *Phys. Rev. Lett.*, Vol.58: 2486–2489.
[3] Joannopoulos, J.D.; Johnson, S.G.; Winn J.N.; & Meade, R.D. (2008). *Photonic Crystals: Molding the Flow of Light*, 2nd Ed. Princeton University Press, Princeton and Oxford, USA.
[4] Fox, D.L. (1976). *Animal Biochromes and Structural Colours*, University of California Press, Berkeley, USA.
[5] Srinivasarao, M. (1999). Nano-Optics in the biological world: beetles, butterflies, birds, and moths. *Chem. Rev.*, Vol.99: 1935–1961.

[6] Vukusic, P. & Sambles, J.R. (2003). Photonic structures in biology. *Nature*, Vol.424: 852–855.

[7] Parker, A.R. (2005). A geological history of reflecting optics. *J. R. Soc. Interface*, Vol.2: 1–17.

[8] Seago, A. Brady, P. Vigneron, J-P. & Schultz, T.D. (2009). Gold bugs and beyond: a review of iridescence and structural colour mechanisms in beetles (Coleoptera). *J. R. Soc. Interface*, Vol.6: S165–S184.

[9] Shawkey, M.D. Morehouse, N.I. & Vukusic, P. (2009). A protean palette: colour materials and mixing in birds and butterflies. *J. R. Soc. Interface*, Vol.6: S221–S231.

[10] Kinoshita, S. & Yoshioka, S. (2005). Structural colors in nature: the role of regularity and irregularity in the structure. *ChemPhysChem*, Vol.6: 1442–1459.

[11] Parker, A.R. & Townley, H.E. (2007). Biomimetics of photonic nanostructures. *Nat. Nanotech.*, Vol.2: 347–353.

[12] Pulsifer, D.P. Lakhtakia, A. (2011). Background and survey of bioreplication techniques. *Bioinsp. Biomim.*, Vol.6(No. 031001).

[13] Chung, W-J. Oh, J-W. Kwak, K. Lee, B.Y. Meyer, J. Wang, E. Hexemer, A. & Lee, S-W. (2011). Biomimetic self-templating supramolecular structures. *Nature*, Vol.478: 364–368.

[14] Fudouzi, S. (2011). Tunable structural color in organisms and photonic materials for design of bioinspired materials. *Sci. Technol. Adv. Mater.*, Vol.12(No.064704).

[15] Zhao, Y. Xie, Z. Gu, H. Zhu, C. & Gu, Z. (2012). Bio-inspired variable structural color materials. *Chem. Soc. Rev.*, Vol.41: 3297–3317.

[16] (b) Li, Y. Zhang, J. & Yang, B. (2010). Antireflective surfaces based on biomimetic nanopillared arrays. *Nano Today*, Vol.5: 117–127.

[17] Land, M.F.; & Nilsson, D.E. (2001). *Animal Eyes.*, Oxford University Press, Oxford, UK.

[18] Yoshida, A. Motoyama, M. Kosaku, A. & Miyamoto, K. (1997). Antireflective nanoproturberance array in the transparent wing of a hawkmoth Cephanodes hylas. *Zool. Sci.*, Vol.14: 737–741.

[19] Zhang, G. Zhang, J. Xie, G. Liu, Z. & Shao, H. (2006). Cicada wings: a stamp from nature for nanoimprint lithography. *Small*, Vol.2(No.12): 1440–1443.

[20] Born, M. Wolf, E. (1999). *Principles of Optics*, 7th Ed. Cambridge University Press, Cambridge, UK.

[21] Xi, J.-Q. Schubert, M.F. Kim, J.H. Schubert, E.F. Chen, M. Lin, S. Liu, W. & Smart, J.A. (2007). Optical thin-film materials with low refractive index for broadband elimination of Fresnel reflection. *Nat. Photon.*, Vol.1: 176–179.

[22] Martín-Palma, R.J. Pantano, C.G. & Lakhtakia, A. (2008). Replication of fly eyes by the conformal-evaporated-film-by-rotation technique. *Nanotechnology*, Vol.19 (No.355704).

[23] Yang, S-M. (2006). Nanomachining by colloidal lithography. *Small*, Vol.2(No.4): 458–475.

[24] Pearton, S.J. & Norton, D.P. (2005). Dry etching of electronic oxides, polymers, and semiconductors. *Plasma Process. Polym.*, Vol.2: 16–37.

[25] Hsu, C. Lo, H. Chen, C. Wu, C.T. Hwang, J. Das D. Tsai, J. Chen, L. & Chen K. (2004). Generally applicable self-masked dry etching technique for nanotip arrayfabrication. *Nano Lett.*, Vol.4 (No.3): 471–475.

[26] Hsu, C. Huang, Y.F. Chen L.C. Chattopadhyay, S. Chen, K.H. Lo, H.C. & Chen, C.F. (2006). Morphology control of silicon nanotips fabricated by electron cyclotron resonanceplasma etching. *J. Vac. Sci. Technol. B*, Vol.24(No.1): 308–311.

[27] Huang, Y. Huang, Chattopadhyay, S. Jen, Y. Peng, C. Liu, T. Hsu, Y. Pan, C. Lo. H. Hsu, C. Chang Y. Lee, C. Chen, K. & Chen L. (2007). Improved broadband and quasi-omnidirectional anti-reflection properties with biomimetic silicon nanostructures. *Nat. Nanotech.*, Vol.2: 770–774.

[28] Gao, H. Liu, Z. Zhang, J. Zhang, G. & Xie, G. (2007). Precise replication of antireflective nanostructures from biotemplates. *Appl. Phys. Lett.*, Vol.90(No. 12): 1–3.

[29] Huang, J. Wang, X. & Wang, Z.L. (2008). Bio-inspired fabrication of antireflection nanostructures by replicating fly eyes. *Nanotechnology*, Vol.19(No.025602).

[30] Green, M.A. (1987). *Higher Efficiency Silicon Solar Cells,* Trans Tech Pub, Aedermannsdorf, Switzerland.

[31] Zhao, J. & Green, M.A. (1991). Optimized antireflection coatings for high-efficiency silicon solar cells. *IEEE Trans. Electron Devices*, Vol.38(No.8): 1925–1934.

[32] Zhu, J. Hsu, C. Yu, Z. Fan, S. & Cui, Y. (2010). Nanodome Solar Cells with Efficient Light Management and Self-Cleaning. *Nano Lett.*, Vol.6(No.6): 1979–1984.

[33] Chen, H.L. Chuang, S.Y. Lin, C.H. & Lin, Y.H. (2007). Using colloidal lithography to fabricate and optimize sub-wavelength pyramidal and honeycomb structures in solar cells. *Opt. Express*, Vol.15: 14793–14803.

[34] Wang, K.X. Yu, Z. Liu, V. Cui, Y. & Fan S. (2012). Absorption enhancement in ultrathin crystalline silicon solar cells with antireflection and light-trapping nanocone gratings. *Nano Lett.*, Vol.12(No.3): 1616–1619.

[35] Ishimori, M. Kanamori, Y. Sasaki, M. Hane, K. & Xie, G. (2002). Subwavelength antireflection gratings for light emitting diodes and photodiodes fabricated by fast atom beam etching. *Jpn. J. Appl. Phys.*, Vol.41: 4346–4349.

[36] (a) Li, Y. Li, F. Zhang, J. Wang, C. Zhu, S. Yu, H. Wang, Z. & Yang, B. (2010). Improved light extraction efficiency of white organic light-emitting devices by biomimetic antireflective surfaces. *Appl. Phys. Lett.*, Vol. 96(No.153305).

[37] Song, Y.M. Choi, E.S. Park, G.C. Park, C.Y. Jang, S.J. & Lee Y.T. (2010). Disordered antireflective nanostructures on GaN-based light-emitting diodes using Ag nanoparticles for improved light extraction efficiency. *Appl. Phys. Lett.*, Vol.97(No. 093110).

[38] Lee, C. Bae, S.Y. Mobasser, S. & Manohara, H. (2005). A novel silicon nanotips antireflection surface for the micro Sun sensor. *Nano Lett.*, Vol.5(No.12): 2438–2442.

[39] Cong, H. Yu, B. & Zhao, X.S. (2011). Imitation of variable structural color in *paracheirodon innesi* using colloidal crystal films. *Opt. Express*, Vol.19(No.13): 12799–12808.

[40] Hadley, N.F. (1979). Wax secretion and color phases of the desert Tenebrionid beetle *Cryptoglossa verrucosa* (LeConte). *Science*, Vol.203: 367–369.

[41] Mäthger, L.M. Land, M.F. Siebeck, U.E. & Marshall, N.J. (2003). Rapid colour changes in multilayer reflecting stripes in the paradise whiptail, *Pentapodus paradiseus*. *J. Exp. Biol.*, Vol.206: 3607–3613.

[42] McClain, E. Seely, M.K. Hadley, N.F. & Fray, V. (1985). Wax blooms in Tenebrionid beetles of the Namib desert: correlations with environment. *Ecology*, Vol.66: 112–118.

[43] Jolivet, P. (1994). Phsiological colour changes in tortoise beetles, In: *Novel Aspect of the Biology of Chrysomelidae*, Cox, M.L. & Petitpierre, E., (Eds.), page numbers (331-335), Kluwer Academic, Netherland.

[44] Vigneron, J.P. Pasteels, J.M. Windsor, D.M. Vertésy, Z. Rassart, M. Seldrum, T. Dumont, J. Deparis, O. Lousse, V. Biró, L.P. Ertz, D. & Welch, V. (2007). Switchable reflector in the Panamanian tortoise beetle *Charidotella egregia* (Chrysomelidae: Cassidinae). *Phys. Rev. E.*, Vol.76(No. 031907).

[45] Mason, C.W. (1929). Transient color changes in the tortoise beetles (Coleoptera: Chrysomelidae). *Entomol. News*, Vol.45: 52–56.

[46] Eliason, C.M. & Shawkey, M.D. (2010). Rapid, reversible response of iridescent feather color to ambient humidity. *Opt. Express*, Vol.18(No.20): 21284–21292.

[47] Kasukawa, H. Oshima, N. & Fujii, R. (1986). Control of chromatophore movements in dermal chromatic units of blue damselfish-II. The motile iridophore. *Comp. Biochem. Physiol. C*, Vol.83: 1–7.

[48] Kasukawa, H. Oshima, N. & Fujii, R. (1987). Mechanism of light reflection in blue damselfish motile iridophore. *Zool. Sci.*, Vol.4: 243–257.

[49] Oshima, N. & Fujii, R. (1987). Mobile mechanisms of blue damselfish (*Chrysiptera cyanea*) iridophores *Cell Motil. Cytoskel.*, Vol.8: 85–90.

[50] Liu, F. Bong, B.Q. Liu, X.H. Zheng, Y.M. & Zi, J. (2009). Structural color change in longhorn beetles *Tmesisternus isabellae*. *Opt. Express*, Vol.17(No.18): 16183–16191.

[51] Lythgoe, J.N. & Shand, J. (1982). Changes in spectral reflexions from the iridophores of the neon tetra. *J. Physiol.*, Vol.325: 23–34.

[52] Oshima, N. (2005). Light reflection in motile iridophores of Fish, In: *Structural Colors in Biological SystemsąłPrinciples and Applications*, Kinoshita, S. & Yoshioka, S., (Eds.), page numbers (211), Osaka University Press, Japan.

[53] Yoshioka, S. Matsuhana, B. Tanaka, S. Inouye, Y. Oshima, N. & Kinoshita, S. (2011). Mechanism of variable structural colour in the neon tetra: quantitative evaluation of the Venetian blind model. *J. R. Soc. Interface*, Vol.8: 56–66.

[54] Hinton, H.E. & Jarman, G.M. (1972). Physiological colour change in the Hercules beetle. *Nature*, Vol.238: 160–161.

[55] Hinton, H.E. & Jarman, G.M. (1973). Physiological colour changes in the elytra of the Hercules beetles, Dynastes hercules. *J. Insect Physiol.*, Vol.19: 533–549.

[56] Rassart, M. Colomer, J-F. Tabarrant, T. Vigneron, J.P. (2008). Diffractive hygrochromic effect in the cuticle of the hercules beetle *Dynastes hercules*. *New J. Phys.*, Vol.10(No. 033014).

[57] Kim, H. Ge, J. Kim, J. Choi, S. Lee, H. Lee, H. Park, W. Yin, Y. & Kwon, S. (2009). Structural colour printing using a magnetically tunable and lithographically fixable photonic crystal. *Nat. photon.*, Vol.3: 534–540.

[58] Shim, T.S. Kim, S-H. Sim, J.Y. Lim, J-M. & Yang, S-M. (2010). Dynamic Modulation of Photonic Bandgaps in Crystalline Colloidal Arrays Under Electric Field. *Adv. Mater.*, Vol.22: 4494–4498.

[59] Puzzo, D.P. Arsenault, A.C. Manners, I. & Ozin, G.A. (2009). Electroactive Inverse Opal: A Single Material for All Colors. *Angew. Chem., Int. Ed.*, Vol.47: 943–347.

[60] Arsenault, A.C. Puzzo, D.P. Manners, I. & Ozin, G.A. (2007). Photonic-crystal full-colour displays. *Nat. Photon.*, Vol.1: 468–472.

[61] Walish, J.J. Kang, Y. Mickiewicz, A. & Thomas E.L. (2009). Bioinspired electrochemically tunable block copolymer full color pixels. *Adv. Mater.*, Vol.21: 3078–3081.

[62] Hwang, K. Kwak, D. Kang, C. Kim, D. Ahn Y. & Kang, Y. (2011). Electrically tunable hysteretic photonic gels for nonvolatile display pixels. *Angew. Chem., Int. Ed.*, Vol.50: 6311–6314.

[63] Fudouzi, H. & Sawada, T. (2006). Photonic rubber sheets with tunable color by elastic deforonmation. *Langmuir*, Vol.22(No.3): 1365–1368.

[64] Sumioka, K. Kayashima, H. & Tsutsui, T. (2002). Tuning the Optical Properties of Inverse Opal Photonic Crystals by Deformation. *Adv. Mater.*, Vol.13(No.18): 1284–1286.

[65] Fudouzi, S. Kanai, T. & Sawada, T. (2011). Widely tunable lasing in a colloidal crystal gel film permanently stabilized by an ionic liquid. *Adv. Mater.*, Vol.23(No.33): 3815–3820.

[66] Arsenault, A.C. Kitaev, V. Manners, I. Ozin, G.A. Mihi, A. & Míguez, H. (2005). Vapor swellable colloidal photonic crystals with pressure tunability. *J. Mater. Chem.*, Vol.15: 133–138.

[67] Arsenault, A.C. Clark, T.J. Freymann, G.V. Cademartiri, L. Sapienza, R. Bertolotti, J. Vekris, E. Wong, S. Kitaev, V. Manners, I. Wang, R.Z. John, S. Wiersma, D. & Ozin, G.A. (2006). From colour fingerprinting to the control of photoluminescence in elastic photonic crystals. *Nat. Mater.*, Vol.5: 179–184.
[68] (a) Fudouzi, H. & Xia, Y. (2003). Photonic papers and inks: color writing with colorless materials. *Adv. Mater.*, Vol.15(No.11): 892–896.
[69] (b) Fudouzi, H. & Xia, Y. (2003). Colloidal crystals with tunable colors and their use as photonic papers. *Langmuir*, Vol.19(No.23): 9653–9660.
[70] Convertino, A. Capobianchi, A. Valentini, A. & Emilio, N.M.C. (2003). A new approach to organic solvent detection: high-reflectivity Bragg reflectors based on a gold nanoparticle/Teflon-like composite material. *Adv. Mater.*, Vol.15(No.13): 1103–1105.
[71] Weissman, J.M. Sunkara, H.B. Tse, A.S. & Asher, S.A. (1996). Thermally Switchable Periodicities and Diffraction from Mesoscopically Ordered Materials. *Science*, Vol.274(No.5289): 959–963.
[72] Holtz, J.H & Asher, S.A. (1997). Polymerized colloidal crystal hydrogel films as intelligent chemical sensing materials. *Nature*, Vol.389: 829–832 .
[73] Hu, Z. Lu, X. & Gao, J. (2001). Hydrogel opals. *Adv. Mater.*, Vol.13(No.22): 1708–1712.
[74] Gates, B. Park, S.H. & Xia, Y. (2000). Tuning the photonic bandgap properties of crystalline arrays of polystyrene beads by annealing at elevated temperatures. *Adv. Mater.*, Vol.12(No. 9): 653–655.
[75] Kim, J.H. Moon, J.H. Lee, S-Y. & Park, J. (2010). Biologically inspired humidity sensor based on three-dimensional photonic crystals. *Appl. Phys. Lett.*,Vol.97(No.103701).
[76] Burgess, I.B. Mishchenko, L. Hatton, B.D. Kolle, M. Loncar M. & Aizenberg, J.(2011). Encoding complex wettability patterns in chemically functionalized 3D photonic crystals. *J. Am. Chem. Soc.*, Vol.133: 12430–12432.
[77] Kubo, S. Gu, Z.Z. Takahashi, K. Ohko, Y. Sato, O. & Fujishima, A. (2002). Control of the optical band structure of liquid crystal infiltrated inverse opal by a photoinduced nematic-isotropic phase transition. *J. Am. Chem. Soc.*, Vol.124: 10950–10951.
[78] Kubo, S. Gu, Z.Z. Takahashi, K. Fujishima, A. Segawa, H. & Sato, O. (2004). Tunable photonic band Gap crystals based on a liquid crystal-infiltrated inverse opal structure. *J. Am. Chem. Soc.*, Vol.126: 8314–8319.
[79] Kubo, S. Gu, Z.Z. Takahashi, K. Fujishima, A. Segawa, H. & Sato, O. (2005). Control of the optical properties of liquid crystal-infiltrated inverse opal structures using photo irradiation and/or an electric field. *Chem. Mater.*, Vol.17: 2298–2309.
[80] Jeong, U. & Xia, Y.N. (2005). Photonic crystals with thermally switchable stop bands fabricated from Se@Ag_2Se spherical colloids. *Angew. Chem., Int. Ed.*, Vol.44: 3099–3103.
[81] Li, B. Zhou, J. Li, L. Wang, J. Liu, X.H. & Zi, J. (2003). Ferroelectric inverse opals with electrically tunable photonic band gap . *Appl. Phys. Lett.*, Vol.83: 4704–4706.
[82] Kuai, S-L. Bader, G. & Ashrit, P.V. (2005). Tunable electrochromic photonic crystals. *Appl. Phys. Lett.*, Vol.86(No.221110).
[83] Khalack, J. & Ashrit, P.V. (2006). Tunable pseudogaps in electrochromic WO_3 inverted opal photonic crystals. *Appl. Phys. Lett.*, Vol.89(No.211112).
[84] Kim, S. Jeon, S. Jeong, W.C. Pank, H.S. & Yang, S. (2008). Optofluidic Synthesis of Electroresponsive Photonic Janus Balls with Isotropic Structural Colors. *Adv. Mater.*, Vol.20(No.21): 4129–4134.
[85] Ge, J. Lee, H. He, L. Kim, J. Lu, Z. Kim, H. Goebl, J. Kwon, S. & Yin, Y. (2009). Magnetochromatic microspheres: rotating photonic crystals. *J. Am. Chem. Soc.*, Vol.131(No.43): 15687–15694.

[86] Kim, J. Song, Y. He, L. Kim, H. Lee, H. Park, W. Yin, Y. & Kwon, S. (2011). Real-time optofluidic synthesis of magnetochromatic microspheres for reversible structural color patterning. *Small*, Vol.7: 1163–1168.

[87] Schultz, T.D. & Rankin, M.A. (1985). The ultrastructure of the epicuticular interference reflectors of tiger beetles(Cicindela). *J. Exp. Biol.*, Vol.117: 87–110.

[88] Schultz, T.D. & Rankin, M.A. (1989).Schultz, T.D. & Bernard, G.D. (1989). Pointillistic mixing of interference colours in cryptic tiger beetles. *Nature*, Vol.337: 72–73.

[89] Berthier, J. Boulenguez, J. & Balint, Z. (2007). Multiscaled polarization effects in *Suneve coronata* (Lepidoptera) and other insects: application to anti-counterfeiting of banknotes. *Appl. Phys. A*, Vol.86(No.1): 123–130.

[90] Vukusic, P. Sambles, J.R. & Lawrence, C.R. (2000). Colour mixing in wing scales of a butterfly. *Nature*, Vol.404: 457.

[91] Liu, F. Yin, H.W. Dong, B.Q. Qing, Y.H. Zhao, L. Meyer, S. Liu, X.H. Zi, J. & Chen, B. (2008). Inconspicuous structural coloration in the elytra of beetles *Chlorophila obscuripennis* (Coleoptera). *Phys. Rev. E*, Vol.77(No.012901).

[92] Liu, F. Wang, G. Jiang, L.P. & Dong, B.Q. (2010). Structural colouration and optical effects in the wings of *Papilio peranthus*. *J. Opt.*, Vol.12(No.065301).

[93] Vinther, J. Briggs, D.E.G. Clarke, J. Mayr, G. & Prum, R.O. (2010). Structural coloration in a fossil feather. *Biol. Lett.*, Vol.6: 128–131.

[94] Parker, A.R. (2000). 515 million years of structural colour. *J. Opt. A: Pure Appl. Opt.*, Vol.2: R15–R28.

[95] Kolle, M. Salgard-Cunha, P.M. Scherer, M. Huang, F.M. Vukusic, P. Mahajan, S. Baumberg, J.J. & Steiner, U. (2010).Mimicking the colourful wing scale structure of the Papilio blumei butterfly. *Nat. Nanotechnol.*, Vol.5: 511–515.

[96] Vukusic, P., Kelly R. & Hooper I. (2009). A biological sub-micron thickness optical broadband reflector characterized using both light and microwaves. *J. R. Soc. Interface*, Vol.6: S193–S201.

[97] Huang, J. Wang, X. & Wang, Z.L. (2008). Controlled replicaton of butterfly wings for achieving tunable photonic properties. *Nano Lett.*, Vol.6(No.10): 2325–2331.

[98] Biro, L.P. Kertesz, K. Horvath, E. Mark, G.I. Molnar, G. Vertesy, Z. Tsai, J.-F. Kun, A. Bailint,Z. & Vigneron, J.P.(2009). Bioinspired artificial photonic nanoarchitecture using the elytron of the beetle Trigonophorus rothschildi varians as a 'blueprint'. *J. R. Soc. Interface*, Vol.7(No.47): 887–894.

[99] Liu, F. Shi, W.Z. Hu, X.H. & Dong, B.Q. (2012). Hybrid structures and the optical effects in *Morpho* scales with thin and thick coatings using an atomic layer deposition method. Unpublished data.

[100] Zhang, W. Zhang, D. Fan, T.X. Gu, J.J, Ding, J. Wang, H. Guo, Q. & Ogawa, H. (2009). Novel Photoanode Srucure Templated from Butterfly Wing Scales. *Chem. Mater.*, Vol.21(No.1): 33–40.

[101] Potyrailo, R.A. Ghiradella, H. Vertiatchikh, A. Dovidenko, K. Cournoyer, J.R. & Olson, E. (2007). Morpho butterfly wing scales demonstrate highly selective vapour response. *Nat. photon.*, Vol.1: 123–128.

[102] Liu, F. Dong, B.Q. Zhao, F, Hu, X. Liu, X. Zi, J. (2011). Ultranegative angular dispersion of diffraction in quasiordered biophotonic structures *Opt. Express*, Vol.19(No.8): 7750–7755.

Chapter 7

Optical Devices Based on Symmetrical Metal Cladding Waveguides

Lin Chen

Additional information is available at the end of the chapter

http://dx.doi.org/10.5772/48146

1. Introduction

Controlling and guiding light with planar waveguide has a great potential to fabricate attractive optical devices such as modulators [1], filters [2] and sensors [3]. Although many studies use planar waveguide made of dielectric materials or semiconductors, metals also play an important role in this field. Metals have been usually used as mirror in the visible or infrared regions. By constructing the dielectric layer sandwiched by two metal layers and forming the metal-dielectric-metal (MDM) structure, we can obtain the unique optical properties which the dielectric planar waveguides do not have. Recently, Shin *et al*. reported that this structure can function as the negative refraction lens for surface plasmon waves on a metal surface. This structure provides a new way of controlling the propagation of surface plasmons, which are important for nanoscale manipulation of optical waves [4].

This type of MDM can be expected to have the interesting optical features not only in the negative refraction index but also in the mode properties. The above mentioned waveguide structure can commonly accommodate only surface plasmon mode. When the thickness of guiding layer is increased to millimeter-scale, such waveguide can accommodate thousands of guided modes, and ultrahigh-order modes (UHM) can be excited. In this case the MDM is commonly called symmetrical metal-cladding waveguide (SMCW) [5]. To our knowledge, however, there have been few investigations about the UHM properties of the SMCW. In this chapter, we have reported the UHM properties of the SMCW and their applications on optical devices. The UHM of SMCW can be excited by free space coupling [6] in small incident angle. In section 2, we present some details of the UHM properties such as large mode spacing, sensitive to the change of waveguide parameters, and slow wave effect. Section 3 introduces applications on optical devices such as modulators, filters and sensors, which are closely related to the UHM in guiding layer.

2. Properties of SMCW

The SMCWs are the millimeter-scale guiding layers (dielectric constant ε_1, thickness d) sandwiched between two metal films (Fig.1). The upper metal film (dielectric constant ε_2, thickness h) of several dozens of nanometer acts as coupling layer as well as a metal cladding; and the base metal layer (dielectric constant ε_2) acts as a substrate of the waveguide, which is thick enough to prevent the influence of the glass flat on guiding layer. For modulation applications, the upper and base metal films also serve as electrodes of the device. If the light incidents from air to the guiding layer directly, it can meet the coupling condition and couple the mode with effective index less than 1. The allowed range of the effective index in SMCW can be reached from 0 to n_1 for the guided modes, where $n_1 = (\varepsilon_1)^{1/2}$ is the refractive index of the guiding slab. This range is much wider than that available to the usual waveguides. The principle model of free-space coupling is shown in Fig.2. When the beam incomes from the air to the metal surface, it will generate evanescent field in the metal. Since the upper metal film is thin (dozens of nanometer), the tail of evanescent field can reach the interface of guided wave layer of the metal. It will generate an opposite evanescent field on the interface. Due to the interaction between two evanescent fields from opposite directions, the incident light can be coupled into waveguide.

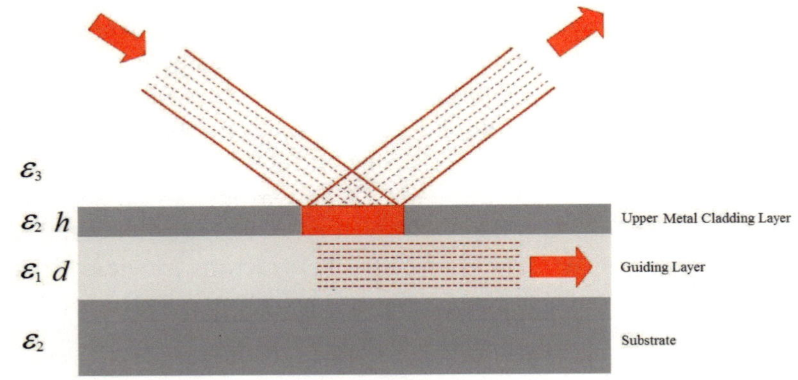

Figure 1. Symmetrical metal-cladding waveguide structure

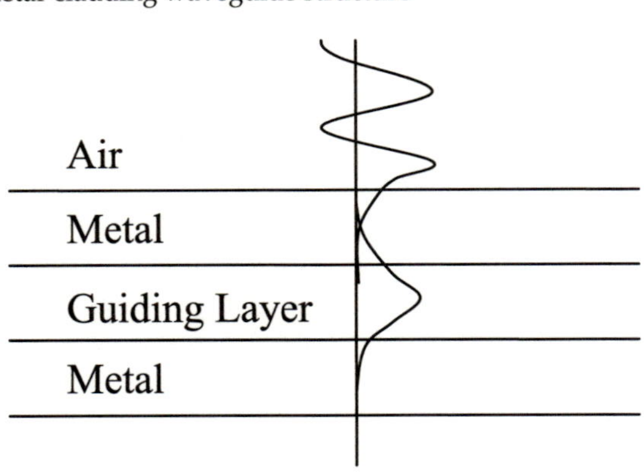

Figure 2. Principle model of free-space coupling

2.1. The attenuated total reflection spectrum

When a laser beam cast on the upper metal layer with resonant conditions, a large part of the light energy is transferred in the guiding layer, resulting in the attenuated total reflection (ATR) spectrum, which describes the relation between reflectivity and incident angle or wavelength of the reflected light. We take TE mode for example. The reflection coefficient of the four-layer optical system for TE mode is written as [7]

$$r=\frac{r_{32}-r_{32}r_{21}^{2}\exp(2i\kappa_1 d)+r_{21}\exp(2i\kappa_2 h)-r_{21}\exp(2i\kappa_1 d)\exp(2i\kappa_2 h)}{1-r_{21}^{2}\exp(2i\kappa_1 d)+r_{32}r_{21}\exp(2i\kappa_2 h)-r_{32}r_{21}\exp(2i\kappa_1 d)\exp(2i\kappa_2 h)} \tag{1}$$

here $r_{ij}=(\kappa_i-\kappa_j)/(\kappa_i+\kappa_j)$, is the complex Fresnel reflection coefficient for the boundary between media i and j, in which the normal components of the wave vectors and propagation constant of guided modes are, respectively, expressed as follows:

$$\kappa_j=\sqrt{k_0^2\varepsilon_j-\beta^2},(j=1,2,3) \tag{2}$$

$$\beta=k_0\sqrt{\varepsilon_3}\sin\theta \tag{3}$$

here $k_0=2\pi/\lambda$ is the wavenumber in vacuum; ε_3 is the dielectric constant at medium in which the light incident and reflected; β is the propagation constant of the guided modes; θ and λ are the incident angle and wavelength, respectively. As shown in Fig.3, when the energy of the incident light is coupled into the guided modes, the intensity of the reflected light $R=|r|^2$ decreases dramatically, and a series of reflection dips in the ATR spectrum are produced.

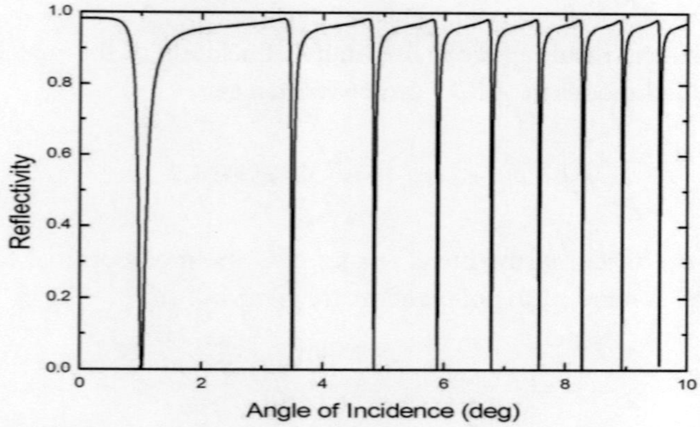

Figure 3. Simulated ATR curve with the following parameters: ε_2=-28+i1.8, ε_1=2.278, λ= 859.8nm, *h*=30nm, and *d*= 0.38mm.

The experimental arrangement for measuring ATR spectrum is shown schematically in Fig.4. The SMCW is fixed on a computer controlled $\theta/2\theta$ goniometer. After passing through the polarizer and mirror, a collimated laser light beam is incident into the structure. Angular scan is carried out by a computer-controlled $\theta/2\theta$ goniometer. As the goniometer rotates, the incident angle will change, while the photodiode keeps consistently monitoring the reflected light intensity to scan the reflectance curve. The reflected light intensity is captured by a photodiode fixed on the 2θ plate of the goniometer and converted into a voltage signal. At small incident angles, UHM resonance will occur and couple the incident light energy into the SMCW by the free space coupling technique. Then in the reflective curve, a series of resonance dips take place in the angular spectrum, as illustrated in Fig.3.

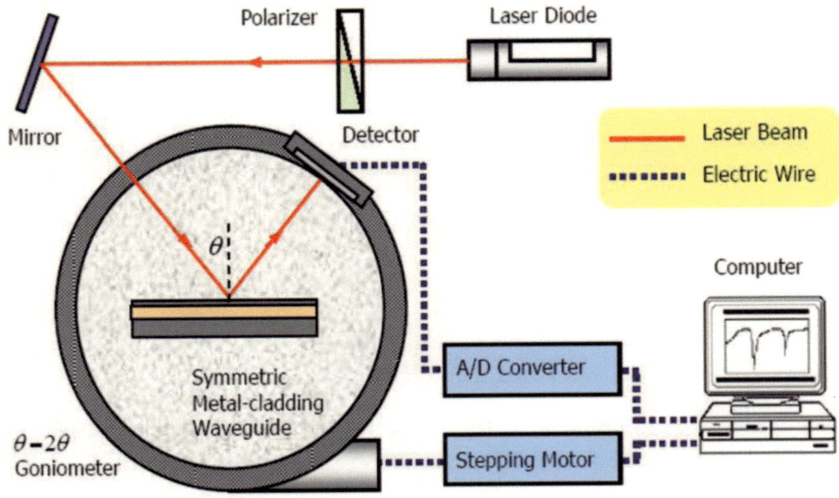

Figure 4. Experimental arrangement for measuring ATR spectrum

2.2. Ultrahigh-order mode

Disregarding the effects resulting from the limited thickness of the metal film, dispersion equation of the guided modes in SMCW can be written as

$$\kappa_1 h = m\pi + 2\tan^{-1}(\rho\frac{\alpha_2}{\kappa_1}), \quad m = 0,1,2,...... \tag{4}$$

where attenuated coefficient in the metal $\alpha_2 = i\kappa_2$, m is the mode ordinal number of guided mode. The parameters relevant to polarization are given by:

$$\rho = \begin{cases} 1 & \textit{TE Polarization} \\ \varepsilon_1/\varepsilon_2 & \textit{TM Polarization} \end{cases} \tag{5}$$

According to Eq. (4), we can deduce an approximate formula:

$$\Delta m \propto \sin 2\theta \Delta\theta \tag{6}$$

Since Δm=1, when the incident angle is small, a bigger $\Delta\theta$ can be obtained. So the mode spacing effect of UHMs is evident. Such property is convenient to comb filter design for optical communication applications.

Furthermore, when d reaches millimeter-scale, the waveguide can accommodate thousands of guided modes. For example, use the parameters: $\varepsilon_2 = -28 + i1.8$, $\varepsilon_1 = 2.278$, $\lambda = 859.8$nm, $h = 30$ nm, and $d = 0.38$ mm, m is 1333 for the highest mode. However, the maximum of the absolute value of the second term on the right in Eq. (4) is π. It will not generate much error if ignoring it, and the approximate dispersion equation of the UHMs for both TE and TM modes is [5]

$$\frac{2\pi}{\lambda} h \cdot \sqrt{\varepsilon_1 - N^2} = m\pi \qquad (m = 0,1,2,......) \tag{7}$$

where $N = \beta / k_0$ is the effective index of the UHM. Eq. (7) implies that when N is close to zero, the UHM exhibits polarization insensibility.

Because UHMs have a short retention time in waveguide layer, any tiny change of λ, n_1 and d will cause the sensitive change of N. If we define sensitivity S_N as the derivative of effective index to certain characteristic parameter, that is

$$S_N = \frac{\partial N}{\partial \xi} \tag{8}$$

where ξ represents λ, n_1 or d. By the total differential of Eq. (7), we can get:

$$\frac{\partial N}{\partial n_1} = \frac{n_1}{N} \tag{9}$$

$$\frac{\partial N}{\partial \lambda} = \frac{n_1^2 - N^2}{N\lambda} \tag{10}$$

$$\frac{\partial N}{\partial d} = \frac{n_1^2 - N^2}{Nd} \tag{11}$$

From above mentioned three equations, sensitivity is in inverse proportion to effective index N. Therefore, we can obtain high sensitivity when UHM is excited. According to ray optics theory, with the same propagation distance, the small incident angle means that UHM will experience more times reflection and light propagation distance will be longer, resulting in a series of special features different from low-order modes. This property is extremely useful to design sensors and modulators.

Finally, according to Eq. (4), tiny change of wavelength can generate great change of effective index, illustrating that UHM has strong dispersion property and consequent slow light effect. Using Eq. (7), we can also obtain the group velocity of UHM:

$$v_g = \frac{d\omega}{d\beta} = \frac{N}{n_1} \cdot \frac{c}{n_1 + \omega dn_1 / d\omega} \tag{12}$$

In the equation, group velocity expression is totally different from those conventional slow light schemes, which is composed of two contributions that are shown in Eq. (12): one originates from the first-order dispersion resembling the conventional slow light system, and factor N/n_1 can be called slow light factor which is related to the effective index of the UHMs. The deduced results offer us a new physical mechanism for realizing slow light that may not rely on the existence of a sharp single resonance or multiple resonances, but the effective index of the UHM for the small incident angle (θ→0).

2.3. Propagation loss

Once SMCW is used to achieving optical devices, the important concern relating to the SMCW is that metallic structures exhibit high losses at optical wavelengths. An issue arising is whether the UHMs could be efficiently confined to the guiding layer over a long distance transmission. To see this clearly, four types of metal cladding waveguides have been considered as shown in Fig.5. Fig.5(a) is a structure of three-layer metal cladding waveguide without considering metal and radiative loss. By using three-layer waveguide theory, for TE mode, propagation constant, β^a, of guided modes for three-layer metal cladding waveguide without considering the metal absorption, can be expressed as:

$$\beta^a = \frac{\pi}{\lambda d}\sqrt{4\varepsilon_1 d^2 - (m_0\lambda)^2} \tag{13}$$

where m_0 is mode order.

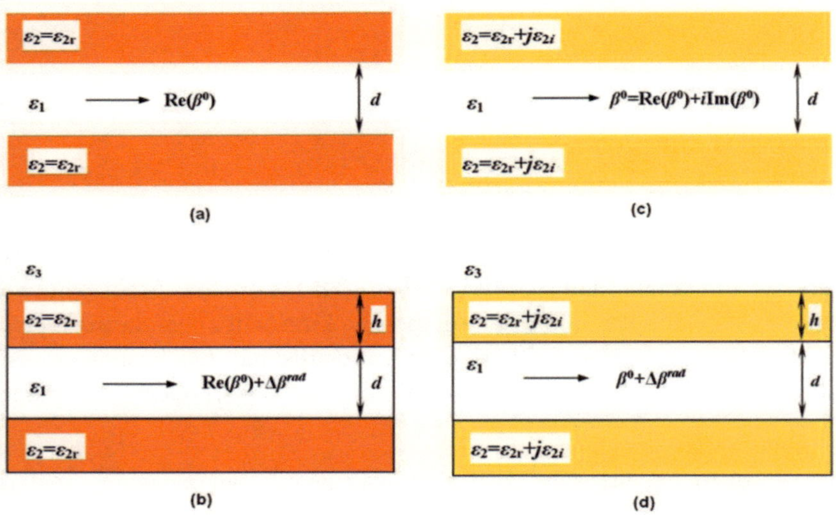

Figure 5. (a) three-layer metal cladding waveguide with $\varepsilon_2=\varepsilon_{2r}$ and h→∞, (b) SMCW with $\varepsilon_2=\varepsilon_{2r}$ and $h \neq \infty$, (c) three-layer metal cladding waveguide with $\varepsilon_2=\varepsilon_{2r}+i\varepsilon_{2i}$ and $h\to\infty$, (d) SMCW with $\varepsilon_2=\varepsilon_{2r}+i\varepsilon_{2i}$ and $h\neq\infty$.

Fig.5(b) is a structure of SMCW only with considering radiative loss. Comparing Fig.5(b) with Fig.5(a) by using weak-coupling condition, which is satisfied with four-layer system [8], the radiative damping, $\mathrm{Im}(\Delta\beta^{rad})$, can be expressed as,

$$\mathrm{Im}(\Delta\beta^{rad}) = \frac{4\alpha_2^a(\kappa_1^a)^2\kappa_3^a\exp(-2\alpha_2^a h)}{((\kappa_1^a)^2+(\alpha_2^a)^2)((\kappa_3^a)^2+(\alpha_2^a)^2)\beta^a d_{eff}} \tag{14}$$

with

$$\kappa_1^a = \sqrt{k_0^2\varepsilon_1 - (\beta^a)^2} \tag{15}$$

$$\alpha_2^a = \sqrt{(\beta^a)^2 - k_0^2\varepsilon_{2r}} \tag{16}$$

$$\kappa_3^a = \left(k_0^2\varepsilon_3 - \left(\beta^a\right)^2\right)^{1/2} \tag{17}$$

and

$$d_{eff} = d + 2/\alpha_2^a \tag{18}$$

Where α_2^a is the absorption constant of the cladding layer, κ_1^a is the vertical guiding wave vector of three-layer metal cladding waveguide without considering the metal absorption, ε_{2r} and ε_{2i} are real and imaginary parts of dielectric constant ($\varepsilon_2=\varepsilon_{2r}+j\varepsilon_{2i}$), and d_{eff} is the effective thickness. $\Delta\beta^{rad}$ represents the difference of the eigen-propagation constant between three-layer waveguide and radiative waveguide coupling system. Because the radiative damping is inversely proportional to the exponential function of h as shown in Eq. (14), the radiative damping can be adjusted by changing the parameter h.

Fig.5(c) is a structure of three-layer metal cladding waveguide only with considering metal loss. Comparing Fig.5(a) with Fig.5(c) under the condition of $|\varepsilon_{2r}|>>\varepsilon_{2i}$ at visible and near infrared wavelength, we can obtain that metal loss only affects the imaginary part of propagation constant in three-layer metal cladding waveguide. Then the intrinsic damping, $\mathrm{Im}(\beta^0)$, can be written as [8]:

$$\mathrm{Im}(\beta^0) = \frac{\varepsilon_{2i}k_0^2(\kappa_1^a)^2}{\alpha_2^a(\kappa_1^a+(\alpha_2^a)^2)\beta^a d_{eff}} \tag{19}$$

where β^0 is the eigen-propagation constant of the guided mode for three-layer waveguide with semi-infinite-thick coupling layer. Table1 lists the propagation loss of two kinds of SMCW which thickness of guiding layer d is 0.5 mm and 1 mm, corresponding to UHM order m is 1421 and 2843, separately. The dielectric constant of Au and Ag at wavelength λ= 1053 nm is -40+2.5i and -48+1.6i, respectively. η presents the ratio of the remaining power after the guided mode transmits for z= 1 mm to the guided mode initial power and can be expressed as $\exp(-2\Delta\beta z)$. Then the propagation loss ζ can be expressed as $-10/z\times lg\eta$. As shown in table 1, in sub-millimeter scale, the propagation losses of the guided modes are both less than 3 dB/mm, which is benefit for the design of optical devices. In addition, the propagation loss property has little relation to polarization.

h/mm	λ/nm	ε	m	mode	|Δβ|	η/%	ζ/(dB/nm)
0.5	1053	εAn=-40+2.5i	1421	TE	0.31904	52.8	2.78
0.5	1053	εAn=-40+2.5i	1421	TM	0.32289	52.4	2.81
0.5	1053	εAg=-48+1.6i	1421	TE	0.15689	73.1	1.36
0.5	1053	εAg=-48+1.6i	1421	TM	0.15834	72.9	1.38
1	1053	εAn=-40+2.5i	2843	TE	0.15953	72.7	1.38
1	1053	εAn=-40+2.5i	2843	TM	0.16144	72.4	1.4
1	1053	εAg=-48+1.6i	2843	TE	0.078446	85.4	0.69
1	1053	εAg=-48+1.6i	2843	TM	0.079167	85.3	0.7

Table 1. Guided-mode propagation loss with different parameters

2.4. The enhanced Goos–Hänchen effect

When the weak coupling condition $|\exp(2i\kappa_2 h)|<<1$ is satisfied, Eq. (1) can be rewritten as

$$r=|r|e^{i\phi}=r_{32}\frac{\beta-\left[\mathrm{Re}\left(\beta^0\right)+\mathrm{Re}\left(\Delta\beta^{rad}\right)\right]-i\left[\mathrm{Im}\left(\beta^0\right)-\mathrm{Im}\left(\Delta\beta^{rad}\right)\right]}{\beta-\left[\mathrm{Re}\left(\beta^0\right)+\mathrm{Re}\left(\Delta\beta^{rad}\right)\right]-i\left[\mathrm{Im}\left(\beta^0\right)+\mathrm{Im}\left(\Delta\beta^{rad}\right)\right]} \tag{20}$$

where φ is the phase difference between the reflected and incident waves. $\mathrm{Re}(\beta^0)$, $\mathrm{Im}(\beta^0)$ and $\mathrm{Re}(\Delta\beta^{rad})$, $\mathrm{Im}(\Delta\beta^{rad})$ are the real and imaginary parts of the parameters β^0 and $\Delta\beta^{rad}$, respectively.

According to stationary phase method, the Goos-Hänchen (GH) shift L is expressed as:

$$L(\lambda)=-\frac{\lambda}{2\pi\sqrt{\varepsilon_3}}\cdot\left.\frac{d\phi}{d\theta}\right|_{\theta=\theta_0} \tag{21}$$

where θ_0 is the fixed incident angle. Using (20), L at the resonance wavelength of ATR curve can be written as [9]

$$L(\lambda_{res})=-\cos\theta_0\cdot\frac{2\,\mathrm{Im}(\Delta\beta^{rad})}{\mathrm{Im}(\beta^0)^2-\mathrm{Im}(\Delta\beta^{rad})^2} \tag{22}$$

With the parameters ε_3=1, ε_1=2.278, ε_2=-28+1.8j, θ_0=8.11, d=0.38μm, and h=22nm, the calculated GH shifts with respect to wavelength is shown in Fig.6. The reflectivity is also shown in Fig.6 for comparison. It is found from Fig.6 that the enhancement of the lateral shift of reflective light is closely relevant to the coupling of waveguide power. Reflectivity corresponds to the excitation of a guided mode with the change of the incident wavelength. When the incident wavelength gradually meets the condition of resonance, the reflectivity decreased sharply, most of the energy of the incident angle is coupled into waveguide, greatly enhance the reflective GH shift and forms a peak, and this peak of lateral shift corresponds to the minimum of the reflectivity. Moreover, as the radiative damping is inversely proportional to the exponential function of h from Eq. (14), we can adjust radiative damping by varying h. When h is small, the radiative damping is larger than the intrinsic damping, positive lateral shift is obtained. The negative GH shift corresponds to the reverse case. Larger GH shift can be obtained when intrinsic damping approaches the radiative damping. The critical thickness can be determined from Eq. (22) by letting denominator is equal to zero [8]:

$$h_{cr} = \frac{\lambda}{4\pi\kappa}\ln\frac{2}{n} \tag{23}$$

where n and κ are the refractive index of the metal ($n+j\kappa=(\varepsilon_2)^{1/2}$). The critical thickness should be about 31nm using the parameters above.

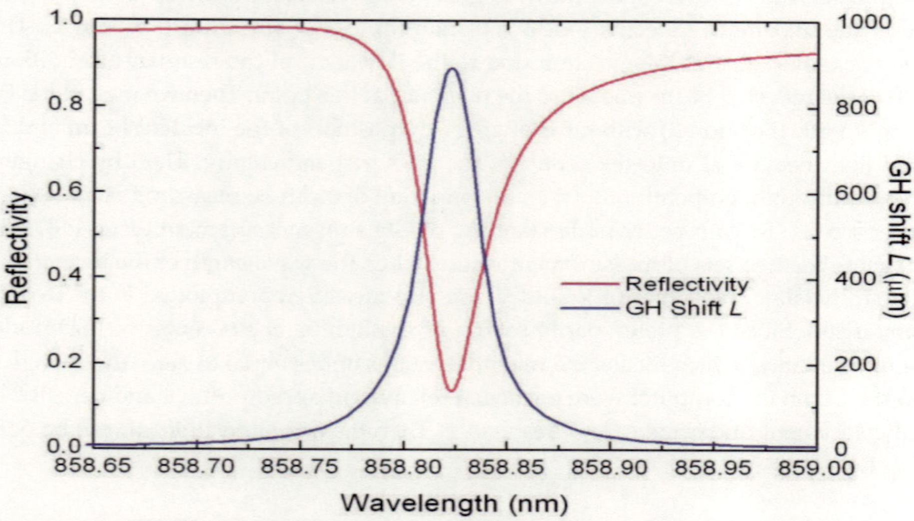

Figure 6. Reflectivity and lateral beam shift with respect to the incident wavelength

To measure the shift while avoiding small spurious displacement, we use the wavelength interrogation-based method by combining a tunable laser and a one-dimensional position sensitive detector (PSD). The position of the incident light can be determined by the PSD through the photocurrents from the output electrodes x_1 and x_2 (see Fig.7):

$$\Delta L = \frac{a}{2}\delta \tag{24}$$

where ΔL is the displacement from the center of the PSD, a is the length of the PSD, δ is defined as

$$\delta = \frac{I_1 - I_2}{I_1 + I_2} = \frac{V_1 - V_2}{V_1 + V_2} \tag{25}$$

here I_1 and I_2 are the photocurrents of the output electrodes x_1 and x_2, respectively. V_1 and V_2 are the voltages converted from I_1 and I_2 after amplifier circuit. The analog voltages V_1 and V_2 are further converted into digital signals and collected by the computer (PC). Light displacement measurement using PSD can achieve high sensitivity and accuracy and will not be affected by the change of the light intensity.

The experimental arrangement for measuring GH shift is shown in Fig.7 [10]. After passing through two apertures (A_1, A_2) and a splitter (S_1), a large part of the Guassian beam from a tunable laser was introduced onto the SMCW. Another part of the beam, which is reflected from S1, irradiated the second splitter (S_2) and is detected by a wavemeter connected to a computer. We choose to excite the UHMs, because of the polarization independence of the UHMs, TE and TM incidences have nearly the same characteristics. The reflected light from the SMCW was first detected by a photodiode (PD). Angular scan was performed by rotating the goniometer and the ATR spectrum was generated. We selected the operation angle to be located at the maximum reflectivity near a certain dip of the spectrum (Position 1). The GH effect is not remarkable at this position due to the deviation of the resonance condition. The position of the reflected beam was set as the reference at this point. Then we moved the PD out of the light path (Position 2) without changing any position of the incident beam and let the reflected light beam cast onto the center of the PSD perpendicularly. Then by changing the wavelength through temperature tuning, the variation of δ can be measured obviously on the computer screen. To gain better understanding of GH shift measurement using PSD, Fig.8(a) and (b) show the two sets of measurement results when the wavelength of the incident light is changed artificially. The values of V_1 and V_2 are also measured and plotted in the two figures for comparison. Since the preliminarily setting of the light is at the center of PSD under the case of unresonance, which means the reference values of δ is equal to zero, the change of the value δ shown on the computer were the actual relative enhanced positive and negative lateral beam displacement under the case of resonance. Then the actual position ΔL can be obtained by using Eq. (24).

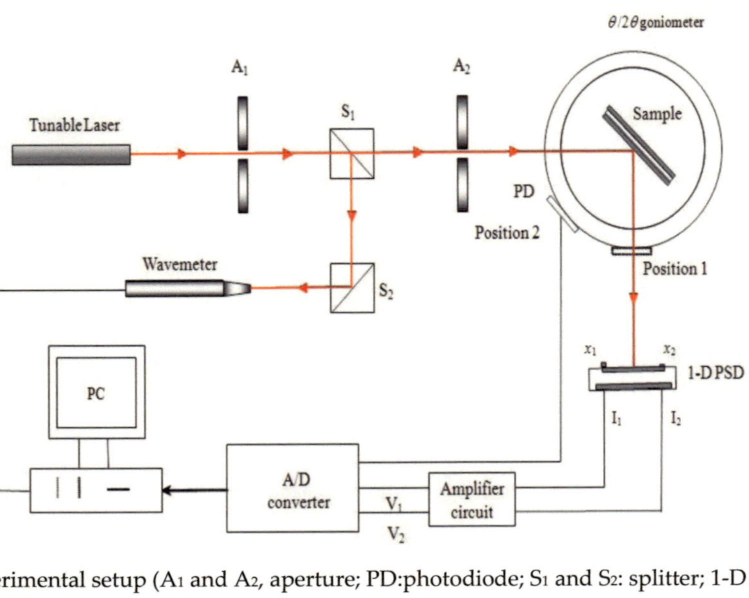

Figure 7. Experimental setup (A_1 and A_2, aperture; PD:photodiode; S_1 and S_2: splitter; 1-D PSD: one-dimensional position sensitive detector and PC: computer).

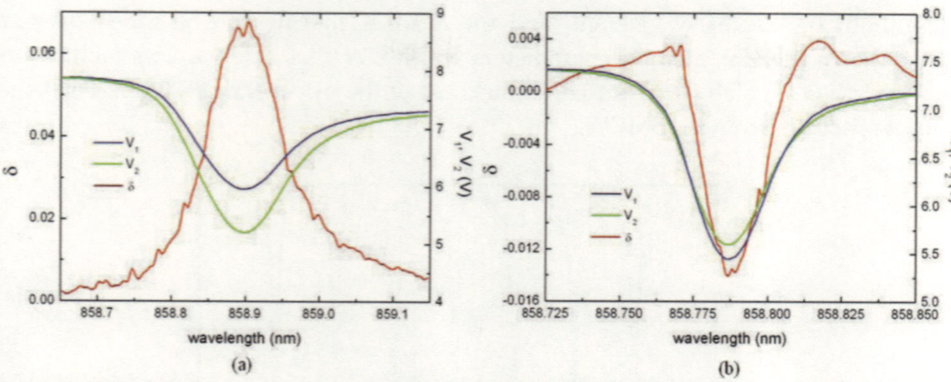

Figure 8. Measured analog voltages V_1 and V_2 and calculated positive value of δ as a function of wavelength with (a) h=8nm and (b) h=50nm. The incident angle θ is 8.11°.

3. Optical devices based on SMCW

The above mentioned properties of UHM in SMCW have been widely used to applications on optical devices. Here, we list several practical application examples, such as filter, sensor, modulator and slow wave device. Due to its excellent performance and simple structure, optical devices based on SMCW have the potential applications in many fields.

3.1. Tunable comb filters

As mentioned above, a sub-millimeter SMCW is capable of coupling the incident light with a fixed wavelength from free space into the glass slab directly. As the excitation of the guided modes results in resonant transfer of energy from the incident light, guided modes manifest themselves by a series of resonant dips in the reflectivity when the incident angle is varied. On the other hand, if a polychromatic light is used instead of a single-wavelength light, resonant dips at a fixed incident angle can also be achieved in a spectral plot of the reflectivity. In this way, a comb filter is built [11]. For a better understanding of how it works, we consider the UHMs of the sub-millimeter SMCW in the case of free space coupling. Eq. (7) can be rewritten as

$$\frac{2\pi v}{c} n_1 d \cos\theta = m\pi \tag{26}$$

where v is the frequency of the light in free space. Without considering the material dispersion of the waveguide, we can express the channel spacing in frequency separated by two neighboring guided modes as

$$\Delta v = \frac{c}{2n_1 d \cos\theta} \tag{27}$$

It is found that θ is a constant as the incident angle of the light is fixed, which means that the channel spacing is equal in frequency for this comb filter. Moreover, according to Eq. (26)

and (27), both the center wavelength and the channel spacing can be tuned by simply changing the incident angle. The parameters of SMCW that achieve comb filter are as follows: a glass (ZF7) slab (d=900μm); the thickness of the upper film h=20 nm. Fig.9 shows the output spectrum from the SMCW.

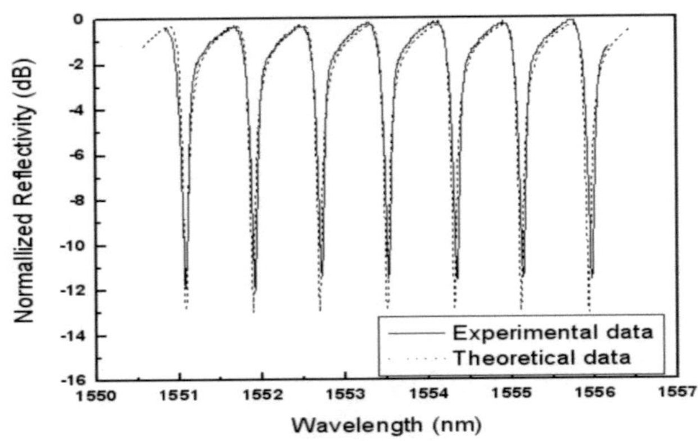

Figure 9. Experimental and theoretical reflective spectrum of SMCW, with parameters $\varepsilon_2 = -132 + i12.56$, $n_1 = 1.765$, $h = 21$ nm, $d = 0.85$ mm and $\theta = 6.1°$

From Fig.9, we can see equally spaced loss peaks with a 3 dB line-width of 0.1 nm appear in the spectral plot from 1551 to 1556 nm. The wavelength spacing of these peaks is just 0.8 nm. The channel isolation, which is closely related to the thickness of the upper gold film of the waveguide, has been found to be 12 dB. Insertion loss is characterized by the maximum reflectivity, which is greater than 95% (~0.2 dB). The tunability of the filter can be obtained simply by slightly rotating the waveguide with respect to the incident light. Tuning of the center wavelength in the range of channel spacing can be easily obtained by changing several degrees of the incident angle according to Eq. (26).

3.2. Optical sensors

In this section, we propose oscillating wave sensors using SMCW structure. The thickness of guiding layer can be expanded to millimeter scale, which it can contain very sensitive UHM. The high sensitivity can be detected by measuring the intensity variation of the reflected light due to the movement of the corresponding synchronous angle by an optical sensor. In addition, an alternative approach is presented via measuring the enhanced GH shifts at excitement of UHM. This approach enables the possibility to obtain a higher resolution and prevent the disturbance caused by the power fluctuation of the light source.

3.2.1. Sensitivity analysis

If we using the intensity measurement interrogation to observe the response of the sensor, the sensitivity in SMCW can be defined as the rate of change of the reflectivity to center characteristics parameters which is expressed as:

$$S_I = \frac{dR}{d\xi} = \left(\frac{\partial R}{\partial \theta}\right)\left(\frac{\partial \theta}{\partial N}\right)\left(\frac{\partial N}{\partial \xi}\right) \tag{28}$$

Using the phase matching condition of resonance energy transfer, the effective refractive index N can be expressed as

$$N = \sqrt{\varepsilon_3}\sin\theta \tag{29}$$

Combined (28) with (29), the sensitivity S_I can be written by

$$S_I = \left(\frac{\partial R}{\partial \theta}\right)\left(\frac{\partial \theta}{\partial N}\right)\left(\frac{\partial N}{\partial \xi}\right) = \frac{1}{\sqrt{\varepsilon_3}\cos\theta}\left(\frac{\partial R}{\partial \theta}\right)\left(\frac{\partial N}{\partial \xi}\right) \equiv \frac{1}{\sqrt{\varepsilon_3}\cos\theta} S_R S_N \tag{30}$$

where S_N is defined as the rate of change of the effective refractive index N with respect to certain characteristic parameter and determined by Eq.(9)-(11). As described in section 2.2, optical waveguide oscillating field sensor exhibits the substantial improvement: the UHMs of the SMCW with millimeter scale is selected to act the sensing probe, so $N \to 0$ and S_N enhancement has been achieved. S_R represents the slope of the dip at the operation angle θ_0, which was usually set at the fall-off side. The absolute maximum values of the slopes are related to the width of the ATR reflection dips and the mode number. The larger the mode number, the wider the corresponding ATR dip, and the smaller the steepest slope. So the divergence angle of the incident light should be taken into consideration. To calculate the maximum value of sensitivity S_I, Lorentzian function is used to approximate the reflection dip in the ATR spectrum. The FWHM is assumed to be larger than the divergence of the incident light in order to prevent profile distortion. According to Ref [12], S_R can be expressed as

$$S_R = \frac{\partial R}{\partial \theta} = \frac{4(R_0 - R_m)}{\alpha\sqrt{2\pi}} \int_{-\infty}^{\infty} \left[\frac{2Q^2}{(x - x_0)^2 + 2Q^2}\right] \cdot x\exp(-x^2)dx \tag{31}$$

with

$$Q = W / \alpha \tag{32}$$

$$x = \sqrt{2}(\theta_i - \theta) / \alpha \tag{33}$$

$$x_0 = \sqrt{2}(\theta_i - \theta_{sp}) / \alpha \tag{34}$$

where R_m and R_0 are the values of the reflectivity when guided mode is and is not excited, respectively. θ_{sp} is the angle for exciting the guided mode, W is the half of the FWHM of the dip, and α is the divergence half-angle of the laser beam.

Taking wavelength sensing for example [13], use the parameters: $\varepsilon_2 = -28 + i1.8$, $\varepsilon_1 = 2.278$ at 859.8 nm, h = 30 nm, α = 0.4 mrad and d = 0.38 mm. The sensitivity S_I with respect to the angle of incidence is shown in Fig.10. The UHMs are more sensitive than the low-order modes and are more applicable for sensing.

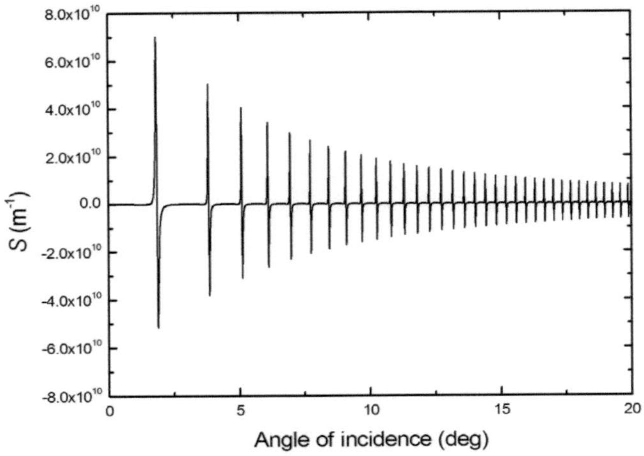

Figure 10. S_I versus incident angle θ

Similarly, in the GH shift interrogation, the sensitivity of the sensor is defined by the change rate of the GH shift (L) with respect to center characteristics parameters and it can be written as

$$S_{GH} = \left(\frac{\partial L}{\partial \theta}\right)\left(\frac{\partial \theta}{\partial N}\right)\left(\frac{\partial N}{\partial \xi}\right) = \frac{1}{\sqrt{\varepsilon_3}\cos\theta}\left(\frac{\partial L}{\partial \theta}\right)\left(\frac{\partial N}{\partial \xi}\right) \equiv \frac{1}{\sqrt{\varepsilon_3}\cos\theta} S_L S_N \tag{35}$$

Compared with the intensity interrogation, the only difference is the replacement of S_R by S_L. According to Eq. (22), if the intrinsic damping of the UHM is close to the radiative damping, a large GH shift can be observed [10]. The dependence of the GH shift on the effective index for a selected mode forms a resonance peak. High sensitivity S_{GH} can be reached at the up and down sides of the peak. In addition, since the magnitude of the GH shift is irrelevant to the incident light intensity, a power fluctuation of the laser brings no disturbance to the resolution of the optical sensor.

Moreover, the thickness of the upper metal cladding is also an important factor that must be considered when analyzing the sensitivity. The thicknesses of upper and bottom metal films are essential parameters in the determination of the sensitivity. If thickness h is excessively thick, it is difficult for the incident light power to be coupled into the planar waveguide structure. However, if thickness h is too thin, it is easy for the incident energy to be coupled into the waveguide. On the other hand, it is also easy for the light energy to be coupled out of the waveguide. As a consequence, there is an optimum solution to thickness h by which the highest sensitivity is obtained. Theoretical and experimental results show that the optimal thickness for the upper metal layer is within the range of about 31–33 nm if the maximum sensitivity is to be achieved. In the following section, refractive index (RI), displacement and light wavelength sensing with extremely high sensitivity are introduced.

3.2.2. *Refractive index sensing*

In the intensity interrogation (as illustrated in Fig. 11(a)), the sample is sealed by an O-ring sandwiched between two gold films that deposited on glass substrates. The two gold films and the sample cell form the SMCW structure. The upper gold film (35 nm in thickness) is deposited on a thin glass slide. The glass slide is 0.178 mm in thickness, with a RI of 1.50. The lower gold film (300 nm) is deposited on a glass plate. The lower glass plate's thickness is 2 mm with an RI of 1.50. The dielectric constants of the gold films are −11.4+i1.50 at the wavelength of 650 nm. The sample cell serves as the guiding layer of the waveguide sensor, and the thickness of the sample cell is governed by the thickness of the O-ring of about 1.99 mm. The aqueous sample could be pumped in and out of the sample cell by a peristaltic pump through the inlet and outlet on the lower substrate. The water sample with an RI of 1.333 can be used as the guiding layer of the waveguide. The guided wave concentrates and propagates in the sample layer as the oscillating field and hence a magnification in sensitivity is expected. Fig.11(b) is the sensor sample [14].

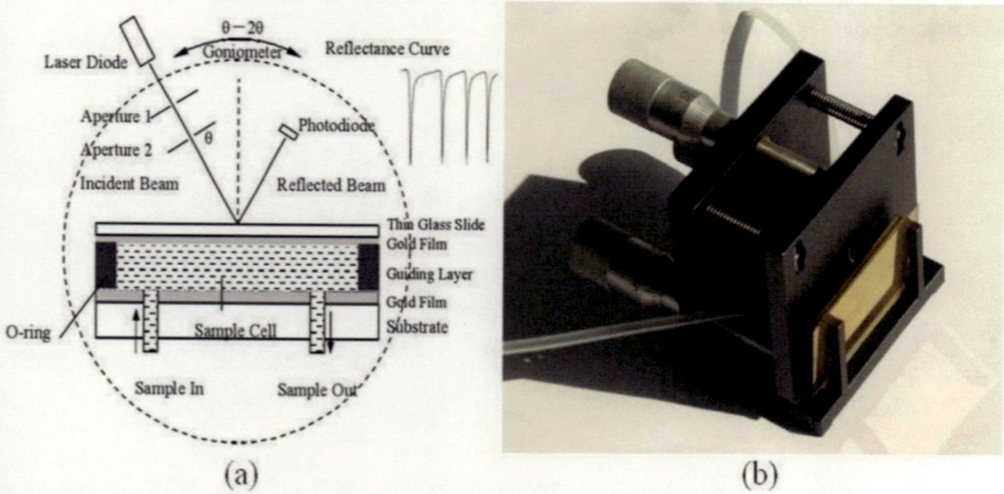

Figure 11. RI sensor of intensity interrogation (a) structural sketch; (b) actual picture

Based on sensitivity analysis, S_I can be cast in the form of

$$S_I = \frac{1}{\sqrt{\varepsilon_3}\cos\theta}\left(\frac{\partial R}{\partial \theta}\right)\left(\frac{\partial N}{\partial n_1}\right) \equiv \frac{S_R n_1}{\cos\theta \cdot \sin\theta} \tag{36}$$

Then we can see that the use of a smaller incident angle (1.69° in the experiment) as a sensing probe can achieve higher sensitivity. The experimental results are shown in Fig.12. In this case, a 20 ppm NaCl concentration change (corresponding to 2.6×10^{-6} RIU) can cause a reflectance change of around 3%.With a standard error of 0.2% for the measurement of the optical intensity[15], its resolution with 1% noise level can reach up to 0.88×10^{-6} RIU for the ideal case.

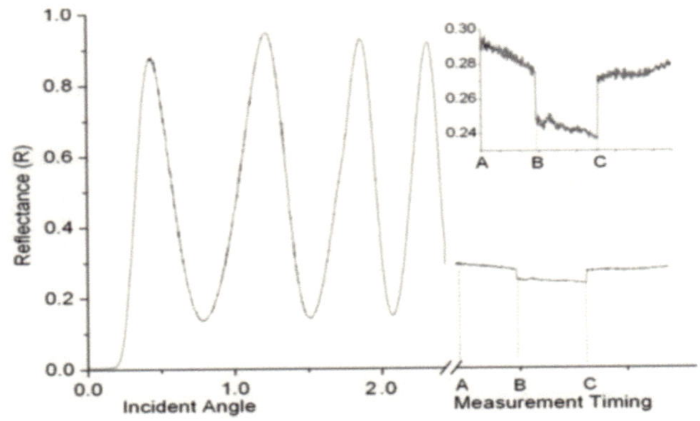

Figure 12. Sensor response for NaCl water solution in the experiment. The left side of the curve is the scanned reflectance curve of the three highest guiding modes. The scan motor stopped at the fall-off of the fourth resonance dip. 20 ppm NaCl water solution is filled in the sample cell and the first measurement starts at A and is then paused. After pure water is pumped in the sample cell and the signal stabilized, measurement starts at B and then back to the 20 ppm NaCl sample measurement at C. The sensor response is enlarged in the inset.

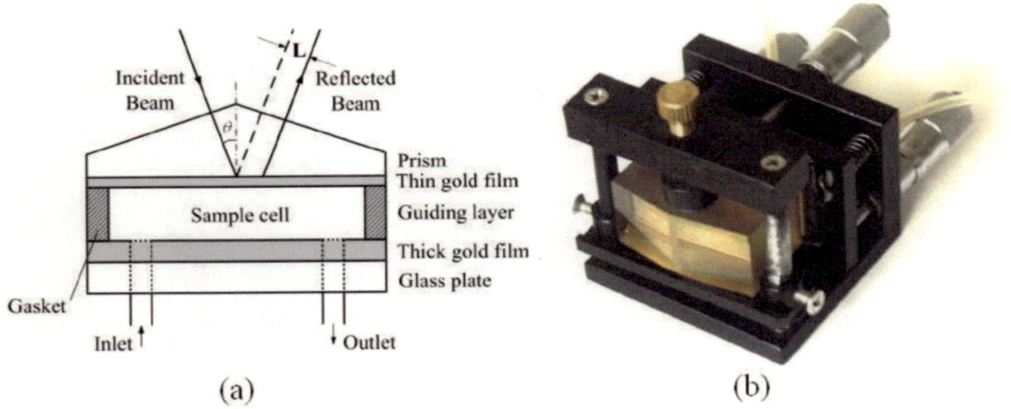

Figure 13. RI sensor of GH shift interrogation (a) structural sketch; (b) actual picture

In the GH shift interrogation (as illustrated in Fig. 13), a glass prism is coated with a 20 nm thick gold film to serve as the coupling layer. A 300 nm thick gold film is sputtered on a glass slab to act as the substrate. The air gap of 0.7 mm sandwiched between two gold films works as the guiding layer, where a gasket is used to form a sealed sample cell. With the help of a peristaltic pump, sample liquids to be detected flow into the cell through the inlet and the outlet tubes embedded in the substrate glass plate.

The expression of sensitivity S_{GH} is similar to Eq. (36) by replacing S_R with S_L. So UHM is also selected as a sensing probe. Experiments are carried out with the waveguide parameters as follows: θ=4.60°, ε_3=2.25, ε_2=−28+1.8i, h=20 nm, and d=0.7 mm, for pure water (solid curve), n_1=1.333 RIU. A series of NaCl solutions with the change step of 20 ppm in concentration is used as sample analyte to be probed. The experimental result is shown in

Fig.14. The step change of 20 ppm NaCl solution in concentration, which corresponds to a variation of 2.64 $\times 10^{-6}$ RIU, induces a GH shift change of at least 20 μm. Considering the noise level in the experiment, the probing sensitivity of 2.0×10^{-7} RIU is resolved since the measurement variation of the GH shift is confined within 1.5 μm for each sample [16].

3.2.3. *Displacement sensing*

In the experimental setup, we propose to use a variable air gap produced by a calibrated piezoelectric translator (PZT) to act as the guiding layer of the optical waveguide. As shown in Fig.15, the sample for minute displacement measurement is composed of two parts: one is a glass prism on its base precoated with a thin gold film; the other is a 500 μm thick LiNbO3 slab sandwiched between two 400 nm thick gold films and serves as a PZT. The two components, separated by an air gap with a thickness of 100 mm, are rigidly mounted on a heavy platform. The gold films deposited on the prism and the upper surface of LiNbO3 slab, together with the air gap form an SMCW.

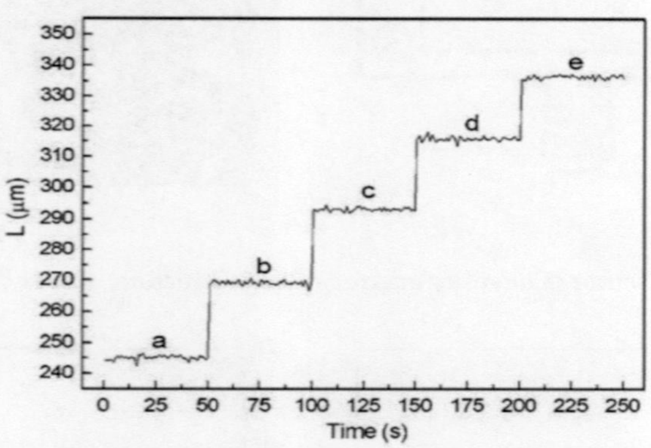

Figure 14. The GH shifts with respect to solutions of different concentrations in the sample cell: (a) pure water, (b) 20ppm NaCl solution, (c) 40 ppm NaCl solution, (d) 60 ppm NaCl solution, and (e) 80 ppm NaCl solution

As soon as applying a dc voltage on the pair electrodes of the PZT, the air gap changes its thickness due to the piezoelectric effect of the LiNbO3 slab. As a result, the reflection dip shifts its peak position and result in a change of the reflectivity. According to the resolution of the reflectivity variation, displacement can be evaluated from the applied voltage and the piezoelectric coefficient of the LiNbO3 slab. In the intensity interrogation, S_I can be cast in the form of

$$S_I = \frac{1}{\sqrt{\varepsilon_3}\cos\theta}\left(\frac{\partial R}{\partial \theta}\right)\left(\frac{\partial N}{\partial d}\right) \equiv \frac{\varepsilon_1 - \varepsilon_3 \sin^2\theta}{\varepsilon_3 \cos\theta \cdot \sin\theta \cdot d} \cdot S_R \tag{37}$$

We can also use UHM as the sensing probe to achieve higher sensitivity. Test experiment has been performed with the waveguide parameters as follows: ε_1=1.0, ε_2=−11+i1.0, ε_3=3.0, d=108 μm, h=40 nm, and λ=650 nm. Displacement sensitivity of proposed configuration is shown in Fig.16. The waveguide thickness d is increased and decreased in steps by increasing and reducing the voltages applied on the electrodes of the PZT. The step-style change of voltage is 50 V. According to the piezoelectric coefficient of a Z-cut LiNbO3 slab, d_{33}=33.45 pm/V, the value of the displacement resolution for the proposed configuration is determined as $S_I = 50\times 33.45\times 10^{-3} = 1.7$ nm, which corresponds to the reflectivity change of R=1%[17].

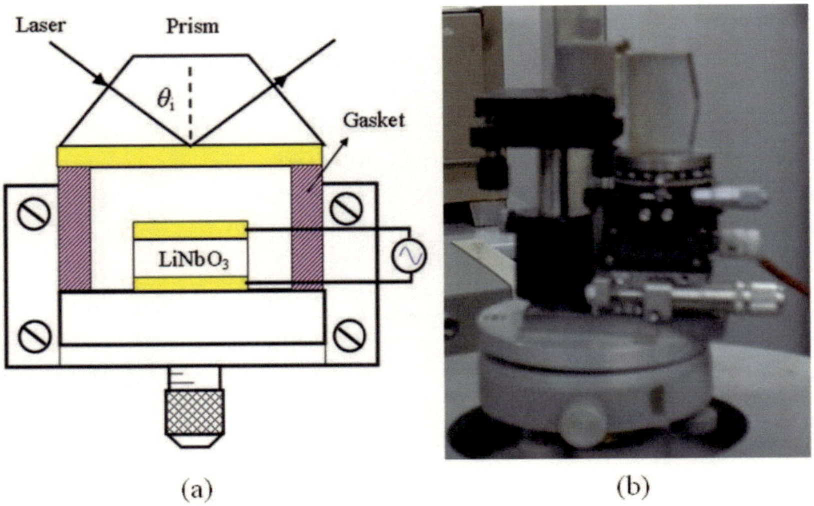

Figure 15. Displacement sensor of intensity interrogation (a) structural sketch; (b) actual picture

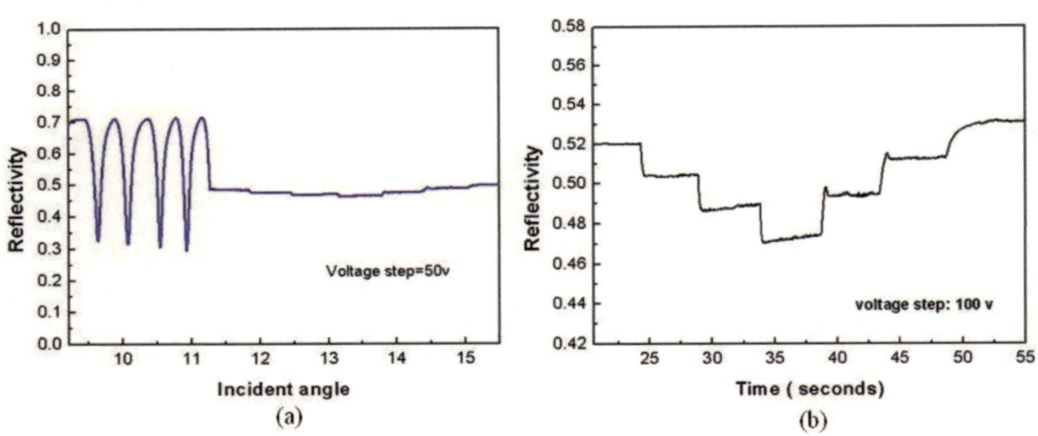

Figure 16. Experimental result of displacement measurement: (a) m-line obtained by angular scanning and displacement experimental result; (b) intensity export by applying step-styled voltages.

In the GH shift interrogation, the sensor structure is the same as Fig. 15. The experiment was performed with the following parameters: ε_2=−28+1.8i, ε_1=1, ε_3=2.25, d=500 μm, and h =18 nm. The incident wavelength is adjusted to be 859.00 nm. The experimental result is shown in Fig.17. The voltage applied on the PZT between each step is 10 V, and the piezoelectric coefficient of the z-cut $LiNbO_3$ is d_{33}=8 ×10^{-12} m/v. Thus, the thickness change per step is determined as Δd=8×10^{-12}×10 m=8×10^{-11} m, which leads to a GH shift change of 2 μm. The experimental ripple of each step is confined to 0.5 μm. With this noise level, the sensing resolution is evaluated to be 40 pm[18].

3.2.4. Wavelength sensing

In the intensity interrogation, the wavelength sensitivity S_I can be written as

$$S_I = \frac{1}{\sqrt{\varepsilon_3}\cos\theta}\left(\frac{\partial R}{\partial \theta}\right)\left(\frac{\partial N}{\partial \lambda}\right) \equiv \frac{\varepsilon_1 - \varepsilon_3 \sin^2\theta}{\varepsilon_3 \cos\theta \cdot \sin\theta \cdot \lambda} \cdot S_R \tag{38}$$

It is found that the effective index is extremely sensitive to λ in the case of $N \to 0$. The waveguide parameters are given as follows: d = 0.38 mm, $\varepsilon_2 = -28 + i1.8$, $\varepsilon_1 = 2.278$, and h= 31 nm. The initial wavelength was set to 859.800 nm. Once this value was stabilized, the reference wavelength was changed subsequently to 859.8005, 859.8010, 859.8015, and 859.8020 nm and decreased back to 859.8000 nm in steps. The step-style change of wavelength is 0.5 pm, with the average 2.5% change in the reflectivity ΔR. Considering a noise level of about 0.05% in Fig.18, a resolution of 0.2pm is finally obtained for the reflectivity change of R=1%[13].

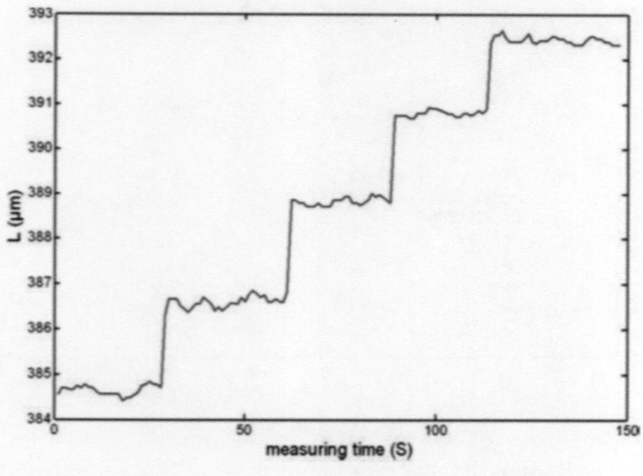

Figure 17. Experimental sensitivity of the proposed configuration. Voltage applied on the PZT between each step is 10 V, which leads to an 8×10^{-11} m change of the air gap thickness.

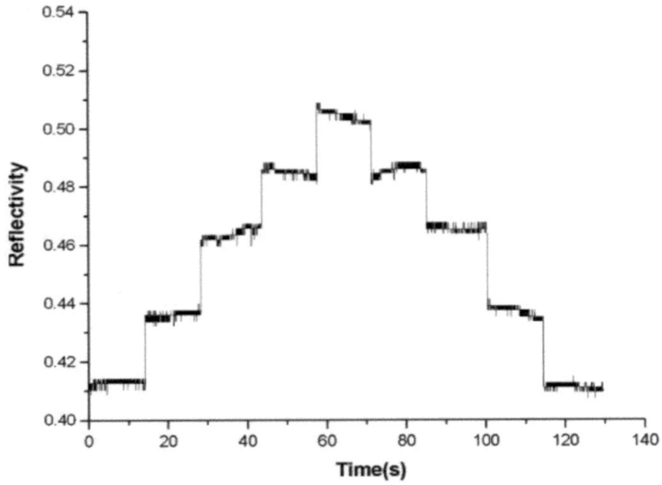

Figure 18. Wavelength sensitivity of the proposed configuration. The shift in the ATR curve was measured by changing the wavelength in steps of 0.5 pm. The angle of incidence is 3.82°.

For the GH shift measurement, the waveguide parameters are given as follows: d = 0.5 mm, ε_2 = −28 + i1.8, ε_1 = 1, ε_3 =2.25, h = 19.8 nm, and θ=4.263°. We first fix the wavelength (λ=859.003 nm). As shown in Fig.19, the change of wavelength 1 pm will cause the average variation of reflective light lateral shift about 10 μm.

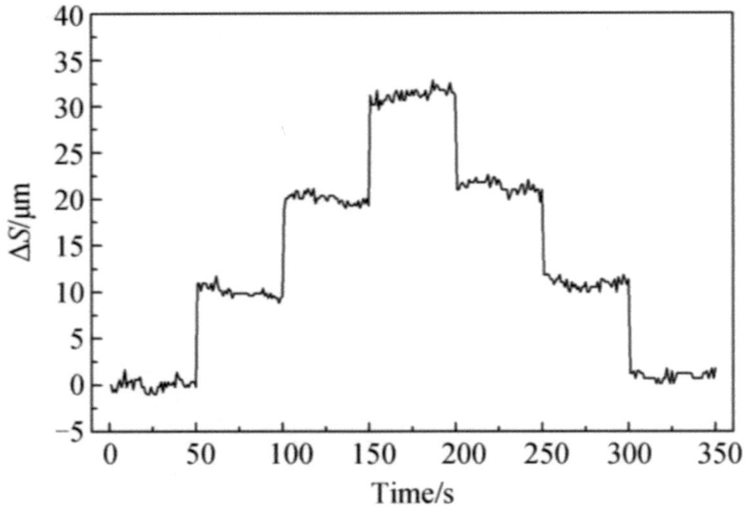

Figure 19. Experiment result using GH shift measurement

3.3. Optical modulators

The configuration of electro-optic (EO) modulator is a SMCW on a glass (K9) flat. The cover and the substrate are both gold film, and the waveguide is EO polymer film. An applied electric field modulates refractive index of the EO polymer, resulting in the change of the effective refractive index for the guided modes, shifting the resonance dips along the angular direction in ATR spectrum. If we define γ_{33} as the EO coefficient of the polymer and E is the applied electric field across the EO polymer film. The refractive index change of guiding layer Δn_1 is written as [19]:

$$\Delta n_1 = -\frac{1}{2} n_1^3 \gamma_{33} E \tag{39}$$

At the midst of the fall-offs of resonance dip, where a considerably good linearity is observed, the change of the light reflectivity ΔR is:

$$\Delta R = \frac{1}{\sqrt{\varepsilon_3}\cos\theta}\left(\frac{\partial R}{\partial \theta}\right)\left(\frac{\partial N}{\partial n_1}\right)\Delta n_1 = \frac{S_R n_1^4 \gamma_{33}}{\sin 2\theta} E \tag{40}$$

Therefore the reflected light is modulated by the applied electric field. Higher sensitivity is obtained at the midst of the fall-off of the resonance dip excited at smaller resonance angle with thicker guiding layer, so that enhanced modulation is realized by enabling the device to operate with stronger modulation depth and lower driving voltage.

In the experiment, the gold film about 300nm was sputtered onto the surface of the K9 glass flat to serve as substrate and one electrode. A PMMA-based second-order nonlinear optical (NLO) side-chain material containing the disperse red chromophore was synthesized through copolymerization for electro-optic device. The polymer was dissolved in toluene, 25% polymer to 75% toluene by weight. A 12-thick polymer film was spin coated onto the gold film substrate, and then was tempered at 40° in a vacuum for 12 hours to remove the residual solvent. The refractive index of polymer is 1.52 at wavelength 832nm. In order to remove the centrosymmetric structure of the chromophores, the film was corona-poled in the air by an applied electric voltage of 4000V at 110° for 25 min with inter-electrode distance being 20mm, and cooled down to room temperature with the field still applied. Finally, the upper gold film about 30nm thick was deposited on the polymer film by sputtering technique to serve as the coupling layer and another electrode. The complex dielectric constant of the gold film is $\varepsilon_2 = -28 + i1.6$. The EO coefficient γ_{33} of polymer measured at 832 nm using improved ATR method is 11.9 pmV. A collimated TM-polarized beam with 832nm wavelength is used. The modulation working angle was chosen at the midst of the fall-off of the resonance dip excited at resonance angle 9.7°. A sinusoidal electrical field was applied across the two electrodes with 10 $V_{P\text{-}P}$ driving voltage at 1MHz, the electro-optic modulation process was obtained. The oscilloscope traces of modulating voltage and reflected intensity versus time are shown, respectively, in Fig.20. The modulation depth was measured to be 8.7%. However, under the same electrical field the modulation depth was only 5.6% when the resonance dip at 28.9° was used, which is attributed to lower sensitivity at larger resonance angle. The total insertion loss of the sample was 1.08 dB[20].

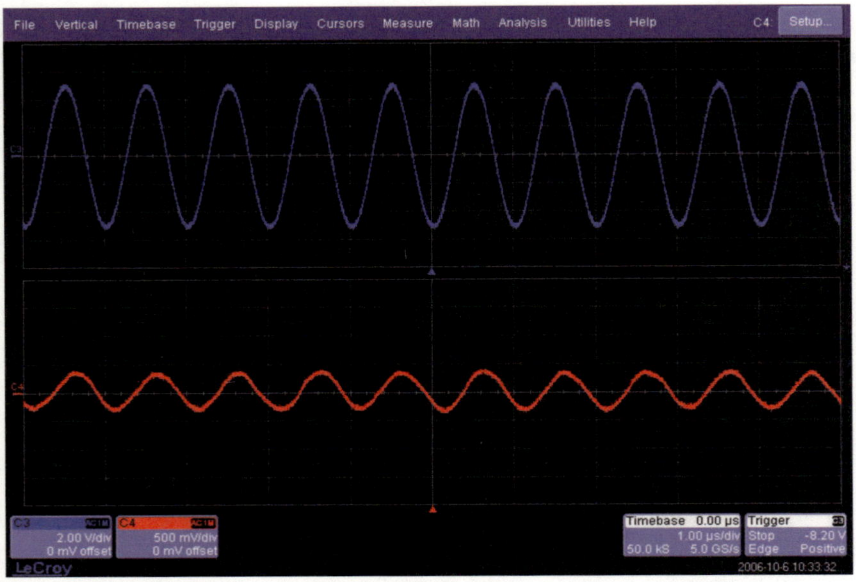

Figure 20. Oscilloscope traces of modulating voltage versus time (upside) and the reflected intensity versus time (downside).

3.4. Slow light devices

The scheme diagram of the SMCW for verifying slow light effect is illustrated in Fig. 21. The advantages of this geometry are that the slow light properties can be tailored to the desired wavelength and the delay is tunable by varying the incident angle and the parameters of the guiding layer. This is important because the applications of slow light require a degree of tunability. These make the proposed slow light scheme useful and practical.

Test experiment has been performed with the waveguide parameters as follows: $\varepsilon_1 = 2.89$, $\varepsilon_2 = -19 + 0.5i$ (silver films), $\varepsilon_3=1.0$, d=2mm (glass slab), h=30 nm, λ=632.8 nm, and $\theta=3°$. An additional silver stripe (about 500nm thick and 1.1 cm wide) is fabricated in the middle of this layer to prevent light leakage. According to Eq.(19), one finds that $\mathrm{Im}(\beta^0)=0.1158526\mathrm{mm}^{-1}$, which shows that the signal power is about 10% after propagating along the guiding layer for about 1 cm. The proposed structure is completely different from the conventional Fabry-Perot cavity, and those similar folded optical delay lines where the incident ray of light bounces in the cavity between the two interfaces (mirrors) before it exits [21]. From Eq. (12), the slow light effect can be observed only when a specific UHM that propagates along the guiding layer is excited, thus the proposed slow light scheme does not rely on obtaining long optical paths. This proposed slow light mechanism can also be interpreted in terms of the anomalous dispersion of the UHMs, which is depicted in Fig.22. It is demonstrated in Fig.22 that anomalous dispersion curves of the proposed structure exhibit an extremely flattened region (slow light region) in the vicinity of zero wave number ($N\rightarrow 0$).

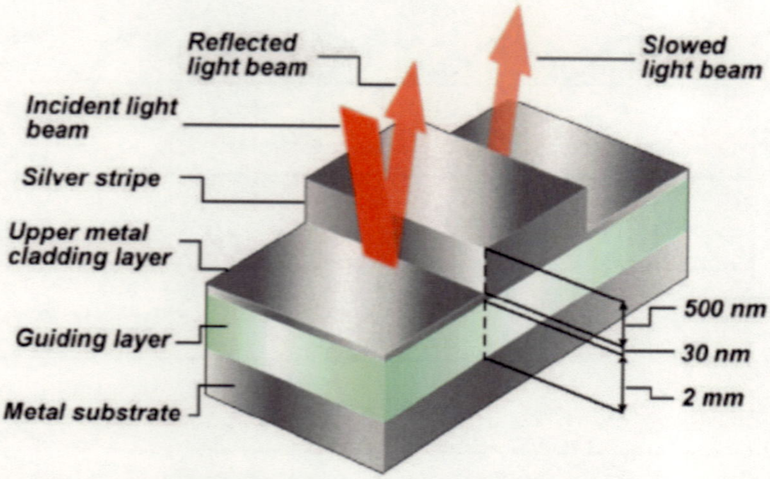

Figure 21. Schematic layout of the SMCW for verifying slow light effect

In the experimental arrangement, the source is a collimated light beam at a wavelength of 650 nm, which is modulated by an EO modulator to produce a signal of 1.0 GHz sinusoidal pulse train. Two photodiodes are setup to detect the light intensity. The first photodiode (PD1) takes in the reflected light beam, which serves as the reference beam. The second photodiode (PD2) is used to measure the time delay of the slowed light beam. The tunable delay is measured by an oscilloscope with the bandwidth of 2 GHz. We measure the delay times Δt at different fixed incident angles, which become larger as the incident angle θ becomes smaller. The experimental results are illustrated in Fig.23. After propagating through a 1.1-cm-long active region, a delay of 2.165 ns was achieved, that corresponds to group velocities less than 0.017c. Further decreasing the incident angle could generate more delay [22].

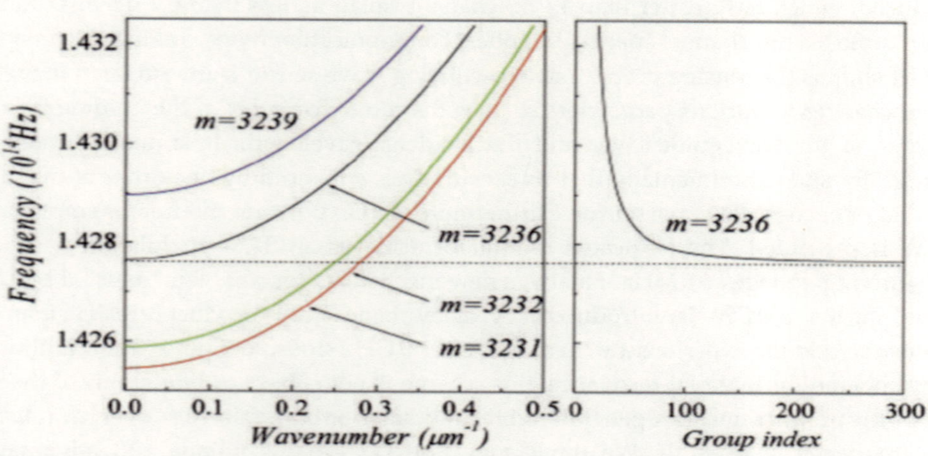

Figure 22. Dispersion curves and group index characteristics of ultrahigh-order modes with parameters $\varepsilon_2 = -19 + 0.5i$, $\varepsilon_1 = 2.89$, and $d = 2$ mm.

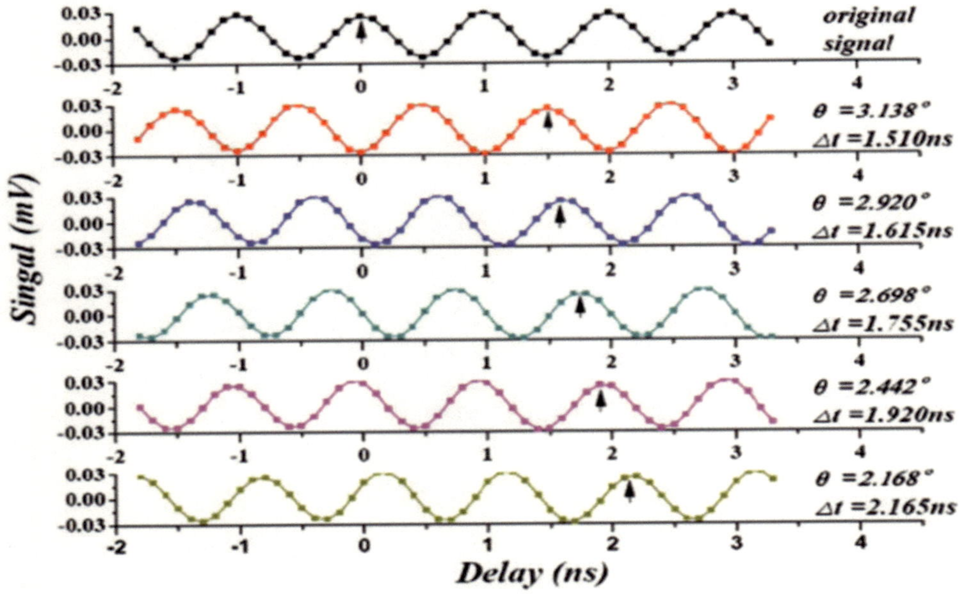

Figure 23. Delay tuning of slow light in an SMCW structure by adjusting the incident angle.

4. Conclusion

The properties of UHM in a SMCW and its applications on optical devices have been demonstrated in this chapter. It is found that the effective refractive index of UHM is sensitive to the refractive index, the thickness of the waveguide layer and the incident wavelength. UHM has also shown strong dispersion and polarization independent effects. Then, a polarization independent and tunable comb filter based on SMCW has been introduced, which has greater than 12 dB channel isolation, less than 0.2 dB insertion loss, and accurate 0.8 nm channel spacing in optical communication range. Taking the reflectivity and GH shift as the sensing probe, a new oscillating wave sensor is investigated to measure minute changes in various parameters such as the refractive index of the guiding layer, the thickness of the waveguide layer and the incident wavelength. It is demonstrated both theoretically and experimentally that its sensitivity is enhanced by one order of magnitude than that of evanescent wave sensor. Furthermore, an EO polymer modulator employing an SMCW is presented. The fabricated modulator achieves an 8.2% modulation depth with $10V_{P-P}$ driving voltage at 1 MHz. Finally, a new mechanism for slow light assisted by UHMs excited in the SMCW is introduced. A delay bandwidth product greater than 2 is demonstrated in the experiment with a signal of 1.0 GHz sinusoidal pulse train. Without use of any coherent or material resonance, this scheme is not subject to limitations of the delay bandwidth product and can generate arbitrarily small group velocities over an unusually large frequency bandwidth. We think such SMCWs possess unique and advantageous properties over the state-of-the-art and may have great potential for next generation optical devices.

Author details

Lin Chen
Engineering Research Center of Optical Instrument and System, Ministry of Education,
Shanghai Key Lab of Modern Optical System,
University of Shanghai for Science and Technology, China

Acknowledgement

The author thanks Ning Yang and YinQi Bao, students from the University of Shanghai for Science and Technology, for editing the manuscript of this chapter. This work is partly supported by the Leading Academic Discipline Project of Shanghai Municipal Government (S30502), "Chen Guang" Research Fund from Shanghai Municipal Education Commission and Shanghai Education Development Foundation (09CG49), and the Basic Research Program of Shanghai from Shanghai Committee of Science and Technology (11ZR1425000).

5. References

[1] O. Solgaard, F. Ho, J. I. Thackara, and D. M. Bloom, High frequency attenuated total internal reflection light modulator, *Appl. Phys. Lett.*, Vol. 61, No. 21, pp. 2500-2502 (1992).

[2] B. Yu, G. Pickrell, and A. Wang, Thermally tunable extrinsic Fabry-Perot filter, *IEEE Photon, Technol. Lett.*, Vol. 16, No. 10, pp. 2296-2298 (2004).

[3] R. Horvath, H. C. Pedersen, and N. B. Larsen, Demonstration of reverse symmetry waveguide sensing in aqueous solutions, *Appl. Phys. Lett.*, Vol. 81, No. 12, pp. 2166-2168 (2002).

[4] H. Shin and S. Fan, All-Angle Negative refraction for surface plasmon waves using a metal-dielectric-metal structure, *Phys. Rev. Lett.*, Vol. 96, No. 7, pp. 3907-3910 (2006).

[5] H. F. Lu, Z. Q. Cao, H. G. Li and Q. S. Shen, Study of ultrahigh-order modes in a symmetrical metal-cladding optical waveguide, *Appl. Phys. Lett.*, Vol. 85, No. 20, pp. 4579-4581 (2004).

[6] H. G. Li, Z. Q. Cao, H. F. Lu, and Q. S. Shen, Free-space coupling of a light beam into a symmetrical metal-cladding optical waveguide, *Appl. Phys. Lett.*, Vol. 83, No. 14, pp. 2757-2759 (2003).

[7] X. M. Liu, P. F. Zhu, Z. Q. Cao, Q. S. Shen, and J. L. Chen, Prism coupling of ultrashort light pulses into waveguides, *J. Opt. Soc. Am. B*, Vol. 23, No. 2, pp. 353-357 (2006).

[8] L. Chen, Y. M. Zhu, D. W. Zhang, Z. Q. Cao and S. L. Zhuang, Investigation of the limit of lateral beam shifts on a symmetrical metal-cladding waveguide, *Chin. Phys. B*, Vol. 18, No. 11, pp. 4875-4880 (2009).

[9] X. B. Liu, Z. Q. Cao, P. F. Zhu, Q. S. Shen, and X. M. Liu, Large positive and negative lateral optical beam shift in prism-waveguide coupling system, *Phys. Rev. E*, Vol. 73, No. 5, pp. 6617-6621 (2006).

[10] L. Chen, Z. Q. Cao, F. Ou, H. G. Li, Q. S. Shen, and H. C. Qiao, Observation of large positive and negative lateral shifts of a reflected beam from symmetrical metal-cladding waveguides, *Opt. lett.*, Vol. 32, No.11, pp. 1432-1434 (2007).

[11] H. Lu, Z. Cao, H. Li, Q. Shen, and X. Deng, Polarization-independent and tunable comb filter based on a free-space coupling technique, *Opt. Lett.*, Vol. 31, No.3, pp. 386-388 (2006).

[12] J. Villatoro and A. Garcia-Valenzuela, Sensitivity of optical sensors based on laser-excited surface-plasmon waves, *Appl. Opt.*, Vol. 38, No. 22, pp. 4837-4844 (1999).

[13] L. Chen, Z. Q. Cao, Q. S Shen, X. X. Deng, F. Ou, and Y. J. Feng, Wavelength sensing with subpicometer resolution using ultrahigh order modes, *J. Lightwave Technol.*, Vol. 25, No.2, pp. 539-543 (2007).

[14] J. H. Gu, G. Chen, Z. Q. Cao, and Q. S. Shen, An intensity measurement refractometer based on a symmetric metal-clad waveguide structure, *J. Phys. D: Appl. Phys.*, Vol. 41, No.18, pp. 5105 (2008).

[15] J. Homola, S. S. Yee, and G. Gauglitz, Surface plasmon resonance sensors: review, *Sens. Actuators B*, Vol. 54, No. 1-2, pp. 3-15 (1999).

[16] Y. Wang, H. G. Li, Z. Q. Cao, T. Y. Yu, Q. S. Shen, and Y. He, Oscillating wave sensor based on the Goos–Hänchen effect, *Appl. Phys. Lett.*, Vol. 92, No. 06, pp. 1117-1119 (2008).

[17] F. Chen, Z. Q. Cao, Q. S. Shen, X. X. Deng, B. M. Duan, W. Yuan, M. H. Sang, and S. Q. Wang, Nanoscale displacement measurement in a variable-air-gap optical waveguide, *App. Phys. Lett.*, Vol. 88, No. 16, pp. 1111-1112 (2006).

[18] T. Y. Yu, H. G. Li, Z. Q. Cao, Y. Wang, Q. S. Shen, and Y. He, Oscillating wave displacement sensor using the enhanced Goos–Hänchen effect in a symmetrical metal-cladding optical waveguide, *Opt. Lett.*, Vol. 33, No. 9, pp. 1001-1003 (2008).

[19] Y. Jiang, Z. Cao, Q. Shen, X. Dou, Y. Chen, and Y. Ozaki, Improved attenuated total reflection technique for measuring the electro-optic coefficients of nonlinear optical polymers, *J. Opt. Soc. Am. B*, Vol. 17, No. 5, pp. 805-808 (2000).

[20] X. X. Deng, P. P. Xiao, X. Zheng, Z. Q. Cao, Q. S. Shen, K. Zhu, H. G. Li, W. Wei, S. X. Xie, and Z. J. Zhang, An electro-optic polymer modulator based on the free-space coupling technique, *J. Opt. A: Pure Appl. Opt.*, Vol. 10, No. 1, pp. 5305 (2008).

[21] D. R. Herriott and H. J. Schulte, Folded optical delay lines, *Appl. Opt.*, Vol. 4, No. 8, pp. 883-889 (1965).

[22] W. Yuan, C. Yin, H. G. Li, P. P. Xiao, and Z. Q. Cao, Wideband slow light assisted by ultrahigh-order modes, *J. Opt. Soc. Am. B*, Vol. 28, No. 5, pp. 968-971 (2011).

Chapter 8

Nano-Plasmonic Filters Based on Tooth-Shaped Waveguide Structures

Xu Guang Huang and Jin Tao

Additional information is available at the end of the chapter

http://dx.doi.org/10.5772/75765

1. Introduction

Along with development of human society and technology, it becomes more dependable on the miniaturization and integration of semiconductor components, circuits and devices. The performance of integrated circuits, such as micro-processor, is in accordance with the famous Moore's law that the number of transistors placed inexpensively on an integrated circuit doubles approximately every two years. However, the integration of modern electronic components and devices for information communication and processing have been approaching its fundamental speed and bandwidth limitation, because the ultra-intensive electrical interconnects have an increased effective resistor-capacitor (RC) time constant that increases the time of charging and discharging [1, 2]. This has caused an increasing serious problem that hinders further development in many fields of modern science and technology. Using light signals instead of electronic is one of the most promising solutions. The speed of optical signal is on the order of 10^8 m/s, which is about 3 orders of the saturation velocity of electrons in a semiconductor such as silicon [3]. However, a major problem with using light as information carrier in conventional optical devices is the poor performance of integration and miniaturization. Dielectric waveguides are basic components and cannot allow the localization of electromagnetic waves into subwavelength-scale regions because of diffraction limit $\lambda_0/2n$, here λ_0 is the wavelength of the light in the free space and n is the refractive index of the dielectric. Photonic crystal (PC) structures and devices have been studied by many researchers since E. Yablonovitch and S. John ' s two milestone published papers in 1987 [4, 5], which confirmed that the light can be confined in the nanoscale. However, the dimensions of the PC system are on the order of the wavelength or even larger, making them less appropriate for nano-scale optical elements integration.

Surface Plasmon polaritons (SPPs) are electromagnetic waves that propagate along the interface of metal and dielectric. In recent years, plasmonics is called the area of nanophotonics under the light diffraction limit that studies the transmission characteristics, localization, guiding of the SPP mode using metallic nanosturctures. The plasmonic waveguides are the most basic components and have been given much attention. Various kinds of metallic nanostructures have been proposed for SPP guiding. Generally these structures could be classified into three big categories: 1) chains of metal nanoparticles [6] and cylindrical metallic nanorods with various geometries [7]. 2) Metal-dielectirc-metal (MDM) or Metal-insulator-metal (MIM) plasmonic waveguides, including groove channel structures in metallic substrates [8], slot waveguide [9]. 3) Dielectric-metal-dielectric (DMD)/ Insulator-metal-Insulator (IMI) waveguide [10]. It should be noted that not all these plasmonic structures can be used for guiding SPP mode to achieve subwavelength localization. Among them, MDM/MIM plasmonic waveguide can propagate SPP mode in the subwavelength scale with relatively low dissipation and large propagation distance. Our following proposed structures are mainly based on the MDM/MIM structures.

Wavelength selection is one of key technologies in fields of optical communication and computing. To achieve wavelength filtering characteristics, plasmonic Bragg reflectors and nanocavities have been proposed. They include the metal hetero-structures constructed with several periodic slots vertically along a metal-dielectric-metal (MDM) waveguide [11], the Bragg grating fabricated by periodic modulating the thickness of thin metal stripes embedded in an insulator [12] and the periodic structure formed by changing alternately two kinds of the insulators [13,14]. Lately, a high-order plasmonic Bragg reflector with a periodic modulation of the core index of the insulators [15], and a structure with periodic variation of the width of the dielectric in MDM waveguide [16] have been proposed. Most of the periodic structures mentioned above, however, have total length of micrometers and relatively high propagation loss of several decibels.

In this chapter, we present our recent work on compact nano-plasmonic waveguide filters based on T-series and nano-capillary structures. In section 2, we introduce the novel nanometeric plasmonic filter in a tooth-shaped MIM waveguide and give an analytic model based on the scattering matrix method. In section 3, we investigate the characteristics of double-side teeth-shaped nano-plasmonic waveguide. In section 4 and section 5, we introduce the multiple multiple-teeth-shaped plasmonic filters and asymmetrical multiple teeth-shaped narrow pass band subwavelength filter. In section 6, we introduce a wavelength demultiplexing structure based on metal-dielectric-metal plasmonic nano-capillary resonators. Finally, we make a conclusion.

2. Single-tooth shaped plasmonic waveguide filter [17]

To begin with the dispersion relation of the fundamental TM mode in a MIM waveguide (shown in the inset of Fig. 1) is given by [18]:

$$\varepsilon_{in} k_{z2} + \varepsilon_m k_{z1} \coth(-\frac{ik_{z1}}{2} w) = 0, \tag{1}$$

with k_{z1} and k_{z2} defined by momentum conservations:

$$k_{z1}^2 = \varepsilon_{in}k_0^2 - \beta^2, \quad k_{z2}^2 = \varepsilon_m k_0^2 - \beta^2. \tag{2}$$

Where ε_{in} and ε_m are respectively dielectric constants of the insulator and the metal, $k_0=2\pi/\lambda_0$ is the free-space wave vector. The propagation constant β is represented as the effective index $n_{eff}=\beta/k_0$ of the waveguide for SPPs. The real part of n_{eff} of the slit waveguide as a function of the slit width at different wavelengths is shown in Fig. 1. It should be noted that the dependence of n_{eff} on waveguide width is also suitable to the region of the tooth waveguide with the tooth width of w_t shown in Fig. 2. The imaginary part of n_{eff} is referred to the propagation length which is defined as the length over which the power carried by the wave decays to 1/e of its initial value: $L_{spps}=\lambda_0/[4\pi\cdot\text{Im}(n_{eff})]$. In the calculation above and the following simulations, the insulator in all of the structures is assumed to be air ($\varepsilon_{in}=1$), and the frequency-dependent complex relative permittivity of silver is characterized by Drude model: $\varepsilon_m(\omega)=\varepsilon_\infty-\omega_p^2/\omega(\omega+i\gamma)$. Here $\omega_p=1.38\times10^{16}\,Hz$ is the bulk plasma frequency, which represents the natural frequency of the oscillations of free conduction electrons. $\gamma=2.73\times10^{13}\,Hz$ is the damping frequency of the oscillations, ω is the angular frequency of the incident electromagnetic radiation, ε_∞ stands for the dielectric constant at infinite angular frequency with the value of 3.7 [16]

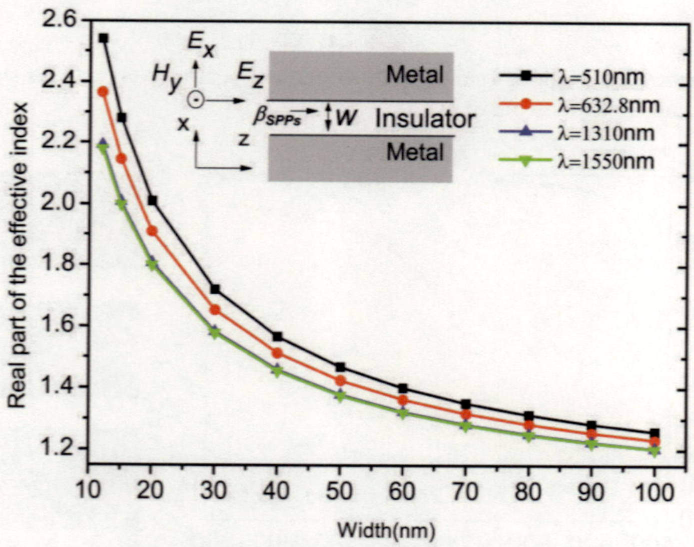

Figure 1. Real part of the effective of refraction index versus the width of a MIM slit waveguide structure.

The tooth-shaped waveguide filter is shown in Fig. 2. In the following FDTD simulations, the grid sizes in the x and z directions are chosen to be 5nm×5nm. Since the width of the waveguide is much smaller than the operating wavelength in the structure, only fundamental waveguide mode is supported. Two power monitors are respectively set at the positions of P and Q to detect the incident and transmission fields for calculating the incident power of P_{in} and the transmitted power of P_{out}. The transmittance is defined to be $T=P_{out}/P_{in}$. The length of L is fixed to be 300nm, while the tooth width and depth are respectively w_t=50nm and d=100nm. The tabulation of the optical constants of silver [19] is used in the simulation. As shown in Fig. 3(a), the tooth-shaped waveguide is of a filtering function: A transmission dip occurs at the free space wavelength nearly 784nm with the transmittance of ~0%. The maximum transmittance at the wavelengths longer than 1700nm is over 90%. The contour profiles of the field distributions around the tooth-shaped area at different wavelengths are shown in Figs. 3(b)-3(d). The filtering structure is distinguished from the Bragg reflectors based on periodical heterostructure.

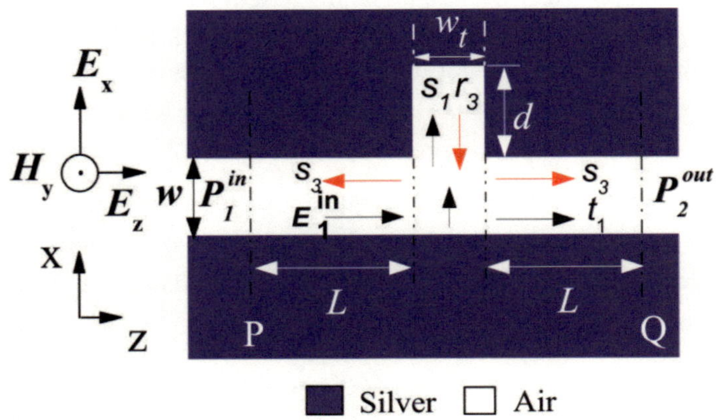

Figure 2. The structure schematic of a single tooth-shaped waveguide filter, with the slit width of w, the tooth width of w_t, and the tooth depth of d.

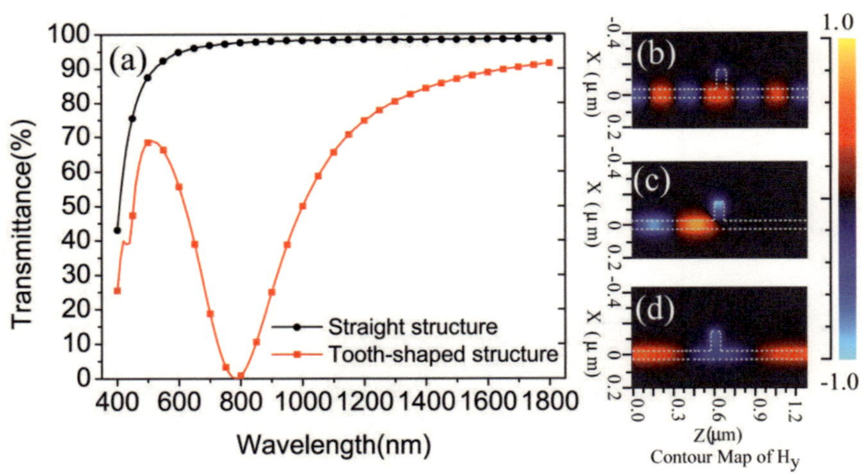

Figure 3. (a) Transmission of the single tooth-shaped MIM waveguide compared with a straight MIM slit waveguide. The width of the waveguide is w=50nm, and the tooth width and depth are respectively w_t=50nm and d=100nm. The contour profiles of field H_y of the tooth-shaped waveguide at different wavelengths of (b) λ=510nm, (c) λ=783nm, and (d) λ=1550nm.

The phenomenon above can be physically explained in the scattering matrix theory [20] as follows:

$$\begin{pmatrix} E_1^{out} \\ E_2^{out} \\ E_3^{out} \end{pmatrix} = S \cdot \begin{pmatrix} E_1^{in} \\ E_2^{in} \\ E_3^{in} \end{pmatrix}, \tag{3}$$

where $S = \begin{bmatrix} r_1 & t_1 & s_3 \\ t_1 & r_1 & s_3 \\ s_1 & s_1 & r_3 \end{bmatrix}$, r_i, t_i and s_i (i=1,2,3) are respectively the reflection, transmission and splitting coefficients of a incident beam from Port i (i=1,2,3), caused by the structure. E_i^{in} and E_i^{out} stand for the fields of incident and output beams at Port i, respectively. Using the fact that $|S| = 1$, one can obtain:

$$r_1^2 r_3 + 2t_1 s_1 s_3 - 2r_1 s_1 s_3 - t_1^2 r_3 = 1. \tag{4}$$

For the case of $E_2^{in} = 0$, one has

$$E_2^{out} = t_1 E_1^{in} + s_3 E_3^{in}, \tag{5}$$

in which E_3^{in} is given as follows:

$$E_3^{in} = s_1 E_1^{in} \exp(i\phi(\lambda))(1 + r_3 \exp(i\phi(\lambda)) + r_3^2 \exp(2i\phi(\lambda)) + ...) = \frac{s_1 E_1^{in}}{1 - r_3 \exp(i\phi(\lambda))} \exp(i\phi(\lambda)), \tag{6}$$

where the phase delay $\phi(\lambda) = \frac{4\pi}{\lambda} n_{eff} \cdot d + \Delta\varphi(\lambda)$, and $\Delta\varphi(\lambda)$ is the phase-shift caused by the reflection on the air-silver surface. Combined Eq. (5) and Eq. (6), the output field at Port 2 is derived as:

$$E_2^{out} = t_1 E_1^{in} + \frac{s_1 s_3 E_1^{in}}{1 - r_3 \exp(i\phi(\lambda))} \exp(i\phi(\lambda)), \tag{7}$$

Therefore, the transmittance T from Port 1 to Port 2 is given by:

$$T = \left| \frac{E_2^{out}}{E_1^{in}} \right|^2 = \left| t_1 + \frac{s_1 s_3}{1 - r_3 \exp(i\phi(\lambda))} \exp(i\phi(\lambda)) \right|^2. \tag{8}$$

It can be seen from Eq. (8) that, if the phase satisfies $\phi(\lambda) = (2m+1)\pi$ (m=0,1,2...), the two terms inside the absolute value sign on the right of the equation will cancel each other (as it can be seen in Fig. 3(c)), so that the transmittance T will become minimum. Therefore, the wavelength λ_m of the transmission dip is determined as follows:

$$\lambda_m = \frac{4 \cdot n_{eff} \cdot d}{(2m+1) - \dfrac{\Delta\varphi(\lambda)}{\pi}}. \tag{9}$$

It can be seen that the wavelength λ_m is linear to the tooth depth d, and depends on tooth width w_t, through the somewhat inverse-proportion-like relationship between n_{eff} and w_t shown in Fig. 1.

Figure 4(a) shows the transmission spectra of the waveguide filters with various tooth widths of w_t. The maximum transmittance can reach 97%. Figure 4(b) shows the wavelength of the trough vs. the tooth width of w_t. The primary dip of the transmission moves very significantly to short wavelength (blue-shift) with the increase of w_t for w_t<20nm. The shift rate rapidly becomes small after w_t>20nm, and tends to be saturated when w_t>200nm. As revealed in the Eq. (9), the above relationship between the dip position and w_t mainly results from the contribution of the inverse-proportion-like dependence of n_{eff} on w_t. The change rate of $\Delta n_{eff} / \Delta w_t$ within the tooth width of 20nm is much higher than that of $\Delta n_{eff} / \Delta w_t$ after w_t>20nm, as shown in Fig. 1, and becomes finally saturated after w_t>200nm. Obviously, tooth width w_t should be chosen within the range of 20-200nm to avoid the critical behavior and the difficulty in fabrication process.

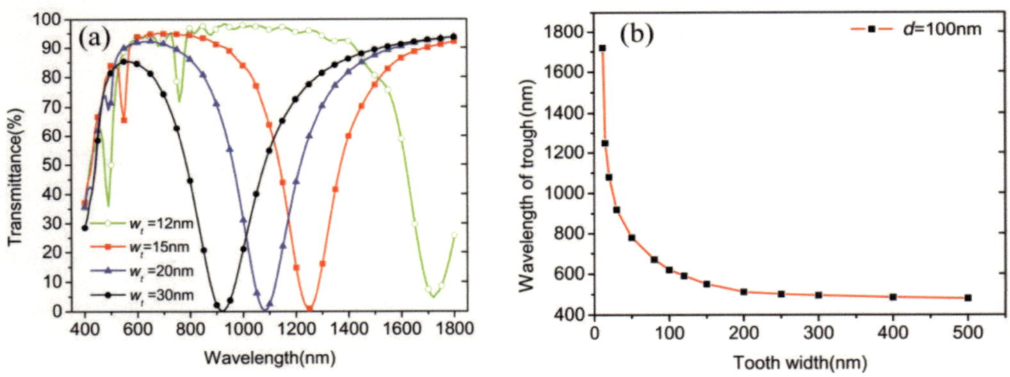

Figure 4. (a) Transmission spectra of the waveguide filters with various tooth widths of w_t, at a fixed tooth depth of d=100nm and the slit width of w=50nm. (b) The wavelength of the trough versus the tooth width of w_t.

Figure 5(a) shows the transmission spectra of the filters with different tooth depths of d. It is found that the wavelength of the transmission dip shifts to long wavelength with the increasing of d. Figure 5(b) reveals that the wavelength of the transmission dip has a linear relationship with the tooth depth, as our expectation in Eq. (9). Therefore, one can realize the filter function in various required wavelength with high performance by changing the width or/and the depth of the tooth. For example, to obtain a filter with a transmission dip at the wavelength of 1550nm, the structural parameters of w_t=w=50nm and d=237.5nm can be chosen.

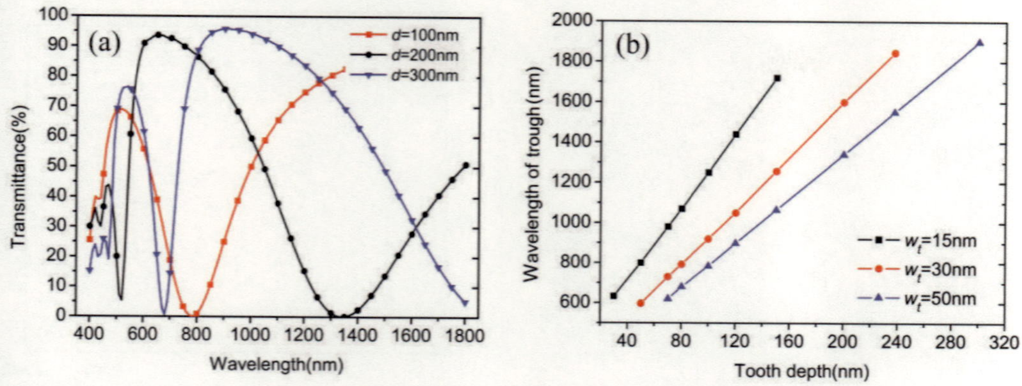

Figure 5. (a) Transmission spectra of the waveguide filters with different tooth depths of d, and with a given tooth width of w_t=50nm and the slit width of w=50nm. (b) The wavelength of the transmission dip vs. the tooth depth of d with w_t=15nm, w_t=30nm and w_t=50nm.

3. Double-side teeth-shaped nano-plasmonic waveguide filters [21]

Figure 6(a) shows the structure of a proposed double-side teeth-shaped waveguide filter. The waveguide width w and the distance L are fixed to be 50nm and 300 nm. d_1 and d_2 are the depths of the teeth on each side of a MIM waveguide. The transmission spectra of the double-side teeth-shaped waveguide filter and the single-side tooth-shaped waveguide filter are shown in Fig. 6(b), which is obtained with the FDTD method. One can see that there is one dip at the free-space wavelength of nearly 1550nm with the transmittance of ~0%. An analytic model to explain the filtering function of single-side tooth-shaped waveguide structure based on multiple-beam-interference and scattering matrix has been given in the section 1

It can be seen that the wavelength λ_m is linear to the tooth depth d and depends on tooth width w through the somewhat inverse-proportionlike relationship between n_{eff} and w shown in Fig. 1. The dip width of the double-side structure is wider compared with that of the single-side tooth-shaped structure. For better understanding the characteristics of double-side teeth structure, we next discuss the dependence of its filtering spectrum on the symmetry of teeth depth.

Figure 7 shows the transmission spectra of the double-side teeth-shaped waveguide filters with different teeth depths of d_1 and d_2. From Fig. 7(a), one can see that there are two transmission dips at the free-space wavelengths of nearly 790nm and 1230nm with d_1=100nm and d_2=180nm. It is found that the wavelength of the second dip shifts to a long wavelength with the increasing of d_2 while the first trough keeps unchanged due to a fixed value of d_1=100nm. From Fig. 7(b), we can see that the first dip shifts to a long wavelength with the increasing of d_1 while the position of the second dip is fixed due to a given d_2=180nm.

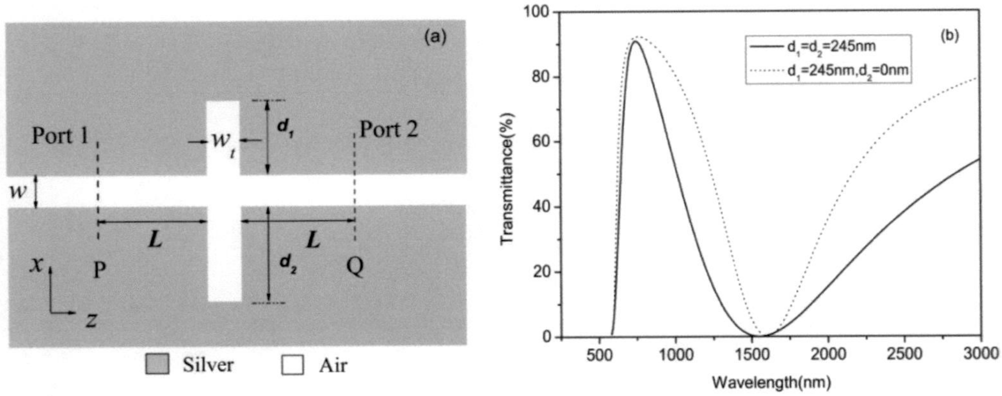

Figure 6. (a) Schematic of a double-side teeth-shaped nano-plasmonic waveguide. The double-side teeth-shaped structure can be asymmetrical, if $d_1 \neq d_2$. (b) The transmission spectra of a symmetrical double-side teeth-shaped waveguide filter with w_t=50nm, d_1=d_2=245nm and a single-side tooth-shaped waveguide filter with w_t=50nm, d_1= 245nm.

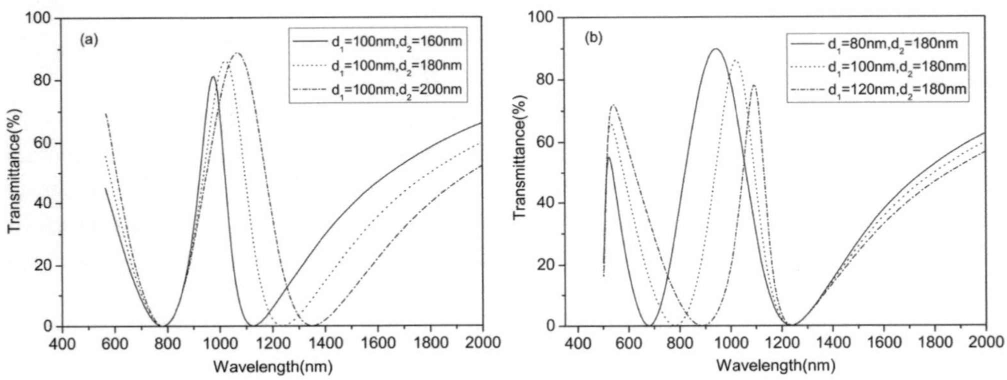

Figure 7. (a)Transmission spectra of the asymmetrical double-side teeth-shaped waveguide filters for different tooth depths of d_2 with a fixed d_1=100nm and w_t=50nm. (b) Transmission spectra of the asymmetrical double-side teeth-shaped waveguide filters for different tooth depths of d_1 with a fixed d_2=180nm and w_t=50nm.

To better understand the origin of the two dips, the transmission spectra of an asymmetrical double-side teeth-shaped waveguide filter with d_1=100nm and d_2=180nm and two single tooth shaped structures with the tooth depths of 100nm and 180nm are all shown in Fig. 8. From it, one can see that the trough positions of the two single tooth waveguide filters with the tooth depths d_1=100nm and d_2=0 as well as d_1=0 and d_2=180nm overlap with the positions of the dips of the asymmetrical double-side teeth-shaped structure, which means the positions of the two dips of the double-side teeth-shaped structure are almost respectively determined by its two different single-tooth parts. Therefore, one can realize the filter function in various required wavelengths by respectively changing the depths of d_1 and d_2 of the two single teeth.

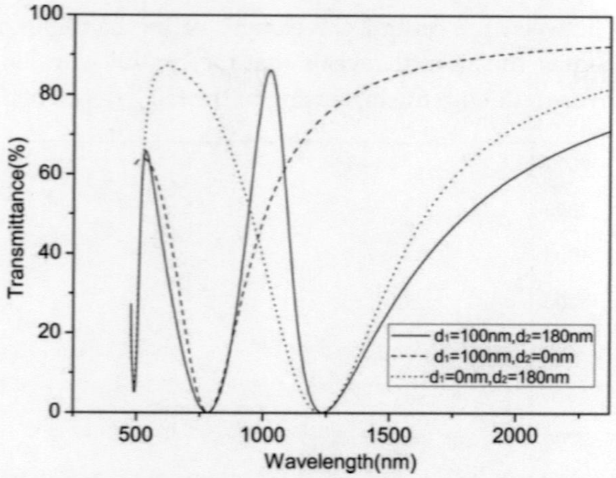

Figure 8. Transmission spectra of the two single-tooth waveguide filters and an asymmetrical double-side teeth-shaped structure with a given tooth width of w_t=50nm and a slit width of w=50nm.

It is also interest to address a staggered double-side teeth-shaped MIM waveguide structure (shown in Fig. 9(a)). L_s stands for the shift length between two single teeth. A typical transmittance of the staggered double-side teeth-shaped waveguide filter with w_t=50nm, d_1=d_2=260.5nm, and L_s =250nm is shown in Fig. 9(b). A wide bandgap occurs around λ=1700nm with the bandgap width (defined as the difference between the two wavelengths at of which the transmittance is equal to 1%) is 660nm, and the transmittance of passband is over 85%. The filter's feature can be attributed to the interference superposition of the reflected and transmitted fields from each tooth of the double-side structure.

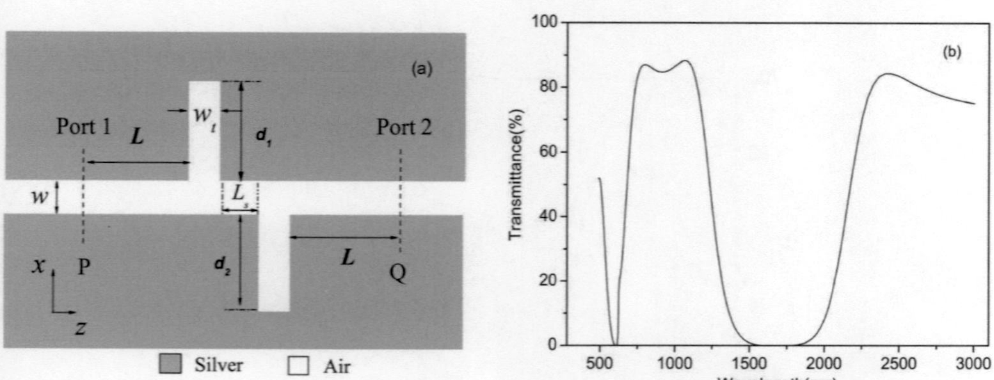

Figure 9. (a) Schematic of a staggered double-side teeth-shaped nano-plasmonic waveguide with a shift length of L_s. (b) The transmittance of the staggered double-side teeth-shaped waveguide filter with w_t=50nm, d_1=d_2=260.5nm, L_s=250nm.

Figure 10 shows the central wavelength of the bandgap as a function of the double-side teeth depth of d. During the simulation, we set d=d_1=d_2. The FDTD simulation results reveal that the relationship between the central wavelength of the bandgap and the double-side teeth depth of *d* is a linear function. It reveals that the central wavelength of the bandgap shifts toward long wavelength with the increasing of the teeth depth of *d*.

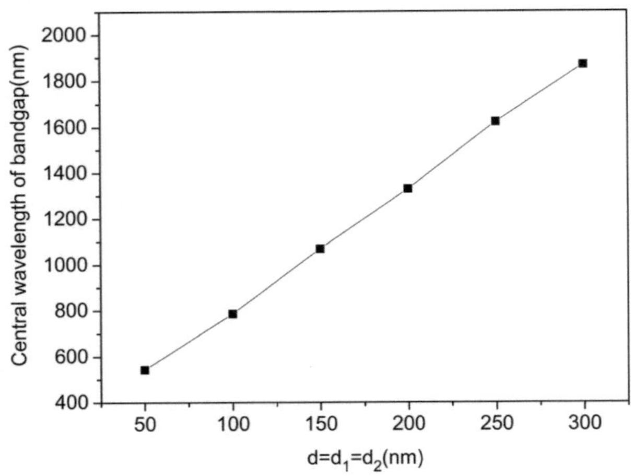

Figure 10. The central wavelength of the bandgap as a function of the double-side teeth depth of d at teeth width of 50nm.

Figure 11 shows how the shift length of the teeth influences or modifies the filtering spectrum of the structure. One can see that central bandgap shifts left and becomes wider with increasing the shift length of L_s. It reveals the filtering characteristics of the structure depend on the phase difference between the plasmon waves passing through the tooth. The optimized filtering response with a sharp left band-edge and high passband-transmittances over 85% can be achieved when the shift length is around 200nm.

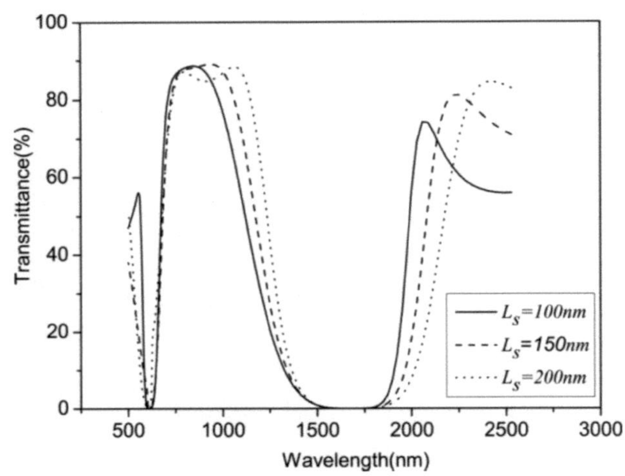

Figure 11. The transmittance of the two-sided staggered teeth-shaped waveguide filter for different shift lengths with w_t=50nm, d_1=d_2=260.5nm.

4. A multiple-teeth-shaped waveguide bandgap filter [22]

It is straight forward and basic interest to expand a single tooth structure to multiple-teeth structure (shown in Fig. 12(a)), and check the difference between them. For the sake of comparison, the waveguide width w and the distance L are fixed to be 50nm and 300nm. Λ and N are the period and the number of rectangular teeth, respectively. w_{gap} stands for the width of the gap that between any two adjacent teeth, and one has $w_t + w_{gap} = \Lambda$. A typical transmittance of the multiple-teeth-shaped waveguide filter with w_t=50nm, Λ=150nm, d=260.5nm and N=5 is shown in the Fig. 12(b), which is obtained with FDTD method. A wide bandgap occurs around λ =1.55μm with the bandgap width 590nm, and the transmittance of passband is over 90%. The filter's feature can be attributed to the superposition of the reflected and transmitted fields from each of the five single tooth-shaped components. Figure 13(a) shows the central wavelength of the transmittance bandgap, while the right y-axis displays the bandgap width of the waveguide filter as a function of teeth depth d at various w_t with the same Λ=150nm and N=5. The FDTD simulation results reveal that the relationship between the central wavelength of the bandgap and the teeth depth d is a linear function for any w_t, which is indeed the one of expectations in Eq. (9). Figure 13(b) shows the central wavelength of the bandgap and the bandgap width as a function of teeth width of w_t at various teeth depths. As revealed in the Eq. (9), the relationship between the bandgap position and w_t mainly results from the contribution of the inverse-proportion-like dependence of n_{eff} on w_t as shown in Fig. 1(a). Obviously, teeth width w_t should be chosen within the range of 50-100nm with a slope of $d\lambda_{center} / dw_t \approx 5$ to avoid a large value of $d\lambda_{center} / dw_t \approx 20$ when w_t<45nm in Fig. 13(b), and to reduce the sensitivity of the central wavelength of bandgap in fabrication process. Therefore, one can realize the filter function at various required wavelengths with high performance, by choosing the width or/and the depth of the teeth.

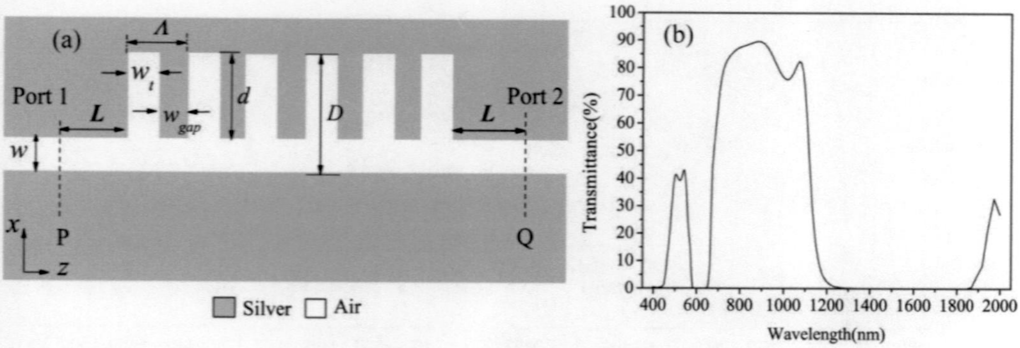

Figure 12. (a) Schematic of a multiple-teeth-shaped MIM waveguide structure. (b) The transmittance of the multiple-teeth-shaped waveguide filter with w_t=50nm, Λ=150nm, d=260.5nm and N=5.

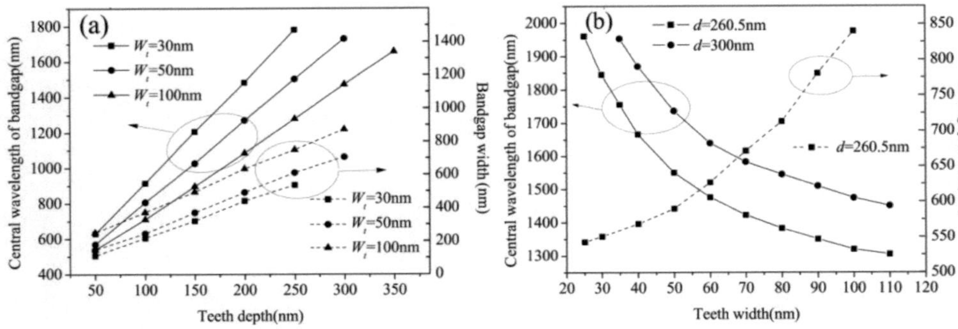

Figure 13. (a) The central wavelength of the bandgap and the bandgap width as a function of the teeth depth of *d* at various teeth widths. (b) The central wavelength of the bandgap and the bandgap width as a function of teeth widths of w_t at various teeth depths.

For the multiple-teeth-shaped structure with the parameters of $n_{eff,teeth} = 1.070$ for the width of *D=d+w*=(260.5+50)nm, $n_{eff,wg} = 1.375$ for *w*=50nm, and the teeth period of $\Lambda = w_t + w_{gap} = 150\text{nm}$ in z-axis direction, one can see $w_t \mathrm{Re}(n_{eff,teeth}) + w_{gap} \mathrm{Re}(n_{eff,wg}) \approx 401.1\text{nm} < 1550\text{nm}/2$. Thus the structure does not follow the Bragg condition in z-axis direction.

Figure 14(a) and (b) show the transmission spectra of a multiple-teeth-shaped waveguide filter at different periods Λ and period numbers *N*. As one can see from Fig. 14(a), when Λ=100nm is chosen, the coupling of the SPPs waves between two adjacent teeth is strong which causes the central bandgap wavelength to shift left and the bandgap to be wider. When the period equals to Λ=200nm, the coupling between any two adjacent teeth becomes very weak. One can see in Fig. 14(b) that, the forbidden bandwidth increases little with the changing of the period number from *N*=3 to 7, while the transmittance of the passband decreases from 93% to 86%. The reason for the decreasing in transmittance can be attributed to the increasing of the propagation loss of the lengthened structure with a large period number. From the simulation results, a tradeoff period number *N*=4 is the optimized number with the transverse filter length of 4×150nm, which is ~5 times shorter than the previous grating-like filter structures.

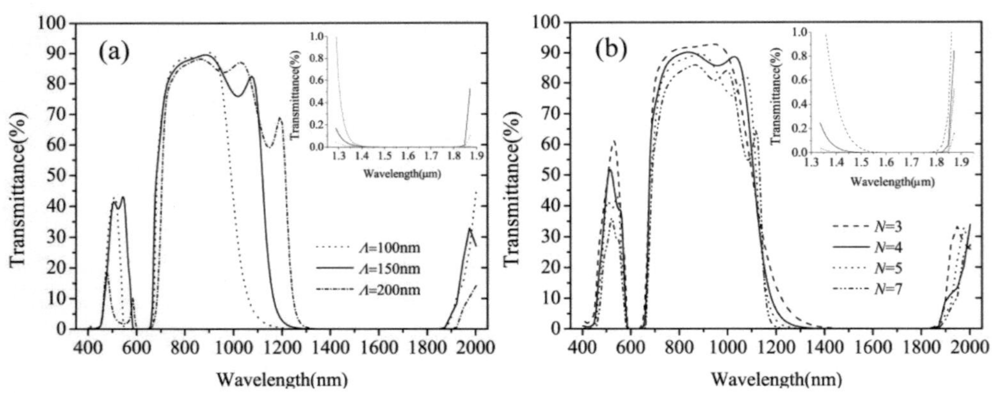

Figure 14. (a) Transmittance spectra of multi-teeth filters with different periods and a fixed *N*=5, (b) Transmittance spectra of multi-teeth filters consisting of 3-7 periods with a fixed Λ=150nm.

5. A narrow band subwavelength plasmonic waveguide filter with asymmetrical multiple teeth-shaped structure [23]

The asymmetrical multiple-teeth-shaped structure is shown in Fig. 15(a), which is composed of two sets of multiple-teeth with two different teeth depths. The short set has three teeth, and the long set has four teeth. Λ, N_1 and N_2, are the period, the numbers of short rectangular teeth and long teeth, respectively. w_{gap} stands for the width of the gap between any two adjacent teeth in multiple-teeth structure, and one has $w_t + w_{gap} = \Lambda$. The separation between the 3rd short tooth and 1st long tooth is w_s. The length of L and the waveguide width w are, respectively, fixed to be 150nm and 50nm. In Fig. 15(a) we set d_1=148nm, d_2=340nm, w_t=50nm, w_{gap}= w_s =84nm. Figure 15(b) shows a typical transmission spectrum of the asymmetrical multiple-teeth-shaped structure using FDTD method.

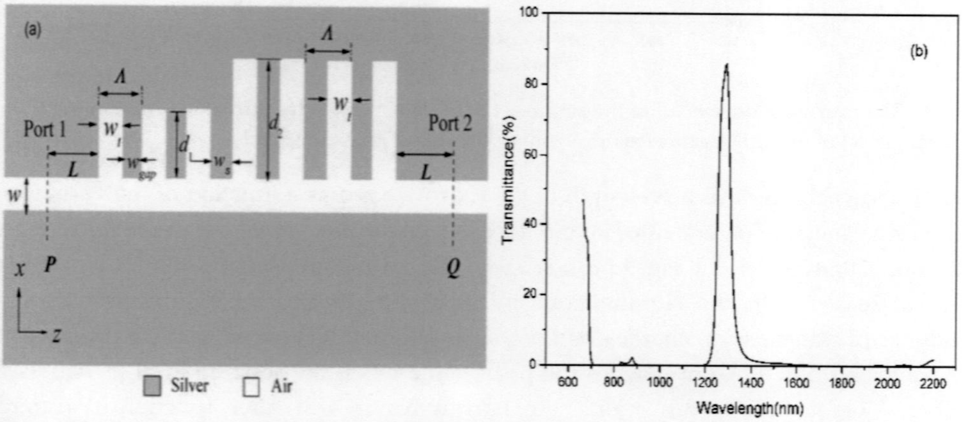

Figure 15. (a) Schematic of an asymmetrical multiple-teeth structure consisted of two sets with different teeth depth. (b) The transmittance of the asymmetrical multiple-teeth-shaped waveguide filter with d_1=148nm, d_2=340nm. w_{gap}= w_s =84nm, N_1=3, and N_2=4.

One can see the maximum transmittance at the wavelength of 1287nm is nearly 90%, and the full-width at half-maximum (FWHM) is nearly 70nm which is much smaller than the bandgap width of 1300nm. The FWHM of the asymmetrical multiple-teeth-shaped structure is also smaller than our previous coupler-type MIM optical filter [24].

In order to understand the origin of the narrow passband of the structure, the spectra of the transmission of a single-set of short three-teeth structure and a single-set of long four-teeth structure are calculated, and shown in Fig. 16. The parameters of the two structures are respectively equal to the parameters of the short teeth part and the long teeth part of the asymmetrical multiple-teeth-shaped structure (shown in Fig. 15(a)). One can see that the passband (or the bandgap) of the long teeth structure and the bandgap (or the passband) of the short teeth are overlapped from 800nm to 1200nm (or from 1450 to 1800nm), and then the transmission of the cascade of the two structure is very low within the two regions. Only the overlapping between the right edge of the passband of the long teeth structure and the left passband of the short teeth is non-zero. This is the reason why the wavelengths around 1300nm have a transmission peak in Fig. 15 (b).

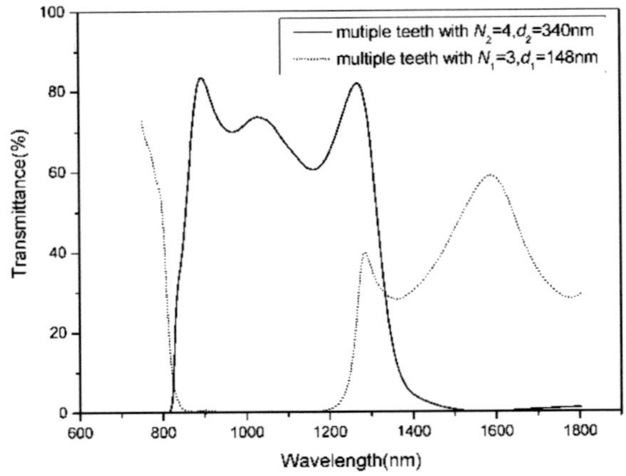

Figure 16. The transmission spectra of the single-set of multiple-teeth structure with d_1=148nm, N_1=3 and the single set of multiple-teeth structure with d_2=340nm, N_2=4, respectively.

Figure 17 shows the central wavelength of the narrow-band as a function of the variation of $\Delta d=\Delta d_1=\Delta d_2$. Δd is the increment of d_1 and d_2. The initial values of d_1 and d_2 are respectively, 128nm and 320nm. From the Fig. 17 can see that the central wavelength of the narrow-band linearly increases with the simultaneous increasing of d_1 and d_2. Figure 18 shows the dependence of transmission characteristic on separation w_s. It is found that the transmission at the wavelength of 1287nm reaches the peak value when the separation of w_s equals the gap of w_t. Therefore, one can realize the narrow-bandwidth filter function at different required wavelengths by means of properly choosing the parameters of the device, such as the teeth-depth, the period or the separation of w_s.

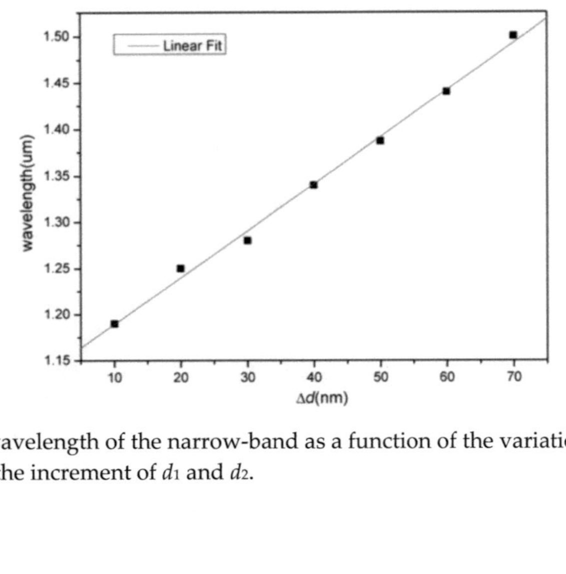

Figure 17. Central wavelength of the narrow-band as a function of the variation of $\Delta d=\Delta d_1=\Delta d_2$, $\Delta d_1$and Δd_2 are respectively the increment of d_1 and d_2.

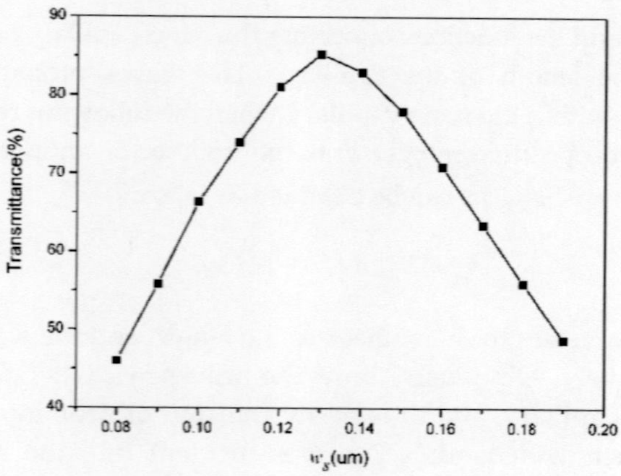

Figure 18. Dependence of transmission characteristic on separation between the 3rd short tooth and the 1st long tooth with d_1=148nm, N_1=3, d_2=340nm, N_2=4, respectively.

6. A wavelength demultiplexing structure based on metal-dielectric-metal plasmonic nano-capillary resonators [25].

The inset of Fig. 19 shows the nano-capillary resonators composed of two parallel metal plates with a dielectric core. Obviously, the structure can be treated as two MDM waveguides with different widths. Because the width of the lower MDM waveguide is much smaller than that of the upper part, here we call the lower (narrower) part as a nano-capillary. When the gap width w of the MDM waveguide is reduced below the diffraction limit, only a single propagation mode TM_0 can exist. The dielectric in the core of the structure is assumed to be air with a permittivity ε_d=1.

To fully understand how the width of the nano-capillary structure influences the SPPs propagation, the dependences of the effective index of SPPs on the width w at various wavelengths of the incident light are calculated and shown in Fig. 19. From the figure 19, one can see that the effective index of the waveguide decreases with increasing of w at the same wavelength. The effective index at short wavelength is larger than that at long wavelength, for a given width w. The effective index n_{eff2} of the nano-capillary can be larger than n_{eff3} of upper MDM part and n_1 of air. As shown in inset of Fig. 19, the waves will flow into the nano-capillary due to its higher effective index, when SPP waves propagate along the interface between metal and air. The wave transmitted into the capillary will be partly reflected at two ends of nano-capillary, because of index differences between n_{eff2} and n_{eff3} as well as n_1. One can expect the nano-capillary operates as a resonator. Resonance waves can be formed only in some appropriate conditions within nano-capillary segment. Defining $\Delta\phi$ to be the phase delay per round-trip in the nano-capillary, one has $\Delta\phi = 4\pi n_{eff} d / \lambda + \phi_r$, where $\phi_r \equiv \phi_1 + \phi_2$, ϕ_1 and ϕ_2 are respectively the phase shifts of a beam reflected on the entrance of the capillary and the junction connecting the nano-capillary and the upper MDM waveguide, and d is the length of the capillary. The waves propagating through the structure will be trapped within the nano-capillary when the following resonant condition is satisfied: $\Delta\phi = m \cdot 2\pi$. Here, positive integer m is the number of antinodes of the standing SPP wave. The resonant wavelengths can be obtained as follows:

$$\lambda_m = 2n_{eff} d / (m - \phi_r / \pi). \tag{10}$$

It can be seen that the wavelength λ_m is linear to the length and the effective index of the nano-capillary, respectively. Obviously, only the waves with the wavelength λ_m can stably exist in the nano-capillary, and thus partly transmit or drop into the output end of the nano-capillary. When wideband SPP waves incident into the structure, only the resonance waves with the wavelength λ_m can be selected and dropped by the nano-capillary. In other words, a transmission peak with the wavelength λ_m will be formed in the output section.

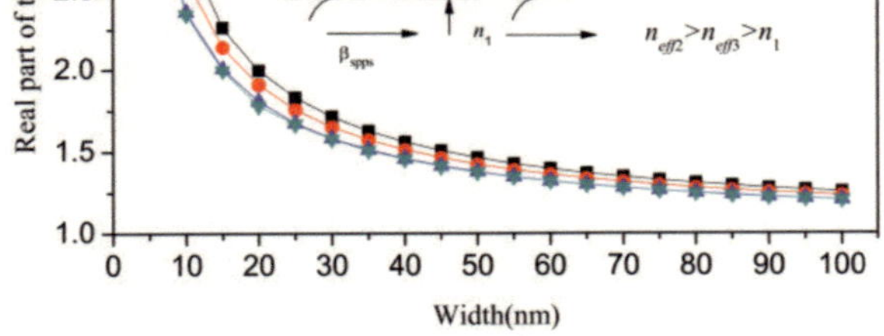

Figure 19. Dependence of real part of the effective index of SPPs in a plasmonic MDM waveguide on wavelength of the incident light and width w. Inset: schematic picture of a MDM nano-capillary resonator.

Fig. 20(a) shows a typical schematic of a 1×3 wavelength demultiplexing structure based on MDM nano-capillary resonators. The wavelength demultiplexing structure consists of three nano-capillary resonators perpendicularly connected to a bus waveguide. w_1 and d_1 stand for the width and the length of the first nano-capillary, respectively. Since the width of the bus waveguide is much smaller than the operating wavelength in the structure, only the excitation of the fundamental waveguide mode is considered. The incident light used to excite SPP wave is a TM-polarized (the magnetic field is parallel to y axis) fundamental mode. In the following FDTD simulation, the grid sizes in the x and the z directions are chosen to be $\Delta x = 5$ nm, $\Delta z = 1.5$ nm. Power monitors are respectively set at the positions of P and Q to detect the incident power of P_{in} and the transmitted power of P_{out}. The transmittance is defined to be $T = P_{out} / P_{in}$. The width w' of the bus waveguide is set to be 250 nm while the length of L_1 and L_2 are fixed to be 50 nm and 500 nm. As an example, three nano-capillaries have been designed to split the first, the second and third optical transmission windows, although more nano-capillaries can be added. The parameters of the structure are set to be $w = 15$ nm, $w_1 = 250$ nm, $d_1 = 202$ nm, $d_2 = 290$ nm, and $d_3 = 347$ nm in calculation. Fig. 20(b) shows the transmission spectra at the outputs of the three channels, and inset of fig. 20(b) shows transmittance and reflectance of the bus waveguide. From it, one can see channels 1-3 can select 980 nm, 1310 nm, 1550 nm bands, respectively, and the maximum transmittance in three bands can exceed 30% (-5.2 dB). And there is also another high transmission in channel 3 around 820 nm wavelength for $m = 2$. Given the total phase shift ϕ_r, one can estimate the resonance wavelength from Eq. (10). Submitting $\lambda_m = 1310$ nm into Eq. (10) gives $\phi_r = 0.35$ for $d = 290$ nm and $n_{eff} = 2.01$. Other resonance wavelengths can be approximately calculated with the formula. For the lengths of the nano-capillaries of 347 nm and 202 nm, resonance wavelengths are simply estimated to be 1559 nm and 926 nm. The deviation between FDTD simulation and the result from Eq. (10) could be partly attributed to the neglecting of wavelength dependence of ϕ_r. And it is partly due to the fact that Eq. (10) is derived based on the effective index approximation that SPP waves with the phase factor of $\exp(i2\pi n_{eff} x / \lambda)$ travel back and forth within a capillary, similar to a 3-dimentional plane wave with $\exp(i2\pi nx / \lambda)$ traveling in a bulk medium with refractive index n.

The FWHM of channel 1-3 are 75 nm, 130 nm, 160 nm, respectively. Obviously, the FWHM of the channel 2 and channel 3 are larger than that of channel 1. The reason is that, from the calculation in Fig. 19, the effective index at short wavelength with a fixed width of nano-capillary is higher compared with the one at long wavelength, thus the waves at short wavelength have a higher reflectivity at two ends of nano-capillary and its Q factor is higher. Cross-talk is defined as the ratio between the power of the undesired and desired bands at the outputs. The cross-talk between channel 1 and channel 2 is around -19.7 dB for the 980 nm branch, and the cross-talk between them is -13.1 dB for the 1310 nm branch. The cross-talk between channel 1 and the whole channel 3 is around -19.2 dB for the 980 nm branch, and is -16.6 dB for the 1550 nm branch, although there is also another high transmission in channel 3 around 820nm wavelength for m=2. Therefore, this structure is suitable for wideband wavelengths demultiplexing.

Equation (10) indicates that the transmission behavior of each nano-capillary (channel of the demultiplexing structure) mainly depends on two parameters: the length of the nano-capillary, and the effective index of SPPs in the nano-capillary, which is determined by its width. Figure 21 shows the central wavelength of the nano-capillary resonator as a function of nano-capillary length d. One can see that the central wavelength of nano-capillary shifts toward longer wavelengths with the increasing of nano-capillary length d, as expected from equation (10). Therefore, one can realize the demultiplexing function at arbitrary wavelengths through the nano-capillary resonator by means of properly choosing the parameters of the structure, such as nano-capillary length and width.

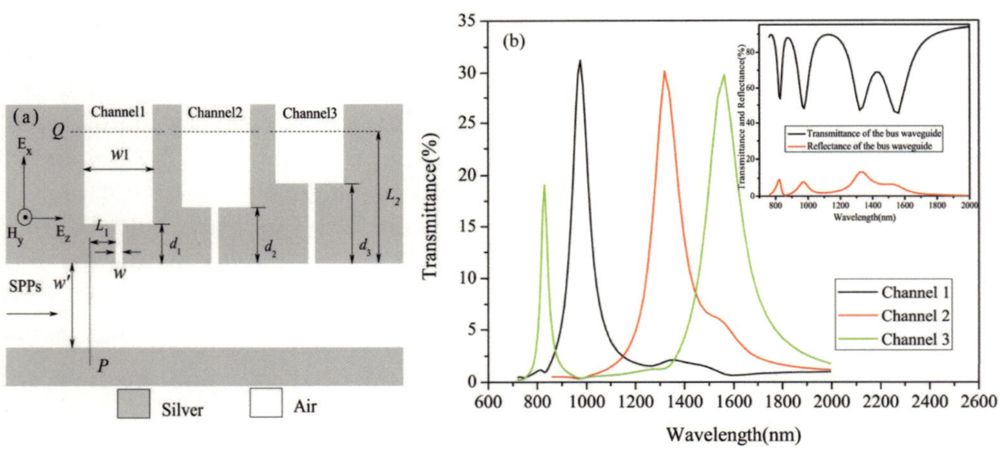

Figure 20. (a) Schematic of a 1×3 wavelength demultiplexing structure based on MDM plasmonic nano-capillary resonators. (b) Transmission spectra of the three channels of the demultiplexing structure with w = 15 nm, w_1 = 250 nm, d_1 = 202 nm, d_2 = 290 nm and d_3 = 347 nm. Inset: Transmittance and reflectance of the bus waveguide.

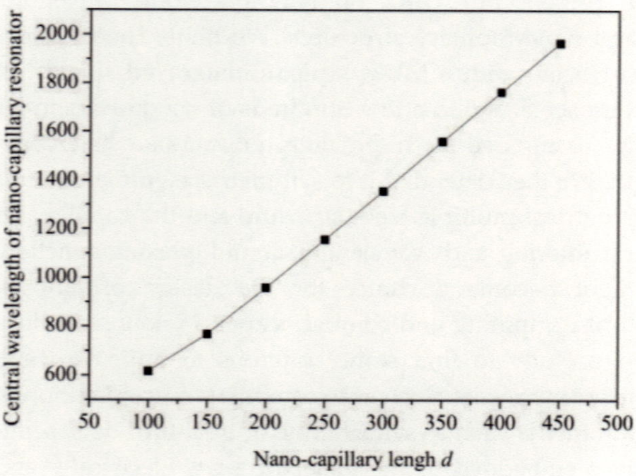

Figure 21. The central wavelength of nano-capillary resonator as a function of nano-capillary length d.

Finally, Figure 22 shows the propagation of field H_y for two monochromatic waves with different wavelengths of 980 nm and 1550 nm launched into nano-capillary resonator demultiplexing structure. The demultiplexing effect is clearly observed. From the figure, one can see the wave with wavelength of 980 nm passing through the first nano-capillary and the wavelength of 1550 nm wave transmitting from the third nano-capillary. This is in good agreement with the transmission spectra shown in Fig. 20(b).

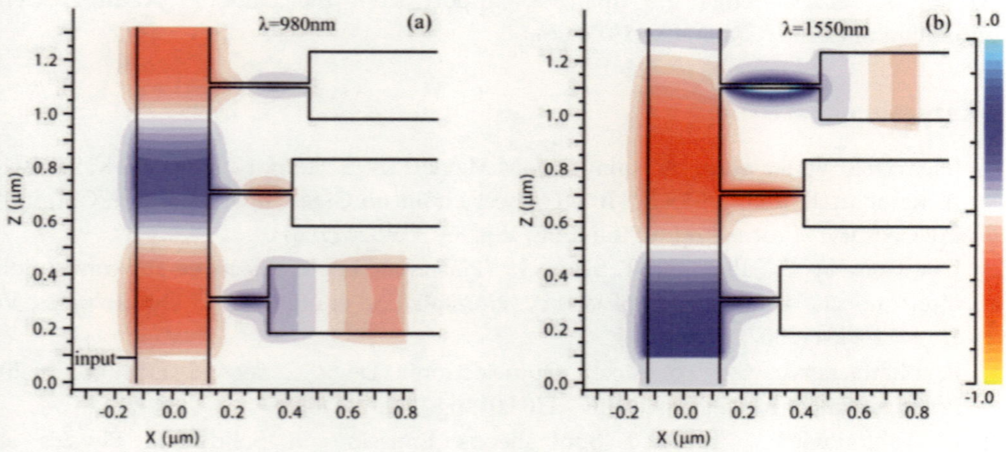

Figure 22. The contour profiles of field H_y of the 1×3 wavelength demultiplexing structure at different wavelengths, (a) $\lambda = 980$ nm, (b) $\lambda = 1310$ nm. All parameters of the structure are same as in Fig. 2(b).

7. Conclusion

In this chapter, we present our work on nano-plasmonic waveguide filters based on tooth/teeth-shaped and nano-capillary structures. We firstly investigated a novel plasmonic waveguide filter constructed with a MDM structure engraved single rectangular tooth. The filter is of an ultra-compact size with a few hundreds of nanometers in length, with reducing fabrication difficulties, compared with previous grating-like heterostructures with a few micrometers in length. We then extended it to symmetric/asymmetric multiple-teeth, capillary structures. The asymmetrical multiple-teeth structure and the capillary structure can achieve selective narrow-band filtering and wavelength demultiplexing functions, respectively. The plasmonic filters might become a choice for the design of all-optical high-integrated architectures for optical computing and communication in nanoscale. In the future, it will be very interesting and usefully to find some solutions to improve the performance of the MIM/MDM plasmonic components. Such as to combine surface plasmons with electrically and optically pumped gain media such as semiconductor quantum dots, semiconductor quantum well, and organic dyes embedded to the dielectric part. Electrically and optically pumped semiconductor gain media and the emerging technology of graphene are also expected to provide loss compensation from visible to terahertz spectra range [26].

Author details

Xu Guang Huang* and Jin Tao

Key Laboratory of Photonic Information Technology of
Guangdong Higher Education Institutes, South China Normal University, Guangzhou, China

Acknowledgement

The authors acknowledge the financial support from the National Natural Science Foundation of China (Grant No. 11977866).

8. References

[1] J. Davis, R. Venkatesam, A. Kaloyeros, M.Meylansky, S. Souri, K. Banerjee, K. Saraswat, A. Rahman, R. Reif, J. Meidl, "Interconnect Limits on Gigascale Integration (GSI) in the 21st Century," Proceedings of the IEEE, Vol. 89, 305-324 (2001).

[2] J. A. Conway, S. Sahni, and T. Szkopek, "Plasmonic interconnects versus conventional interconnects: a comparison of latency, crosstalk and energy costs," Opt. Express, Vol. 15, 4474-4484 (2007).

[3] P. Bhattacharya. Semiconductor Optoelectronic Devices. Second Edition. © 2005 Prentice-Hall of India, New Dehli – 110 001. pg. 102.

[4] E. Yablonovitch, "Inhibited Spontaneous Emission in Solid-State Physics and Electronics," Phys. Rev. Lett 58, Vol. 20, 2059–2062 ,1987.

* Corresponding Author

[5] S. John, "Strong localization of photons in certain disordered dielectric superlattices," Phys. Rev. Lett, Vol. 58 , 2486–2489 ,1987.

[6] Maier, S. A. et al. "Local detection of electromagnetic energy transport below the diffraction limit in metal nanoparticle plasmon waveguides," Nature Mater., Vol. 2, 229–232, 2003.

[7] W. Saj, "FDTD simulations of 2D plasmon waveguide on silver nanorods in hexagonal lattice," Opt. Express, Vol. 13, 4818-4827, 2005.

[8] S. I. Bozhevolyni, V. S. Volkov, E. Devaux, and T. W. Ebbesen, "Channel plasmon–polariton guiding by subwavelength metal grooves," Phys. Rev. Lett., Vol. 95, 046802, 2005.

[9] K. Tanaka, M. Tanaka, and T. Sugiyama, "Simulation of practical nanometric optical circuits based on surface plasmon polariton gap waveguides," Opt. Express, Vol. 13, 256–266, 2005.

[10] J. R. Krenn et al. "Non-diffraction-limited light transport by gold nanowires," Europhys Lett., Vol. 60, 663–669, 2002.

[11] B. Wang and G. Wang, "Plasmon Bragg reflectors and nanocavities on flat metallic surface," Appl. Phys. Lett., Vol. 87, 013107, 2005.

[12] A. Boltasseva, S. I. Bozhevolnyi, T. Nikolajsen, and K. Leosson, "Compact Bragg gratings for Long-Range surface plasmon polaritons," J. Lightwave Technol. Vol. 24, 912-918, 2006.

[13] A. Hossieni and Y. Massoud, "A low-loss metal-insulator-metal plasmonic bragg reflector," Opt. Express, Vol. 14, 11318-11323, 2006.

[14] A. Hosseini, H. Nejati, and Y. Massoud, "Modeling and design methodology for metal-insulator-metal plasmonic Bragg reflectors," Opt. Express, Vol. 16, 1475-1480, 2008.

[15] J. Park, H. Kim, and B. Lee, "High order plasmonic Bragg reflection in the metal-insulator-metal waveguide Bragg grating," Opt. Express, Vol. 16, 413-425, 2008.

[16] Z. Han, E. Forsberg, and S. He, "Surface plasmon Bragg gratings formed in metal-insulator-metal waveguides," IEEE Photon.Technol. Lett, Vol. 19, 91-93, 2007.

[17] Xian Shi Lin and Xu Guang Huang, "Tooth-shaped plasmonic waveguide filters with nanometeric sizes," Opt. Lett, Vol. 33, 2874-2876, 2008.

[18] J. A. Dionne, L. A. Sweatlock, and H. A. Atwater, "Plasmon slot waveguides: Towards chip-scale propagation with subwavelength-scale localization," Phys. Rev. B., Vol 73, 035407 (2006).

[19] E. D. Palik, Handbook of optical constants of *solids* (Academic Press, New York, NY 1985).

[20] H. A. Haus, Waves and Fields in Optoelectronics (Prentice-Hall, Englewood Cliffs, NJ, 1984).

[21] J. Tao, Xu Guang Huang, X. S. Lin, et al, "Systematical research on characteristics of double-sided teeth-shaped nanoplasmonic waveguide filters," J. Opt. Soc. Am. B, Vol. 27, 323-327, 2010.

[22] X. S. Lin and X. G. Huang, "Numerical modeling of a teeth-shaped nanoplasmonic waveguide filter," J. Opt. Soc. Am. B, Vol. 26, 1263-1268, 2009.

[23] J. Tao, X. G. Huang, X. S. Lin, Q. Zhang, X. P, Jin, "A narrow-band subwavelength plasmonic waveguide filter with asymmetrical multiple-teeth-shaped structure," Opt. Express, Vol. 17, 13989-13994, 2009.

[24] Qin Zhang, Xu-Guang Huang, Xian-Shi Lin, Jin Tao, and Xiao-Ping Jin, "A subwavelength coupler-type MIM optical filter," Opt. Express, Vol. 17, 7549-7555, 2009.

[25] Jin Tao, Xu Guang Huang, and Jia Hu Zhu, "A wavelength demultiplexing structure based on metal-dielectric-metal plasmonic nano-capillary resonators," Opt. Express, Vol. 18, 11111-11116, 2010.

[26] A. A Dubinov, V. Y. Aleshkin, V. Mitin, T. Otsuji and V. Ryzhii, "Terahertz surface plasmons in optically pumped graphene structures," J. Phys.Condens. Matter, Vol. 23, 145302 2011.

Section 3

Functional Optical Materials

Chapter 9

Fluidic Optical Devices Based on Thermal Lens Effect

Duc Doan Hong
and Fushinobu Kazuyoshi

Additional information is available at the end of the chapter

http://dx.doi.org/10.5772/48072

1. Introduction

Gordon et al. [1] reported that the beam shape of incident laser light expands after passing through a liquid medium. This phenomenon was termed "the thermal lens effect," and it has become a well-known photo-thermal phenomenon. Phenomenological, optical, and spectroscopic studies of the thermal lens effect have been carried out to describe nonlinear defocusing effect [2-8]. Recent progress in laser technology has revealed the various aspects of the thermal lens effect. Based on these efforts, other mechanisms, such as liquid density, electronic population, and molecular orientation, have been found to play important role as well as thermal lens effect. Recent studies term these effects as "the transient lens effect" [9,10]. The main advantage of using the transient lens effect in Photo-Thermal-Spectroscopy is that the sensitivity is 100 to 1000 greater than a traditional absorptiometry [11].

In this research, a new idea of applying the thermal lens effect in order to develop fluidic optical device is proposed. A schematic of the concept is shown in Fig. 1. A rectangular solid region shown in Fig. 1a represents the liquid medium, which has a temperature field generated by a heater-heat sink system or laser-induced absorption. By controlling the temperature field as well as the refractive index distribution of the liquid medium, the refractive angle of each light ray passing through the liquid medium can be controlled in order to develop fluidic optical devices such as: an optical switching in Fig. 1a to change the direction of the input laser beam, a laser beam shaper in Fig. 1b to transform a Gaussian beam to a flat-top beam and a fluidic divergent lens in Fig. 1c. Merits of these devices include flexibility of optical parameters, versatility and low cost.

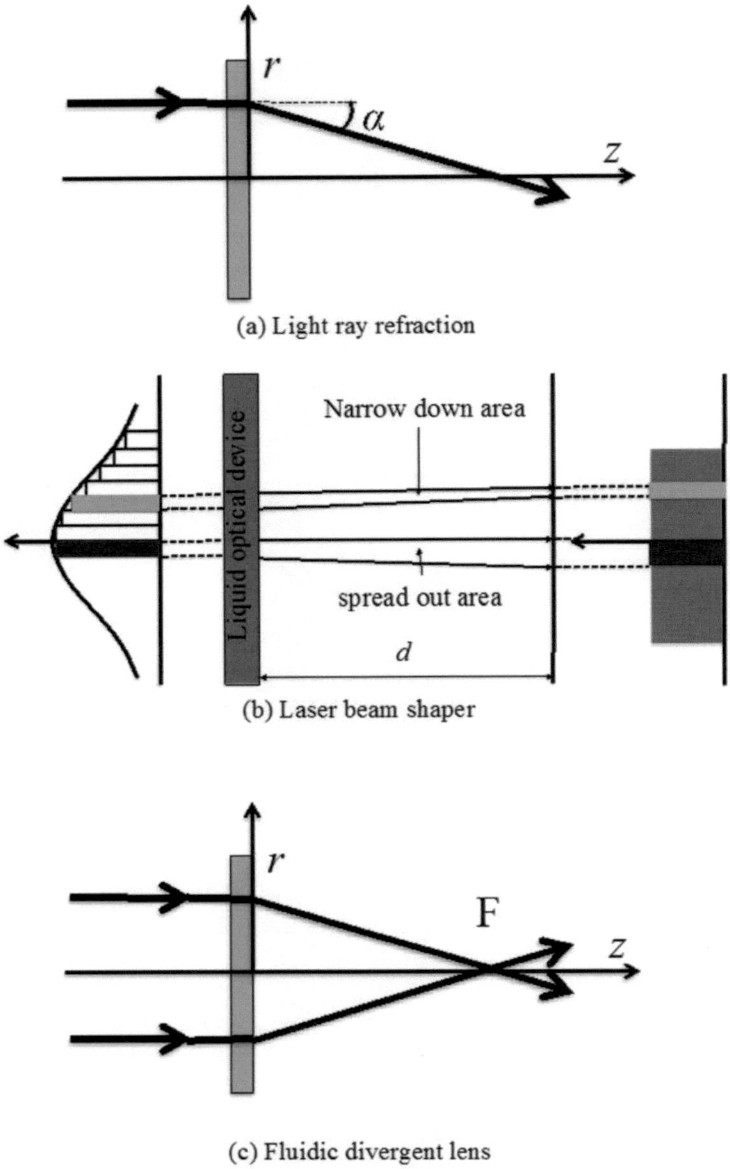

Figure 1. Schematic of the concept of the fluidic optical devices

2. Fundamental light ray transmitted in one-dimensional refractive index medium

In this section, as a first step to develop fluidic optical device, the refractive characteristics of a probe beam, which is transmitted in one-dimensional temperature distribution in a liquid medium is presented.

2.1. Theoretical background

The light ray is modeled in the domain shown in Fig. 2 in order to calculate the refractive angle of the probe beam, which is transmitted in a one-dimensional temperature distribution in the liquid medium. The light ray direction transmitted in a medium having a refractive index dependent only on the *y*-axis, is described by the following form [12]

$$x = \int_0^y \frac{n_0 \sin\theta \cos\varphi}{\sqrt{n^2(y) - n_0^2 \sin^2\theta}} dy \tag{1}$$

$$z = \int_0^y \frac{n_0 \sin\theta \sin\varphi}{\sqrt{n^2(y) - n_0^2 \sin^2\theta}} dy \tag{2}$$

In which, the light ray passes through the medium at coordinate center, θ and ϕ are the incident angle with *y* and *x*-axis respectively, n_0 is the refractive index of the medium at the coordinate center.

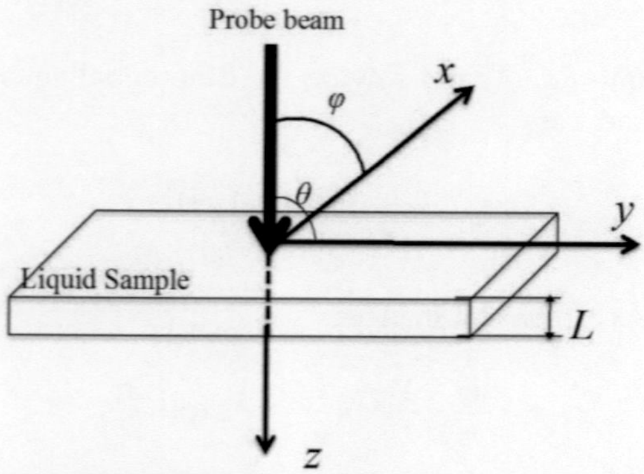

Figure 2. Schematic diagram of computational domain

Figure 2 shows a schematic diagram of the model set up. The rectangular solid medium in the figure represents the domain considered in the calculation which consists of ethylene glycol. In the medium, ethylene glycol has linear temperature distribution in only *y*-axis direction and the probe beam propagate along *z*-axis direction. Therefore, $\theta = \phi = \pi/2$, and Eq. (1), (2) become:

$$x = 0 \tag{3}$$

$$z = \int_0^y \frac{n_0}{\sqrt{n^2(y) - n_0^2}} dy \tag{4}$$

The temperature distribution of the liquid medium is modeled with a linear function of the variable *y* and temperature *T* at point *y* is calculated following:

$$T(y) = T_0 + \frac{dT}{dy} y \tag{5}$$

In which, T_0 is the temperature of the liquid medium at coordinate origin and dT/dy is constant.

Furthermore, between 0°C and 100°C refractive index of ethylene glycol is a linear function of temperature with refractive index change dn/dT = -2.6×10^{-4} 1/K [13]. Therefore, the relationship between refractive index and the variable y can be rewriten as follows:

$$y = \frac{n(y) - n_0}{dn / dy} \tag{6}$$

And

$$dy = dn \times \frac{dy}{dn} = dn \times \frac{dy}{dT} \times \frac{dT}{dn} = \frac{dn}{k} \tag{7}$$

Where,

$$k = -2.6 \times 10^{-4} \times \frac{dT}{dy} \tag{8}$$

By substituting Eq. (7) into Eq. (4) and solving the differential equation, we can obtain the relationship between y and z as:

$$y = \frac{n_0[\exp(kz / n_0) - 1]^2}{2k\exp(kz / n_0)} \tag{9}$$

And refractive angle (RA) can be obtained as:

$$\alpha = \frac{dy}{dz} = \frac{1}{2}\exp(\frac{kz}{n_0}) - \frac{1}{2}\exp(-\frac{kz}{n_0}) \tag{10}$$

Equation (10) shows the expression of the refractive angle as a function of the temperature gradient and the thickness of the liquid medium (optical path length). Figure 3 shows the relationship between refractive angle and temperature gradient at points where the thickness of sample, L, is 1.5, 3.0 and 4.5 mm respectively. As shown in Fig. 3 the relationship between the refractive angle and temperature gradient can be well approximated as linear at a small temperature gradient.

2.2. Experimental set-up

Figure 4 shows the experimental set-up to measure the refractive angle. Fluidic optical device in Fig. 4 is a pyrex vessel (internal size: 21×10×L [mm], thickness of the liquid medium, L, can be varied) filled with ethylene glycol. The vessel is held in an adiabatic material (Mica glass-ceramics (Photoveel®) as shown in Fig. 5. The temperature on both sides of the vessel was controlled by a heater-heat sink system to create a one-dimensional temperature distribution in the liquid medium.

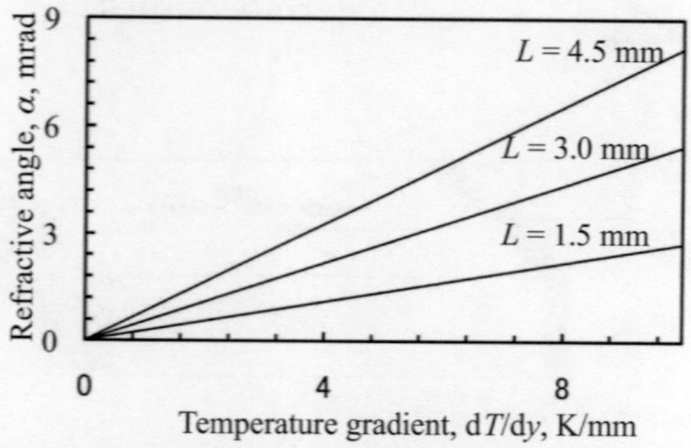

Figure 3. Relationship between the refractive angle and temperature gradient

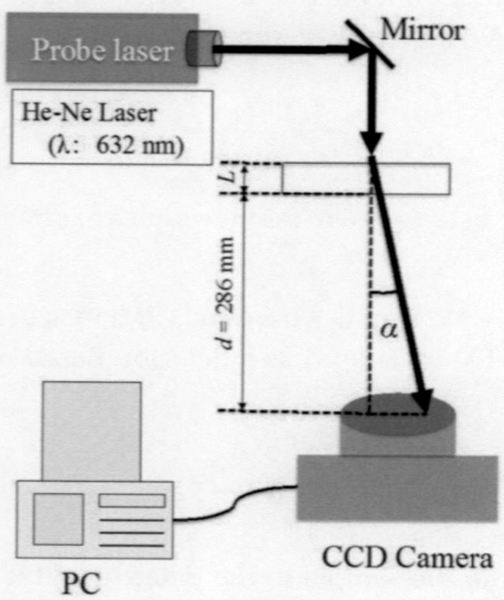

Figure 4. Refractive angle measurement system

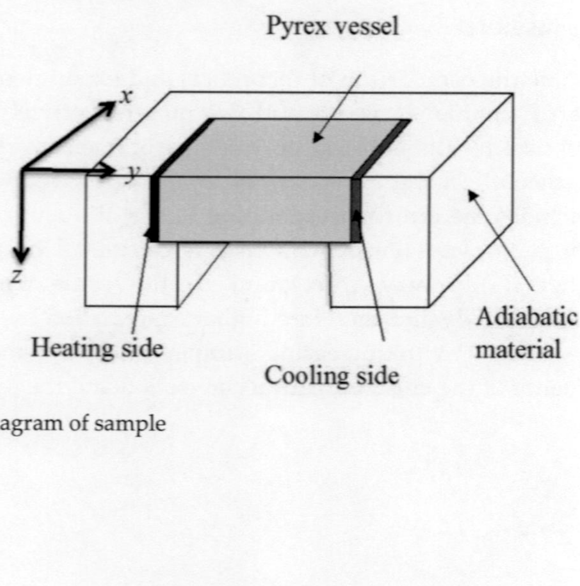

Figure 5. Schematic diagram of sample

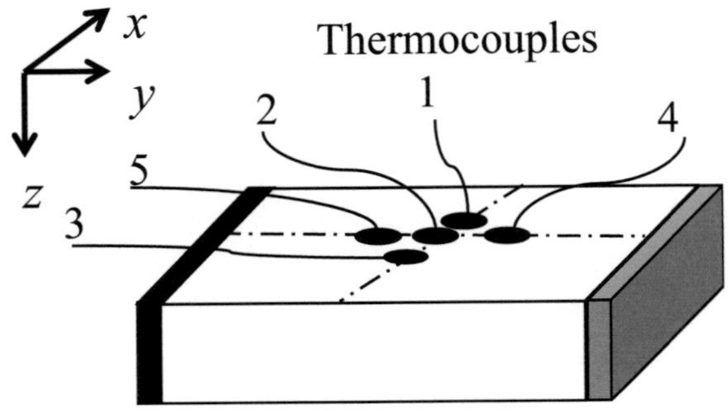

Figure 6. Temperature gradient measurement system

The temperature distribution in the liquid medium is confirmed by measuring the temperatures at 5 points with 2 mm pitch in the vessel using 5 thermocouples as shown in Fig. 6. The temperature gradient in the experiment is given as:

$$\frac{dT}{dy} = \frac{T_4 - T_5}{\Delta y} \tag{11}$$

In which, Δy is the distance between two thermocouples to obtain the temperature gradient, $\Delta y = 4$ mm.

A CW laser (P = 0.6 mW, λ = 632 nm, Φ = 0.8 mm, TEM00) is used as a probe beam. A CCD camera (OPHIR, BeamStar-FX 50) is used as a detector. Based on probe beam position, the refractive angle is estimated with:

$$\alpha = \frac{r}{d} \tag{12}$$

Where d is the distance from the sample to the detector of the camera = 286 mm; r is the beam position.

2.3. Results and discussions

Figure 7(a) and (b) show the comparison of theoretical and experimental results at points where the thickness of sample, L, is 1.5 and 3.0 mm respectively. The temperature gradient in the theoretical results is based on the measurement as described above. As shown in this figure, theoretical and experimental results agree well with each other. The experimental data includes the error corresponding to the difference between the actual temperature gradient at the laser incidence which is calculated by using Eq. (11). The discrepancy at higher dT/dy may correspond to the error where the measured temperature gives higher dT/dy and therefore higher a prediction by using Eq. (10). This discrepancy should increase with increasing sample thickness and the temperature gradient as a consequence of the effect of natural convection and the temperature gradient in the z-axis [14].

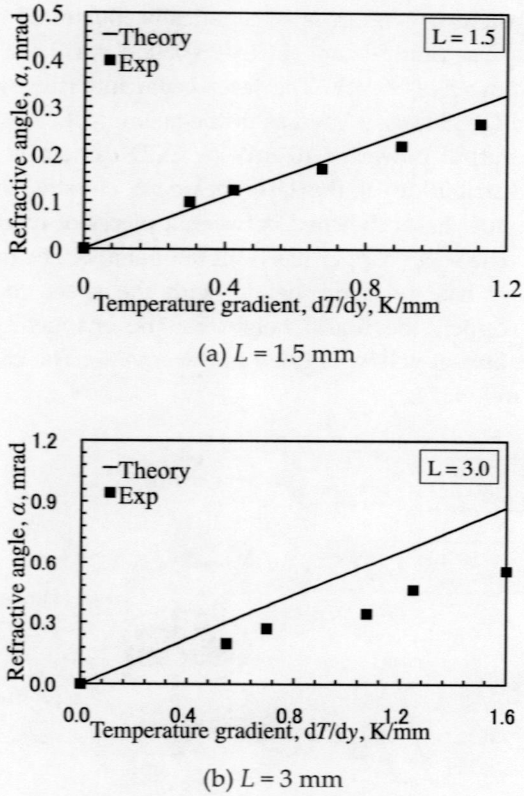

Figure 7. The comparison of theoretical and experimental results

3. Fluidic laser beam shaper

Flat-top laser are well known to present significant advantages for laser technology, such as holographic recording system, Z-scan measurement, laser heat treatment and surface annealing in microelectronics and various nonlinear optical processes [15-19]. For CW beams, several approaches to spatially shape Gaussian beams have been developed, such as the use of aspheric lenses, implement beam shaping or the use of diffractive optical devices [20]. However, these methods have some disadvantages: a refractive beam shaping system lead to large aberration [21] and implemental beam shaping has low energy efficiency and lacks of flexibility [22]; and the use of refractive optical devices requires complex configuration design and high cost [23]. In practice, a low-cost and flexible method to convert a Gaussian beam into a flat-top beam is required. In this section, a novel method to convert a Gaussian beam into a flat-top beam is discussed. The concept is based on the control of the pump power and propagation distance of the probe beam in the thermal lens system.

3.1. Principle of thermal lens effect

The principle of the transient lens effect is schematically illustrated in Fig.8. A CW diode pumped blue laser is used as pump-beam (BCL-473-030, λ = 473 nm, Φ = 0.8 mm, TEM_{00}) with maximum output power of 30mW. The laser beam intensity was adjusted by using ND-filter. A CW infrared DPSS laser is used as probe-beam (MIL, λ = 1064 nm, Φ = 3.0 mm, TEM_{00}) with maximum output power of 10 mW. A CCD camera is used as a detector to measure the intensity distribution of the laser beam. A cuvette, which is a three-layer structure with a sheet copper is sandwiched between 2 pieces of fused silica. The height of the fused silica is 1 mm. The sheet copper has doughnut shape. The liquid that is contained inside the doughnut hole has the same height with the sheet copper. By varying the thickness of the sheet copper, the liquid height can be changed. The ethanol solution dissolved dye termed as Sunset-yellow is filled in the cuvette. The chemical formula of the Sunset-yellow is shown in Ref. 24.

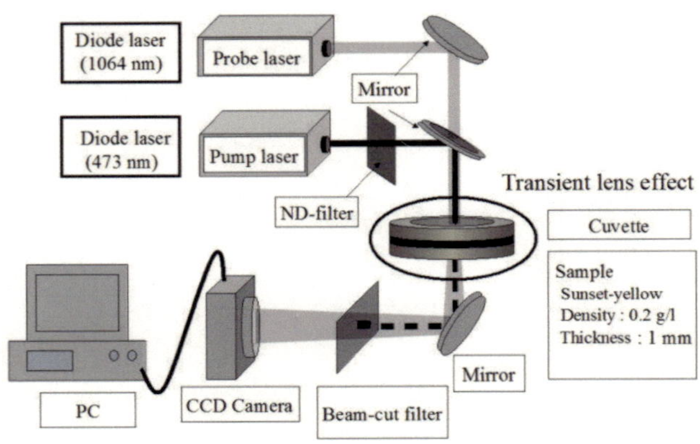

Figure 8. Schematic diagram of a dual-beam thermal lens system

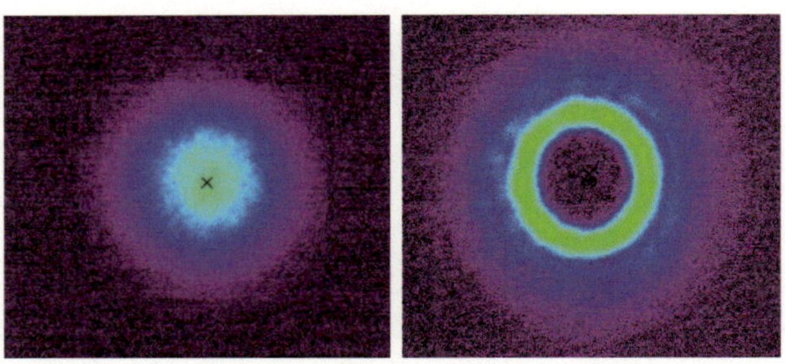

Figure 9. Experiment result: the difference of intensity profile of the probe beam after passing through the thermal lens (left side) and a quart divergent lens (right side)

In this experiment, the absorbance of the pump-beam is 2.776 and that of the probe-beam is negligible small. Figure 9 shows the laser beam profile of the probe beam after propagating through a divergent lens and a thermal lens. It is clear that, the probe beam change its profile from Gaussian to doughnut beam with a hollow center is created.

Theoretical analysis of laser beam profile change in thermal lens effect is done with a model that includes continuity equation, Navier-Stokes equation, energy conservation equation and Helmholtz equation in 2D cylindrical symmetry coordinate. It is assumed that the change of refractive index is caused only by the temperature change of the liquid medium and the thermal coefficient of the refractive index, dn/dT. The concentration is supposed to be constant over the range of the temperature rise induced by the pump beam. When the liquid medium is irradiated, temperature distribution perpendicular to the optical axis is formed due to intensity distribution of laser beam and heat transport. To consider the natural convection effect, the temperature distribution of liquid sample in steady state is calculated numerically following these governing equations [24]:

$$\frac{1}{r}\frac{\partial}{\partial r}(rv_r)+v_z\frac{\partial v_z}{\partial z}=0 \tag{13}$$

$$v_r\frac{\partial v_r}{\partial r}+v_z\frac{\partial v_r}{\partial z}=-\frac{1}{\rho}\frac{\partial p}{\partial r}+\nu\left(\frac{\partial}{\partial r}\left(\frac{1}{r}\frac{\partial}{\partial r}(rv_r)\right)+\frac{\partial^2 v_r}{\partial z^2}\right) \quad (a)$$

$$v_r\frac{\partial v_z}{\partial r}+v_z\frac{\partial v_z}{\partial z}=-\frac{1}{\rho}\frac{\partial p}{\partial z}+\nu\left(\frac{1}{r}\frac{\partial}{\partial r}\left(r\frac{\partial v_z}{\partial r}\right)+\frac{\partial^2 v_z}{\partial z^2}\right)+g\beta(T-T_0) \quad (b) \tag{14}$$

$$v_r\frac{\partial T}{\partial r}+v_z\frac{\partial T}{\partial z}=a\left(\frac{1}{r}\frac{\partial}{\partial r}\left(r\frac{\partial T}{\partial r}\right)+\frac{\partial^2 T}{\partial z^2}\right)+S \tag{15}$$

$$S=\frac{\alpha e^{-\alpha z}I_0(r)}{\rho C_P} \tag{16}$$

Here, $I_0(r)$ is the intensity distribution of the pump laser. The spot sizes of the laser beams are assumed to be constant through the interaction volume within the liquid medium.

The temperature distribution is calculated numerically based on the finite difference method. The 1st order upwind scheme and a 2nd order center differencing are applied to discretize the advection term and the diffusion term respectively. The thermal properties of liquid medium can be found in Ref. 24.

To model the propagation of laser through an inhomogeneous medium, the wave equation which includes an absorption term and an inhomogeneous refractive index term is applied [24]:

$$-\frac{\partial^2 E}{\partial z^2}+2ik_0n_0\frac{\partial E}{\partial z}=\frac{1}{r}\frac{\partial}{\partial r}\left(\frac{\partial(rE)}{\partial r}\right)+k_0^2\left(n^2-n_0^2\right)E-\frac{1}{2}ik_0n_0\alpha E \tag{17}$$

Here, E is the envelope of the oscillating electric field, z is the axis of propagation, r is transverse coordinates, k_0 is the free space wave number and α is the absorption coefficient. The variable n is the refractive index profile depending on medium temperature following:

$$n(T) = n_0 + \frac{dn}{dT}(T - T_0) \tag{18}$$

Here, n_0 = 1.359 is the refractive index of the liquid medium at reference temperature T_0 = 298.15 K, dn/dT is the temperature coefficient of the refractive index. The propagation of laser is calculated based on Pade method. The optical properties parameter can be found in Ref. 24.

3.2. Influences of the pump power and the propagation distance on the change of probe beam profile

Influences of the pump power and the propagation distance to the probe beam profile were investigated numerically using the calculation parameters in Table. 1. In this calculation, both of the pump beam and the probe beam are written as follows.

$$E = E_0 \exp\left(\frac{-r^2}{r_0^2}\right) \exp\left(\frac{-ik_0 n_0 r^2}{2R}\right) \tag{19}$$

$$E_0 = \sqrt{\frac{2P}{\pi r_0^2}} \tag{20}$$

Here P is the power of laser, R is the radius of curvature of the wave front, r and r_0 are distance from laser axis and beam radius respectively.

Parameter	(a)	(b)
Pump power, mW	3	0 ~ 7
Pump beam diameter, mm	0.8	0.8
Probe power, mW	10	10
Probe beam diameter, mm	0.8	0.8
Absorption coefficient, cm^{-1}	2.0	2.0
Distance from experimental section to CCD camera, mm	0 ~ 500	200
Phase front curvature radius, R, mm	∞	∞

Table 1. Calculation conditions

Effects of the pump power and the propagation distance to the probe beam profile are shown in Fig. 10(a) and (b) respectively. The vertical axis and horizontal axis show intensity and distance from laser axis respectively. Plots of 'P = 0 mW' and 'd = 0 mm' represent intensity distribution of the probe beam without thermal lens effect. As shown in Fig. 10(a), the further the propagation distance, the lower intensity at the probe beam center, and higher intensity at the wing. With increasing of the propagation distance, the laser beam profile changes from Gaussian to flat-top and the doughnut beam profile respectively. The profile of the probe beam changes with the same tendency as the increasing of pump power as shown in Fig. 10(b). In particular, when the pump power is 3 mW and propagation distance is 200 mm the probe beam is converted to the flat-top profile approximately. Therefore, by controlling the pump power and the propagation distance the Gaussian beam can be converted into the flat-top beam.

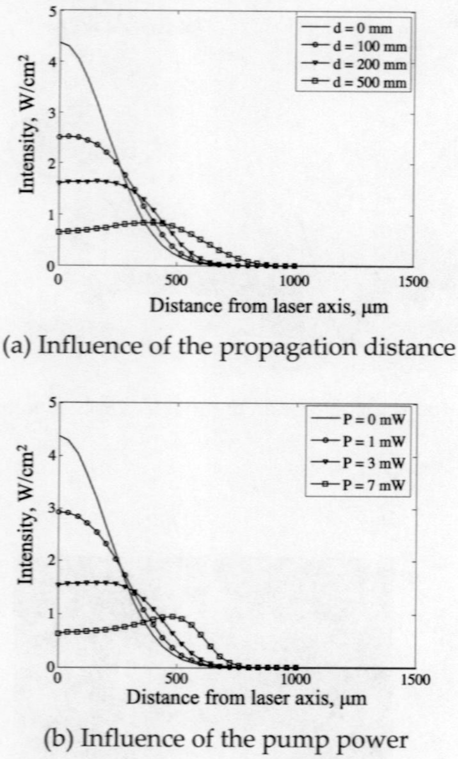

Figure 10. Influence of the propagation distance and the pump power to the probe beam profile

3.3. Experimental set-up to shape spatial profile

In order to confirm the role of the fluidic laser beam shaper, a single-beam experiment is set up as shown in Fig. 11. A CW diode blue laser is used as pump and probe-beam ($P = 10$ mW, $\lambda = 488$ nm, $\Phi = 0.69$ mm, TEM00). In this experiment, the height of the liquid medium is 0.5 mm, the dye concentration is 0.1 g/l and the absorption coefficient is 2.92 cm^{-1} (measured value) respectively. The propagation distance to obtain the flat-top beam profile is measured by changing the distance from the cuvette to the CCD camera. At the propagation distance of 150 mm, the flat-top beam is confirmed as shown in Fig. 12(a).

Figure 12(b) shows the beam profile change from the Gaussian to the flat-top beam. The vertical and horizontal axes show the intensity and distance from the laser axis respectively. The o-line shows the profile of the Gaussian input beam by fitting the laser beam profile measured at the surface of the cuvette. The strange-line shows the profile of the flat-top beam calculated by beam propagation method. The solid-line shows the profile of the flat-top beam measured by CCD camera at propagation distance of 150 mm from the cuvette. Both experimental and calculated results agree well with each other.

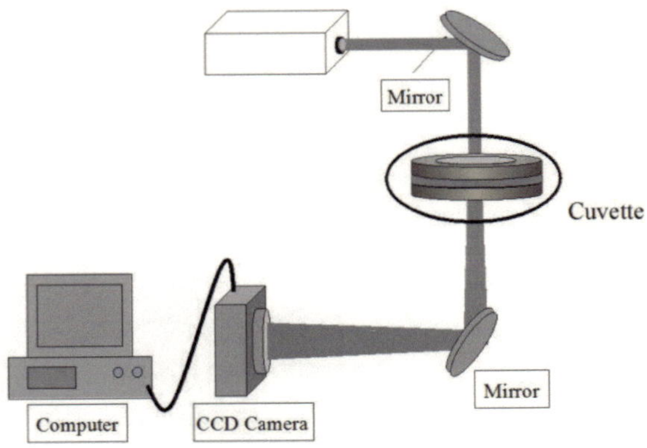

Figure 11. Experimental set up for a single-beam thermal lens system to transfer a Gaussian beam to flat-top beam

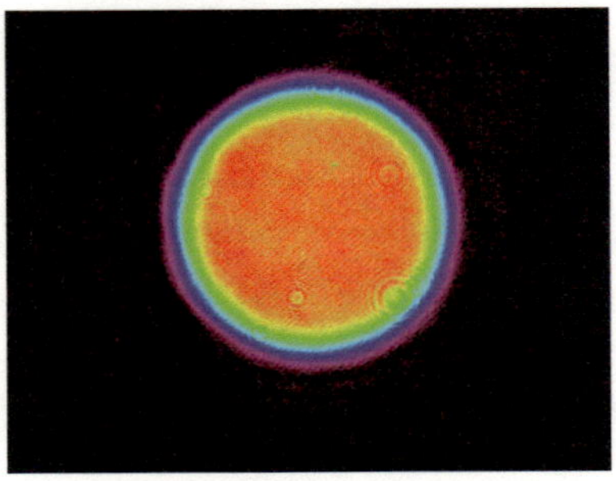

(a) Flat-top beam profile measured by CCD camera

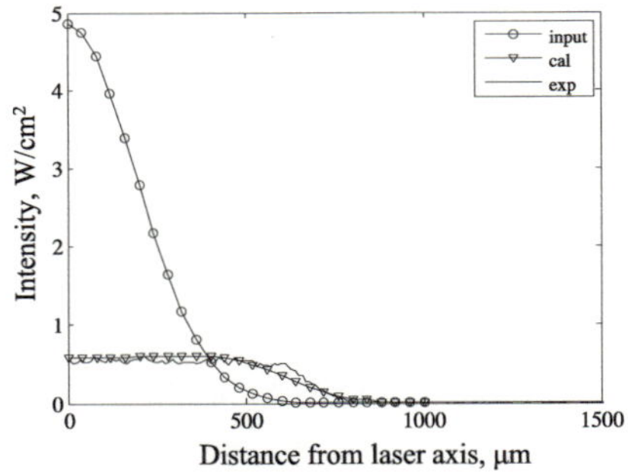

(b) Beam profile change from Gaussian to flat-top

Figure 12. Experimental results

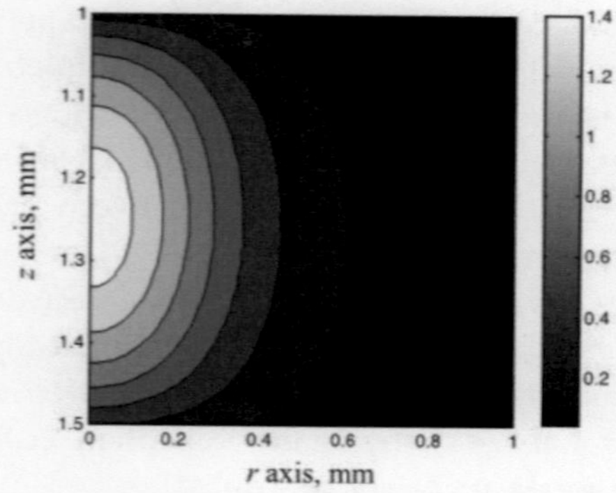

Figure 13. Temperature distribution inside the liquid medium [K]. The calibration shows the difference between temperature inside the liquid medium with the ambient temperature.

In order to explain in more detail about the mechanism of this fluidic beam shaper, the temperature distribution of liquid medium is calculated. As shown in Fig. 13, local heating near the beam axis produces a radially dependent temperature variation, which changes the liquid refractive index in which the lower refractive index is in the region near to the beam center. As a consequence, the radius of curvature of the wave front at the region near the beam center is shorter than one at the beam wing. Therefore the sample liquid locally acts as a micro divergent lens with shorter focal length at beam center. As shown in Fig. 1b, the beam center that passes through shorter focal length is spread out more rapidly than the beam wing. As the probe beam propagates to increasing distance, the intensity in the center region drops rapidly than one in the wing region. At a certain value of propagation distance, the Gaussian beam can be converted into the flat-top beam.

It is noted that, in the case of single-beam shaper, one part of laser beam energy (about 15% in this experiment) is converted into thermal energy in order to change temperature distribution or in other words to change refractive index distribution in the liquid medium. Therefore, in the case of single-beam shaper, the beam shaper has another role, which is as an attenuator. This laser beam shaper/attenuator can be applied in practical laser drilling technology. In the case of applying on only laser beam shaper, the double-beam system is recommended. In this case, it is needed to select dye whose absorbance of the probe-beam is negligible small.

3.4. Relationship between pump power and distance to shape spatial profile

As shown in previous section, the flat-top beam can be obtained only at a fixed distance. In order to control this distance, the influence of pump power is investigated theoretically and experimentally. The calculation parameters are shown in Table. 2. The pump power is changed from 1 to 8 mW. The distance to obtain the flat-top beam is obtained numerically. The relationship between the pump power and the distance to shape spatial profile is shown in Fig. 14(a). The horizontal and vertical axes show pump power and distance to obtain the flat-top beam respectively. As shown in Fig. 14(a), the distance to obtain a flat-top beam is in inverse proportion to the pump power.

In order to validate the numerical prediction, a single beam experiment was carried out. The pump power is changed from 1 to 6 mW and the distance to obtain the flat-top beam was measured. The experimental result shown in Fig. 14(b), shows excellent agreement with calculation prediction. The relationship between pump power and distance to obtain the flat-top beam can be explained by the interaction between energy absorption of liquid medium with the focal length of local micro lens. As the pump power increase, the absorption energy increases. As a consequence, the rate of decreasing of *R* is enhanced. This can be thought as the reason why the distance to obtain a flat-top beam decreases. In other words, the distance to obtain the flat-top beam profile also decreases with the increasing of absorption coefficient. Therefore, by changing the absorption coefficient or the pump power, the distance to obtain a flat-top beam can be controlled.

Pump power, mW	1 ~ 8
Pump beam diameter, mm	0.8
Probe power, mW	10
Probe beam diameter, mm	0.8
Absorption coefficient, cm^{-1}	2.0
Phase front curvature radius, *R*, mm	320

Table 2. Calculation conditions

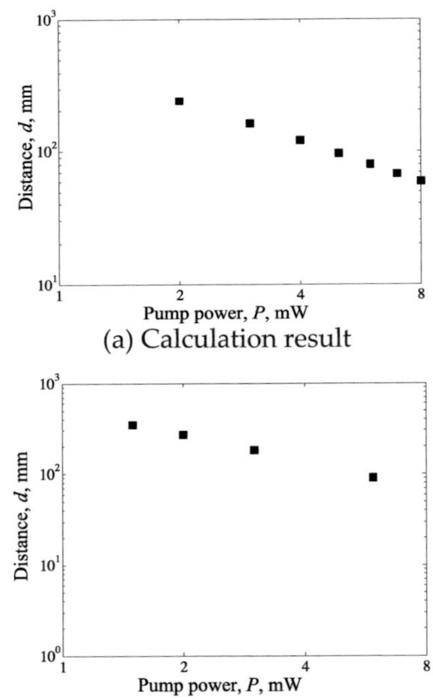

(b) Experimental result

Figure 14. Relationship between the pump power and the distance to obtain the flat-top beam profile. The horizontal and vertical axes show the pump power and the distance to obtain the flat-top beam profile respectively.

4. Tunable fluidic lens

Fluidic lenses are well known to present significant advantages for wide range of applications from mobile phone to laboratory on a chip. Fluidic lenses have a number of apparent advantages such as tunable refractive index and reconfigurable geometry. Several approaches to design the liquid lens have been developed based on the microfluidic techniques to modify the liquid lens shape by using: out-of-plane micro-optofluidic [25-26], in-plane micro-optofluidic [27-28], electron wetting [29], dielectrophoresis [30] and hydrodynamic force [31]. Other approach bases on turning the refractive index of the liquid by different means such as pressure control, optical control, magnetic control, thermo-optic control, and electro-optic control.

4.1. Principle of fluidic lens

When the liquid medium is irradiated, local heating near the beam axis produces a radially dependent temperature variation, which changes the liquid refractive index in which the lower refractive index is in the region near to the beam center. As a consequence, the radius of curvature of the wave front at the region near the beam center is shorter than one at the beam wing. The liquid medium behaviors as a convergence GRIN-L with focal length depends on the radial position of the incident ray relative to the optical axis of the cuvette. The ray equation that is calculated numerically to obtain the path of an incident beam, which is given by:

$$\frac{d}{ds}\left(n\frac{dR}{ds}\right)=\text{grad}(n) \tag{21}$$

Where, d*s* and *R* are the differential element of the path length and the positional vector of the ray respectively. The variable *n* is the refractive index of the liquid sample. The variable *n* is the refractive index profile depending on medium temperature following equation (13-16, 18).

4.2. Influences of the pump beam profile

Influences of the pump beam profile to the focal length of the GRIN-L were investigated numerically with the calculation conditions in Table. 3. The intensity profile of the pump is applied with the Gaussian beam and the quasi-flat-top beam (a super-Gaussian distribution of order *k*) using Eq. 22 and Eq. 23 respectively.

$$I_{\text{Gaussian}}=\frac{2P}{\pi r_0^2}\exp\left(\frac{-2r^2}{r_0^2}\right) \tag{22}$$

$$I_{\text{Flat-top}}=\frac{Pk2^{2/k}}{2\pi r_0^2\Gamma(2/k)}\exp\left(\frac{-2r^k}{r_0^k}\right) \tag{23}$$

Here Γ is the Gamma function, r and r_0 are distance from laser axis and beam radius, respectively. In this calculation, quasi-flat-top beam is the 10 order of the super-Gaussian distribution, two types of the pump intensity profile are shown in Fig.15.

Pump power, mW	10
Pump beam diameter, mm	1.5
Absorption coefficient, cm^{-1}	2.0

Table 3. Calculation conditions

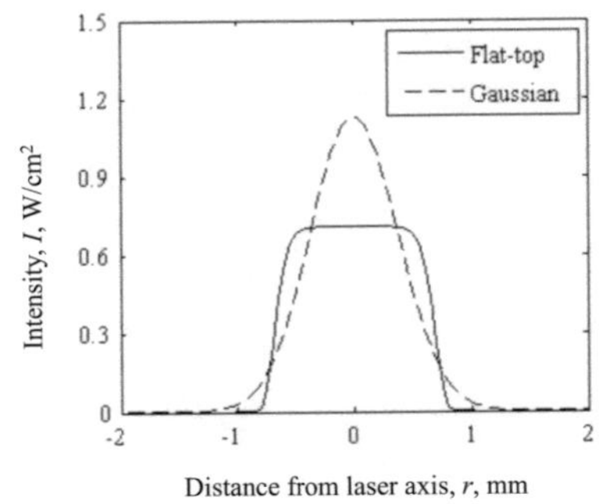

Figure 15. Two types of the pump beam profile

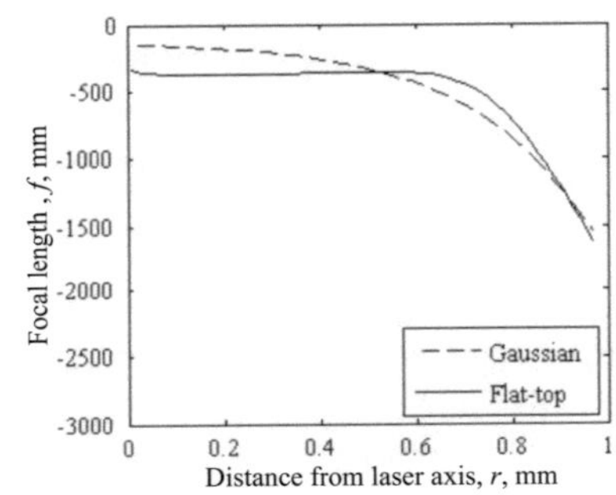

Figure 16. Effect of the pump beam profile to the focal length of the GRIN-L lens

The effect of the pump beam profile to the focal length of the GRIN-L lens is shown in Fig. 16. The vertical and horizontal axes show focal length and distance from laser axis respectively. The solid and dashed lines represent the plot of the focal length again the radial position of the incident ray relative to the optical axis of the cuvette in the case of Gaussian pump beam and quasi flat-top pump beam respectively. As shown in Fig. 16, for the Gaussian pump beam the focal length of the GRIN-L increases sharply with increasing of the distance from laser axis, which means larger spherical aberration. It means that, the beam center which passes through shorter focal length is spread out more rapidly than the beam wing. As a consequence, the further the propagation distance of the probe beam, the laser beam profile changes from Gaussian to the doughnut beam profile [24], which should cause some undesirable results in laser processing [32]. In contrast, with the quasi flat-top pump beam, the focal length of the GRIN-L varies lightly with increasing of the distance from laser axis smaller than beam waist of the flat-top pump beam. The area smaller than the beam waist of the flat-top pump beam acts as a divergent lens with small spherical aberration. Therefore, for the purpose of designing the GRIN-L lens the uniform pump beam shows the advance in reducing the spherical aberration.

4.3. Experimental set-up

In order to confirm the qualities of the GRIN-L, an experiment with the quasi flat-top pump beam is carried out as shown in Fig. 17. A CW diode blue laser is used as pump laser ($P = 10$ mW, $\lambda = 488$ nm, $\Phi = 0.69$ mm, TEM00). In cuvette 1, the height of liquid is 0.5 mm, and the absorption coefficient is 2.92 cm^{-1} (at wavelength of 488 nm). In the cuvette 2, the height of liquid is 1 mm, and the absorption coefficient is 55 cm^{-1} (at wavelength of 488 nm). A CW He-Ne laser is used as probe laser ($P = 0.6$ mW, $\lambda = 632$ nm, $\Phi = 0.8$ mm, TEM00). It is noted that, the absorption of ethanol solution can be ignored at the wavelength of the probe laser. First, the pump beam passes through cuvette1, then the beam profile of pump beam was converted from Gaussian to flat-top during its transmission to cuvette 2 as shown in Fig. 18. Then, the probe laser was adjusted to overlap with pump laser. After propagating through

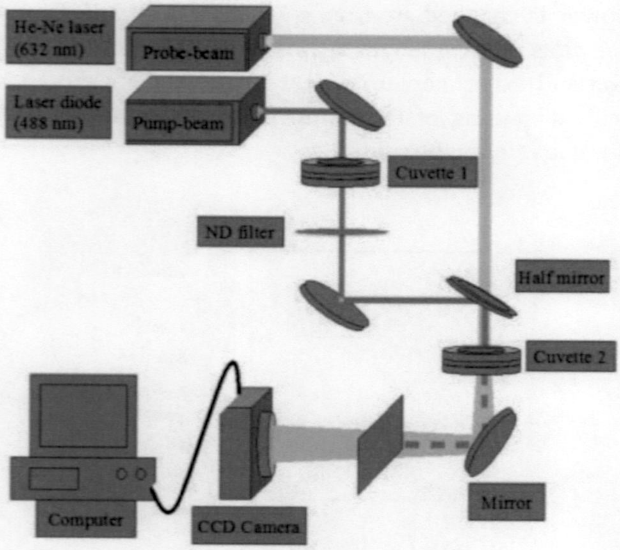

Figure 17. Experimental set-up for the fluidic divergent lens

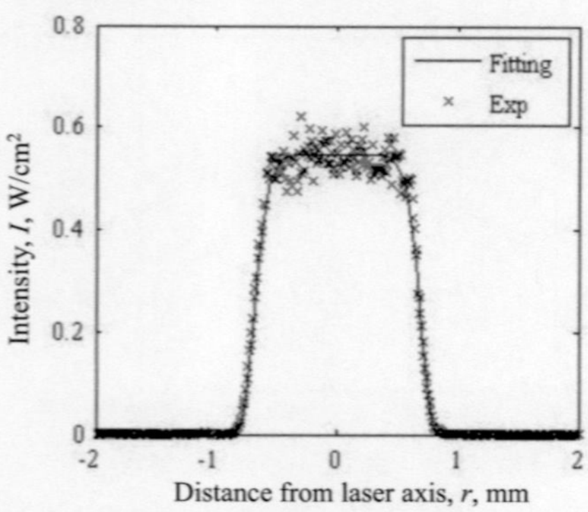

Figure 18. The intensity profile of the pump beam during its transmission to cuvette 2. Dotted and solid lines show the measured result and fitting by super-Gaussian distribution respectively.

the sample the probe laser is directed towards the CCD camera and the pump laser is blocked using filters located at the detection plane. The distance between cuvette 2 and the CCD camera is varied, and the $1/e^2$ diameter of probe laser is measured.

Figure 19(a) shows the change along the propagation direction in the beam profile. The vertical and horizontal axes show the intensity and distance from the laser axis respectively. By using the quasi flat-top pump beam, the beam profile of probe laser can remain in Gaussian distribution during its propagation. Figure 19(b) shows the plot of probe beam waist again propagation distance. As shown in Fig. 19(b), the beam waist of probe laser varies linearly with propagation distance. In other words, cuvette 2 acts as a divergence lens with focal length of f = -424 mm (this value has been calculated by considering the divergence angle of probe laser θ = 1.2 mrad).

Next, the pump power is changed from P_0 = 7.7 mW to $P_0/2$, $P_0/3$ and $P_0/4$ respectively. Figure 20 shows the plots of focal length against the pump power. Square and circle plots show the calculation and experimental result, respectively. As shown in Fig. 20, the focal length increases with increasing of the pump power. This means that, by adjusting the pump power, the focal length can be controlled.

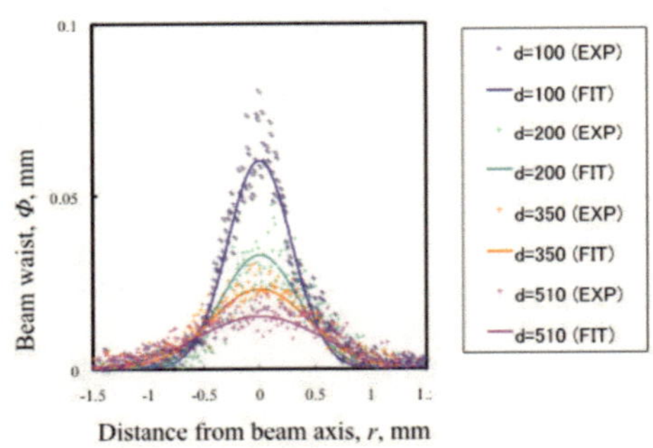

(a) Beam profile

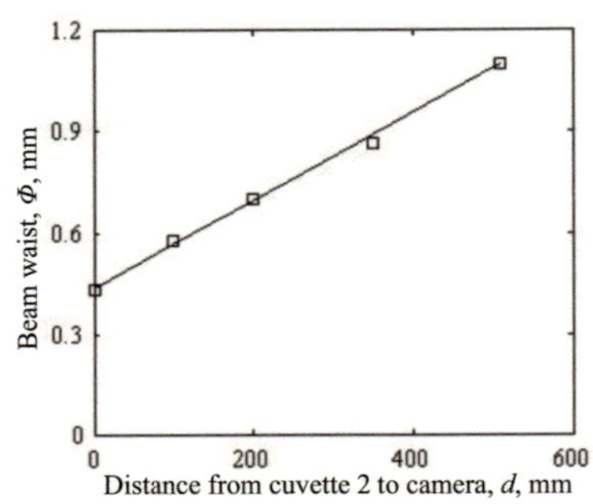

(b) Beam waist

Figure 19. Probe beam changes along the propagation direction

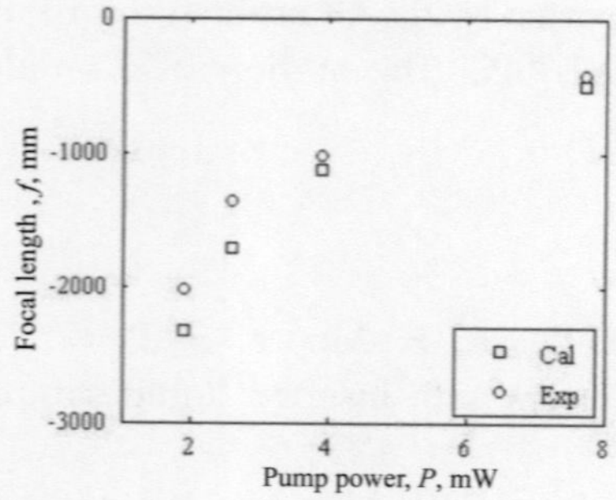

Figure 20. Relationship between the pump power and focal length

5. Conclusion

In this research, a novel idea of fluidic optical devices which includes laser beam shaper and fluidic divergent lens are demonstrated. The fluidic optical devices are based on controlling some parameters in the thermal lens system. The interaction among the intensity distribution, power of the pump beam, the absorption coefficient, the propagation distance and the intensity profile of the probe beam have been investigated experimentally and theoretically. It is found that

- By controlling the pump power and the absorption coefficient, the input Gaussian beam can be converted into a flat-top beam profile. The distance to get the flat-top beam profile can be controlled easily by adjusting the pump power and the absorption coefficient. In actual applications, single-beam shaper has another role, which is as an attenuator. This laser beam shaper/attenuator can be applied in practical laser drilling technology. In the case of applying on only laser beam shaper, the double-beam system is recommended. In this case, it is needed to select a dye whose absorbance of the probe-beam is negligible small.
- The uniform pump beam shows the advance in reducing the spherical aberration. And by adjusting the pump power, the focal length can be controlled

With some merits such as flexiblility, versatility and low cost, these fluidic optical devices will be promising tools in many fields of laser application.

Author details

Hong Duc Doan and Kazuyoshi Fushinobu
Department of Mechanical and Control Engineering,
Tokyo Institute of Technology,
Meguro-ku, Tokyo,
Japan

Acknowledgement

Part of this work has been supported by the Grant-in-Aid for JSPS Fellows and Grant-in-Aid for Scientific Research of MEXT/JSPS. The authors also would like to acknowledge Mr. Akamine Yoshihiko.

6. References

[1] J. P. Gordon, R. C. C. Leite, R. S. Moore, S. P. S. Porto, and J. R. Whinnery, Long-Transient effects in lasers with inserted liquid samples, J. Appl. Phys. 36 (1965) 3-8

[2] S.A. Akhmanov, D.P. Krindach, A.V. Migulin, A.P. Sukhorukov and R.V. Khokhlov, Thermal self-actions of laser beams, IEEE J. Quantum Electronics QE-4 (1968) 568-575

[3] P. M. Livingston, Thermally induced modifications of a high power CW laser beam, Appl. Opt. 10 (1971) 426-436

[4] J. F. Power, Pulsed mode thermal lens effect detection in the near field via thermally induced probe beam spatial phase modulation: a theory, Appl. Opt. 29 (1990) 52-63

[5] P. P. Banerjee, R. M. Misra, and M. Maghraoui, Theoretical and experimental studies of propagation of beams through a finite sample of a cubically nonlinear material, J. Opt. Soc. Am. B 8 (1991) 1072-1080

[6] J. M. Hickmann, A. S. L. Gomes, and C. B. de Araújo, Observation of spatial cross-phase modulation effects in a self-defocusing nonlinear medium, Phys. Rev. Lett. 68 (1992) 3547–3550

[7] Govind P. Agrawal, Transverse modulation instability of copropagating optical beams in nonlinear Kerr media, J. Opt. Soc. Am. B 7 (1990) 1072-1078

[8] C. J. Rosenberg et al., Analysis of the dynamics of high intensity Gaussian laser beams in nonlinear de-focusing Kerr media, Optics Communications 275 (2007) 458–463

[9] M. Sakakura, and M. Terazima, Oscillation of the refractive index at the focal region of a femtosecond laser pulse inside a glass, Opt. lett. 29 (13) (2004) 1548-1550

[10] M. Sakakura, and M. Terazima, Real-time observation of photothermal effect after photo-irradiation of femtosecond laser pulse inside a glass, J. Phys. France 125 (2005) 355-360

[11] M. Terazima, N. Hirota, S. E. Braslavsky, A. Mandelis, S. E. Bialkowski, G. J. Diebold, R. J. D. Miller, D. Fournier, R. A. Palmer, and A. Tam, Quantities, terminology, and symbols in photothermal and related spectroscopies (IUPAC Recommendations 2004), Pure Appl.Chem. , 76, 1083

[12] Kudou Uehara 1990, Basic Optics (Kougaku Kiso) Gendaikougaku, Tokyo, Japan, p. 45-47 (Japanese)

[13] S. K. Y. Tang, B. T. Mayers, D. V. Vezenov, and G. M. Whitesides, Optical waveguiding using thermal gradients across homogeneous liquids in microfluidic channels, *Appl. Phys. Lett.* 88, 06112, 2006

[14] H. D. Doan, K. Fushinobu, and K. Okazaki, Investigation on the interaction among light, material and temperature field in the transient lens effect, transmission characteristics in 1D temperature field, Proc. ITherm 2010, No. 127.

[15] J. Yang, Y. Wang, X. Zhang, C. Li, X. Jin, M. Shui, and Y. Song, Characterization of the transient thermal-lens effect using flat-top beam Z-scan, J. Phys. B: At. Mol. Opt. Phys. 42 (2009) 225404 (5pp)

[16] K. Ebata, K. Fuse, T. Hirai, and K. Kurisu, Advanced laser optics for laser material processing, Proc. SPIE, Vol. 5063, 411 (2003)

[17] E. B. S. Govil, J. P. Longtin, A. Gouldstone and M. D. Frame, Uniform-intensity, visible light source for *in situ* imaging, Journal of Biomedical Optics 14(2) (2009) 024024-024024-7

[18] F. M. Dickey, S. C. Holswade and D. L. Shealy 'Laser Beam Shaping Applications', Taylor & Francis, 2006

[19] M. T. Eismann, A. M. Tai and J. N. Cederquist, Iterative design of a holographic beam former, Appl. Opt. 28 (1998) 2641-1650

[20] F. M. Dickey and S. C. Holswade, Laser beam shaping: Theory and Techniques, Marcel Dekker, New York, 2000

[21] P. Scott, Reflective optics for irradiance redistribution of laser beam design, Appl. Opt. 20 (9) (1981) 1606-1610

[22] S. Zhang, Q. Zhang and G. Lupke, Spatial beam shaping of ultrashort laser pulse: theory and experiment, Appl. Opt.44 (2005) 5818-5823

[23] B. Mercier, J.P. Rousseau, A. Jullien, L. Antonucci, Nonlinear beam shaper for femtosecond laser pulses, from Gaussian to flat-top profile, Optics Communications 283 (2010) 2900-2907

[24] H. D. Doan, Y. Akamine, K. Fushinobu, Fluidic laser beam shaper by using thermal lens effect, Int. J. Heat Mass Transfer (2012) 55 2807–2812

[25] S. H. Ahn, Y. K. Kim, Proposal of human eye's crystalline lens-like variable focusing lens, Sens. Actuators (1999), A 78, 48

[26] D. Y. Zhang, V. Lien, Y. Berdichevsky, J. H. Choi, Y. H. Lo, Fluidic adaptive lens with high focal length tenability, Appl. Phys. Lett. 82 (2003), 3171

[27] S. K. Hsiung, C. H. Lee, and G. B. Lee, Microcapillary electrophoresis chips utilizing controllable micro-lens structures and buried optical fibers for on-line optical detection, Electrophoresis (2008) 29, 1866

[28] V. Lien, Y. Berdichevsky, Y. H. Lo, Microspherical surfaces with predefined focal lengths fabricated using microfluidic capillaries, Appl. Phys. Lett. (2003) 83, 5563

[29] C. B. Gorman, H. A. Biebuyck, G. M. Whitesides, Control of the Shape of Liquid Lenses on a Modified Gold Surface Using an Applied Electrical Potential across a Self-Assembled Monolayer, Langmuir (1995) 11, 2242

[30] C. C. Cheng, C. A. Chang, H. A. Yeh, Variable focus dielectric liquid droplet lens, Opt. Express (2006) 14, 4101

[31] S. K. Y. Tang, C. A. Stan, G. M. Whitesides, Dynamically reconfigurable liquid-core liquid-cladding lens in a microfluidic channel, Lab Chip (2008) 8, 395

[32] D.H. Doan, Y. Yin, N. Iwatani, K. Fushinobu, Laser processing by using fluidic laser beam shaper, Proc. National Heat Transfer Symposium 2012, Inpress

Chapter 10

Novel Optical Device Materials – Molecular-Level Hybridization

Kyung M. Choi

Additional information is available at the end of the chapter

http://dx.doi.org/10.5772/50032

1. Introduction

Silicate glass has been widely used for optical device materials due to its excellent optical transparency. To satisfy our multiple demands in advanced optical device materials, organic/inorganic hybrid composites have been widely prepared by a bulk mixing technique, which is physically mixing multiple components at the bulk scales.

However, conventional glassy materials have shown limitations to modify their physicochemical properties by inserting desired components into glassy hosts. In addition, a significant phase separation occurs during a mixing process of multiple components, especially immiscible phases.

To overcome those limitations, there are growing interests in doping organic components or semiconductor particles into glassy hosts without any phase separation to combine beneficial properties at the molecular scales, and thus to bring desired properties (Figure 1). [1]

Hybrid materials lie at the interface of the organic and inorganic material regimes, where versatility in molecular tailoring approach offers novel molecular modifications in design of new chemical structures. Hybrid materials can also range, depending on the method of formation and domain size, from physical mixtures of inorganic oxides and organics (blends, composites) to nanocomposites and molecular composites that utilize formal chemical linkages between the organic and inorganic domains on the molecular scale.

Hybrid materials are ranged from the bulk-scales to molecular scales as shown in Figure 2 to mix up multiple components. [1-10]

Usually, hybrid materials mixed at the bulk scales retain the original properties of the individual organic and inorganic components. In other words, their final properties are significantly influenced by the characteristics and their domain sizes of individual components after the mixing process at the bulk scales.

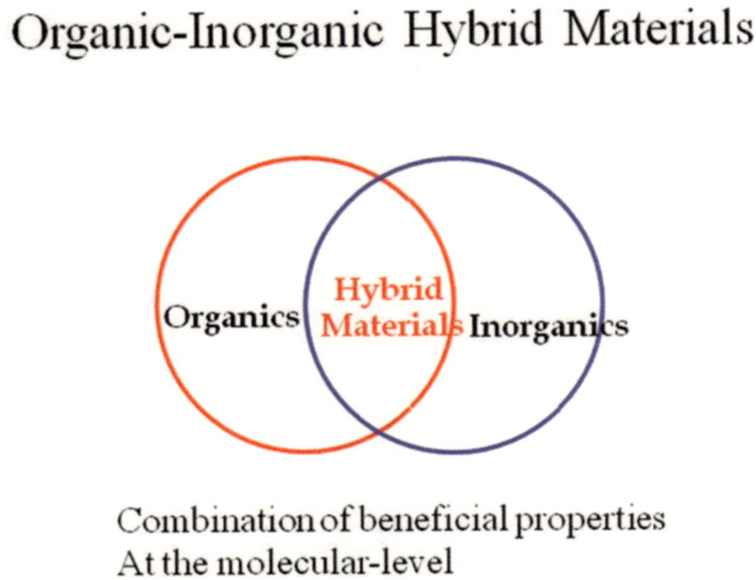

Figure 1. Organic/inorganic hybrid materials. [1]

In addition, a phase separation problem also limits to achieve a uniform mixing of multiple components during the bulk-mixing process.

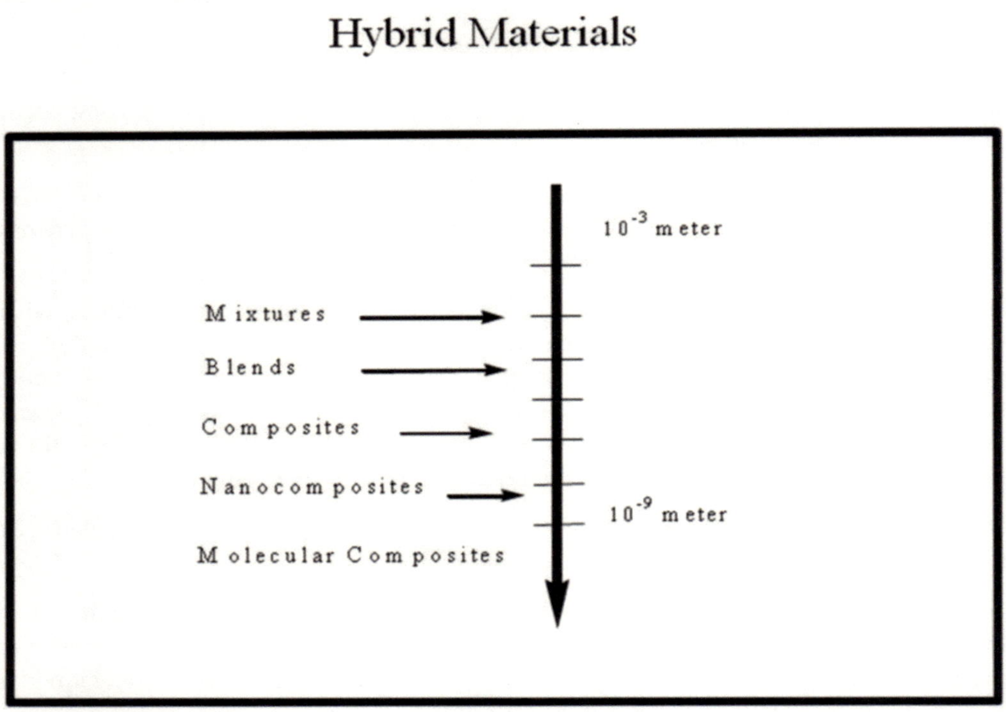

Figure 2. Relative size scale of mixing domains for different types of hybrid materials. [1-10]

To overcome those limitations of the bulk-scale mixing technique, *molecular-level hybridization technique* has been actively investigated. [1-10] This technique results in molecular-level composites, which are their domain sizes in the nanometer scale often create new properties, which would not be expected from those individuals by the loss of individuals' identities after the molecular-level mixing process thereby creating new properties.

Especially for optical device materials, desired properties of hybrid glasses can be chemically designed and then prepared by incorporating functional organic fragments between inorganic oxides.

With this strategy, novel optical device materials with beneficial properties can be obtained by embedding organic spacers into silicate network to create organically modified hybrid glasses as demonstrated in earlier publications (Figure 3). [1-10] Furthermore, the molecular-level mixing technique doesn't show any significant phase separation; because, *the molecular-level hybridization* is based on a microscopically homogeneous mixing and thus the uniform distribution of organic and inorganic moieties in a domain size at the molecular level is provided.

Those organically modified glasses mixed at the molecular scale, were provided by a molecular modification, which inserts desired organic fragments between two inorganic oxides to create entirely new optical properties. Figure 3 shows *the molecular-level hybridization* to produce polysilsesquioxanes.

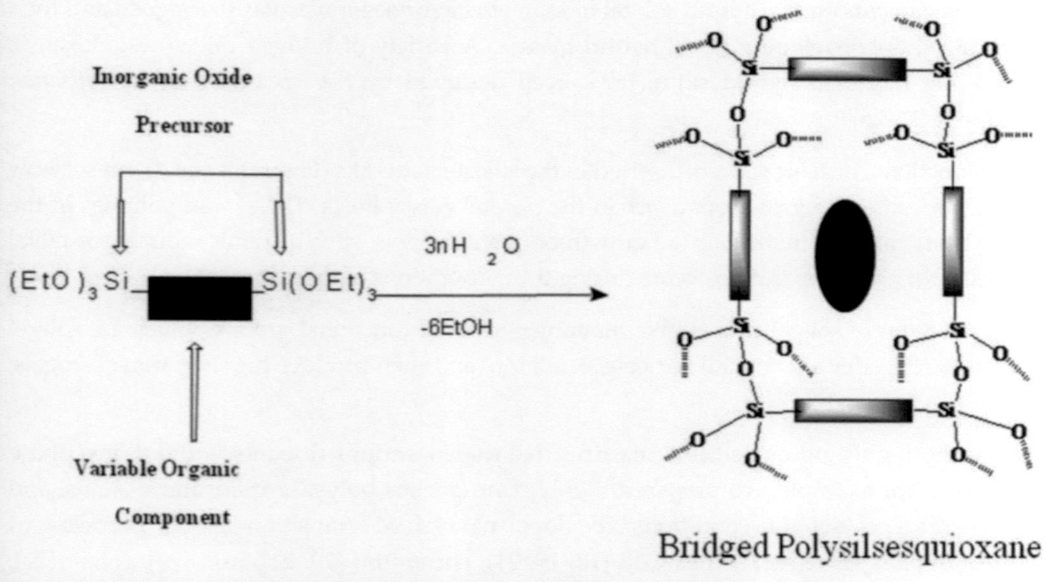

Figure 3. Molecular-level hybridization technique to produce polysilsesquioxanes. [6]

Bridged polysilsesquioxanes are a new family of molecular-level composites, which also a new version of hybrid glasses; these are prepared from hybrid sol–gel processable monomers through a sol-gel polymerization. These are microscopically homogeneous with uniform distribution of the organic spacers and thus also show an excellent optical transparency. There are many publications of novel optical device materials based on organic/inorganic hybrid glasses. [11-17]

In optical device materials, organically modified silicate glasses/transparent polymers have been actively pursued to develop novel optical devices, such as lasers, optical switches, optical fibers, waveguides, laser amplifiers, optical displays, and data storage devices.

[11-17]

In addition, we can also control the porosity of those organically modified silica, polysilsesquioxanes, by inserting different molecules or sizes of organic spacers as shown the void space in Figure 3 (right); due to the insertion of organic spacers, the porosity of the resulting hybrid glass significantly increased. [18-26]

The expanded pores allow us to dope semiconductor or metal particles without any significant mechanical cracks. *The molecular-level hybridization* also solves the phase separation problem during molecular-level mixing process.

The pore size of organically modified silicate materials can be controlled by both choices of organic spacers and sol-gel conditions. Those molecularly designed hybrid glasses have shown high surface areas and a relatively narrow distribution of pore sizes that range from the high micropore to the low mesopore domain (15-100 Å).

Several organic/inorganic hybrid sol-gel monomers have been molecularly designed and then synthesized for developing novel hybrid glasses. A variety of bridged polysilsesquioxanes, organically modified hybrid silica, have been designed by *the molecular-level hybridization* (Figure 4). [18-26]

Those functional organic spacers inserted in the silicate networks (Figures 3 and 4) can serve as dopant precursors to growth particles in the porous glassy hosts. Those void volumes in the glassy hosts can be a matrix for the growth of quantum dots, such as semiconductor or metal particles. No phase separation occurs during the corporations of those dopants.

In Figure 5, those sol-gel processable monomers contain functional groups, which are sol-gel polymerized under either acidic or base condition, and then produce highly porous xerogels. [4]

Due to the highly porous silicate matrices, we doped various dopants without any phase separation; for example, we prepared highly nano-porous polysilsesquioxane systems, and then controlled sol-gel conditions to dope nano-sized transition metal particles or semiconductor particles, such as CdS [18, 19, 21], chromium [20, 21], iron [22], cobalt [27], and platinum [28] into the silicate hosts.

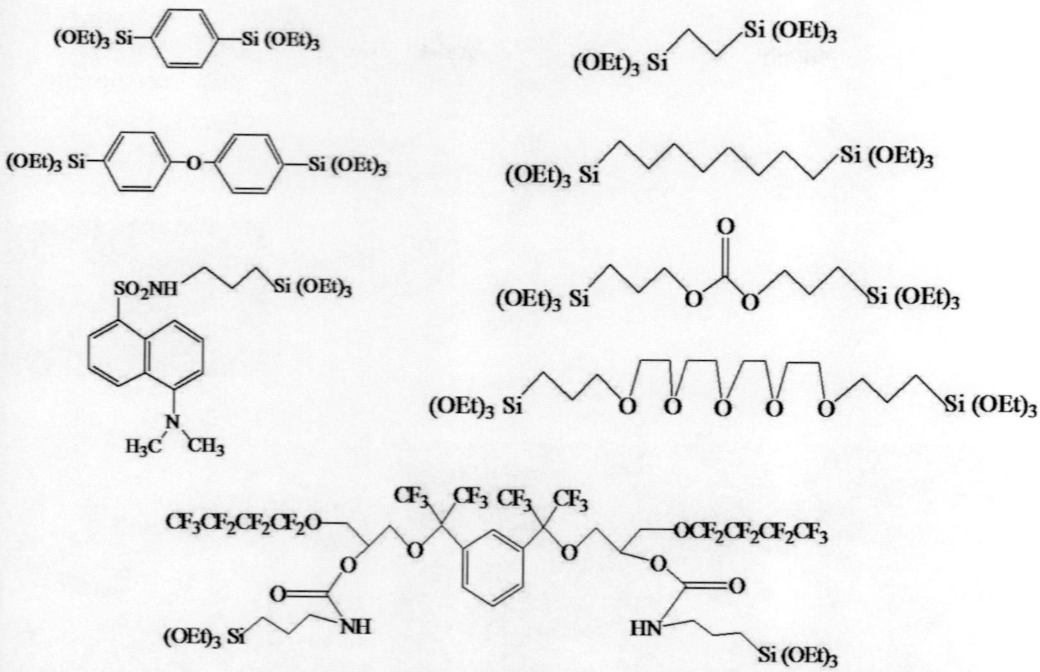

Figure 4. A variety of sol-gel processable monomers prepared by *the molecular-level hybridization.*

2. Novel optical device materials for laser amplifier [25]

We have developed novel laser devices based on those organic/inorganic hybrid glasses prepared by *the molecular-level hybridization.*

2.1. Rare-earth ion doped laser amplifier

Silicate-based optical fibers and planar waveguide amplifiers are widely studied for optoelectronic applications because of the superior chemical resistance and compatibility with other optical devices based on polymeric materials.

Transparent silica doped with rare-earth metal ions has been used for laser amplifiers, such as photonic fiber amplifiers, solid-state lasers, compact laser amplifiers, ultra short pulse lasers, high-power lasers, so on. [29-39]

Planar waveguides and fibers doped with rare-earth metal ions are a key challenge; thus, an enormous amount of publications in rare-earth metal ion doped optical devices has been found. [29-37] Fabrication of optical devices with high resolution offers an efficient approach to minimize the cost and size of optical amplifiers.

High gain laser amplifiers can be achieved by improving several relative factors, such as optical losses, phonon energies, pumping powers and distances, fluorescence life times, and refractive indices of optical medium. The lasing efficiency can be also improved by changing the chemical environment of rare-earth metal ions.

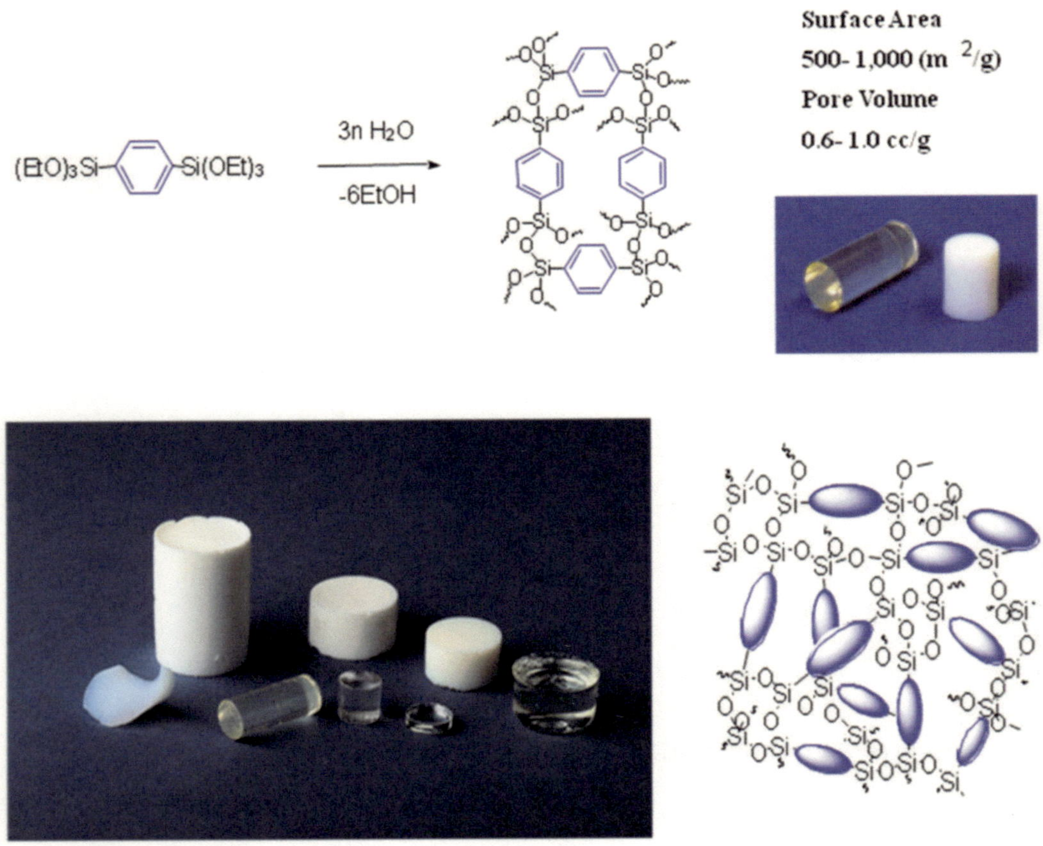

Figure 5. Highly porous polysilsesquioxanes. [4]

In a laser amplification based on rare-earth metal ions, erbium (Er) has been widely used as a gain medium due to its strong fluorescence at 1540 nm, which is a useful wavelength in optical amplifications. Especially, erbium-doped fiber amplifiers (EDFA) dominate this object of high gain optical amplifications. [32-35]

Since the performance of laser amplifiers is significantly influenced by optical media, scientists have been investigating organic/inorganic hybrid silicate hosts doped with Er^{3+} ions, in order to achieve high NIR efficiency and low phonon energy of the matrix to shorten the pumping distance and thus to obtain proper gain/life time. [32-35, 38, 39]

However, inorganic based optical media have shown a limitation to adjust the chemical environment of doped metal ions. For example, conventional silicate-based laser amplifiers often fail to produce high lasing performance because of a strong absorption raised from the OH-group at 1540 nm. The low concentration of erbium-ions in silicate host and the small absorption cross-section of the erbium-ions also limit the performance of normal silicate-based laser amplifiers since the doping level of rare-earth ions in glassy hosts significantly depends upon lasing efficiency.

Organic/inorganic hybrid glasses have been actively pursued as an alternative to conventional silicate glass for fabricating laser devices due to low temperature process and the promise of bringing new optical properties that are not possible from inorganic silica. [1, 40-42]

Furthermore, organic/inorganic hybrid silica is a good candidate to adjust the fluorescence environment of rare-earth metal ions by incorporating desired organic precursors into glassy hosts without any phase separation; usually, in inorganic silicate hosts, rare-earth ions tend to aggregate due to the absence of non-bridging oxygens, which cause a significant deduction of lasing efficiency. The synthesis of chemically modified silica with homogeneous doping of rare-earth ions is a key contributor to improve the performance of laser amplifications.

Optical device materials are required good optical transparency, controllable porosity, chemical purity, tunable refractive index, so on. Our goal for achieving those desired properties is to improve the fluorescence environment of Er^{3+} -ions in glassy hosts. For that, we devoted our attention to achieve an excellent chemical homogeneity of Er^{3+}-ion environment in glassy hosts.

2.2. Design of fluoroalkylene-bridged xerogel doped with Er^{3+}/CdSe

To demonstrate enhanced performance in laser amplifiers, we designed fluoroalkylene-bridged polysilsesquioxanes doped with Er^{3+}/CdSe nano-particles. [25] The fluoroalkylene-bridged silica was initially designed to reduce the phonon energy of the glassy host. Furthermore, CdSe nano-particles were also provided for further manipulation of the photochemical environment of erbium-ions in the matrix.

Use of organic silanes incorporates an organic fragment as an internal component of the silicate network. Sol–gel polymerization involves the hydrolysis of ethoxysilyl groups to yield silanols follows by subsequent condensation to form siloxane (Si–O–Si) linkages. In the sol-state, the condensation is insufficient to form a network, and the solution remains processable.

When sufficient cross-linking occurs, a network is formed and the transition from the sol-state to the gel state occurs. The presence of organic fragments within the 3-D structure imparts organic character to the hybrid glass that changes the microenvironment of additional fluorescence rare-earth ions incorporated in the glassy host.

In this study, fluorocarbon-linkages were designed to achieve high hydrophobicity within the hybrid glassy matrix. Erbium isopropoxide was also employed as the source for the Er^{3+} ions. Furthermore, CdSe nano-particles were also prepared and incorporated into the fluorinated glassy matrix to reduce the phonon energy of the glassy host (the phonon energy of CdSe = 200 cm^{-1}). [43]

In principle, when rare-earth metal ions are excited in transparent glassy matrices, they can behave as a laser, which enables amplification of the incident light intensity. The lasing performance significantly relies on rare-earth metal ion doping level, host materials' physicochemical property, and chemical homogeneity.

2.3. Experimental

A set of three different sol–gel processable monomers were prepared; tetraethoxy-silane (TEOS), 1,6-bis (triethoxysilyl)hexane, and 2,2,3,3,4,4,5,5-octafluoro-1,6-hexanediol bis(3-triethoxysilyl)propyl carbamate.

2.3.1. Tetraethoxysilane (TEOS)

TEOS was purified by drying over 4 Å molecular sieves followed by a vacuum distillation.

2.3.2. Synthesis of 1,6-bis(triethoxysilyl)hexane

1,6-Bis(triethoxysilyl)hexane was synthesized by a 'hydrosilylation' of the corresponding α, ω-alkyldienes with triethoxysilane employing chloroplatinic acid (H_2PtCl_6) as a catalyst. 1,5-Hexadiene (12.3 g, 0.15 mol), triethoxysilane (54.1 g, 0.33 mol), and chloroplatinic acid (1mL of 7.5×10^{-5} mol in isopropanol) were placed in a round bottle flask. After 10 hours of simple stirring process at a room temperature, the reaction mixture darkened.

The reaction was monitored by GC. The crude product was purified by a vacuum distillation with a resulting purity of 99.87 % by GC analysis: bp 130 °C/0.1 mmHg. The final product was verified by NMR analysis and mass spectroscopic analysis and was consistent with data previously report. [11]

2.3.3. Synthesis of 2,2,3,3,4,4,5,5-octafluoro-1,6-hexanediol bis(3-triethoxysilyl)propyl carbamate

2,2,3,3,4,4,5,5-Octafluoro-1,6-hexanediol (1g, 3.8 mmol) and 3-isocyanatopropyl triethoxy-silane (1.9 g, 4.3 mmol) were placed in a round bottom flask with a magnetic stirrer. The flask was sealed, purged with nitrogen, and 10 mg of dibutyl-tin-dilaurate was injected into the vial using a syringe as a catalyst.

The reaction was kept at room temperature under a nitrogen flow for several hours and monitored for the disappearance of the isocyanate peak at 2270 cm^{-1} (CN) in FT-IR. As the isocyanate group was converted to urethane group, identical peaks of 3400 cm^{-1} (NH) and 1720 cm^{-1} (CO) were observed. The product was dissolved in methanol and the tin catalyst was completely removed using a separated funnel. Methanol was then removed by a rotary evaporator. The product was used without further purification.

2.3.4. Er^{3+}-ion/CdSe doping procedure

A sol-gel processible monomer, erbium isopropoxide (Chemat Technology), was used as a source of Er^{3+}-ions. CdSe nano-particles were synthesized by a previously reported procedure. [44, 45] Those dopants were incorporated by mixing with the appropriate sol–gel mixtures (Table 1); xerogels-a5 and -a10 denote higher erbium concentrations (5 and 10 times higher) than that of xerogel-a system.

We also examined the homogeneity of three sol–gel mixtures (Figure 6). Ethanol was used as a solvent. A visual inspection was carried out to determine the comparative homogeneity of those sol–gel mixtures containing erbium isopropoxide (Figure 6-1) or both of erbium isopropoxide/CdSe nano-particles (Figure 6-2). In a mixing test (Figure 6), erbium isopropoxide was indicated as a bright pink color in the photographs. The CdSe nano-particles also show a characteristic bright orange-color in those mixtures.

Xerogels	Silicate matrices	Dopants
T-Xerogel	TEOS	None
H-Xerogel	Hexylene-	None
F-Xerogel	Fluoroalkylene-	None
Xerogel-a	TEOS	Erbium ions
Xerogel-b	TEOS	Erbium ions/CdSe
H-Xerogel-a	Hexylene-	Erbium ions
H-Xerogel-b	Hexylene-	Erbium ions/CdSe
F-Xerogels-a, -a5, and -a10	Fluoroalkylene-	Erbium ions
F-Xerogels-b, -b5, and -b10	Fluoroalkylene-	Erbium ions/CdSe

Table 1. List of xerogels doped with different components.

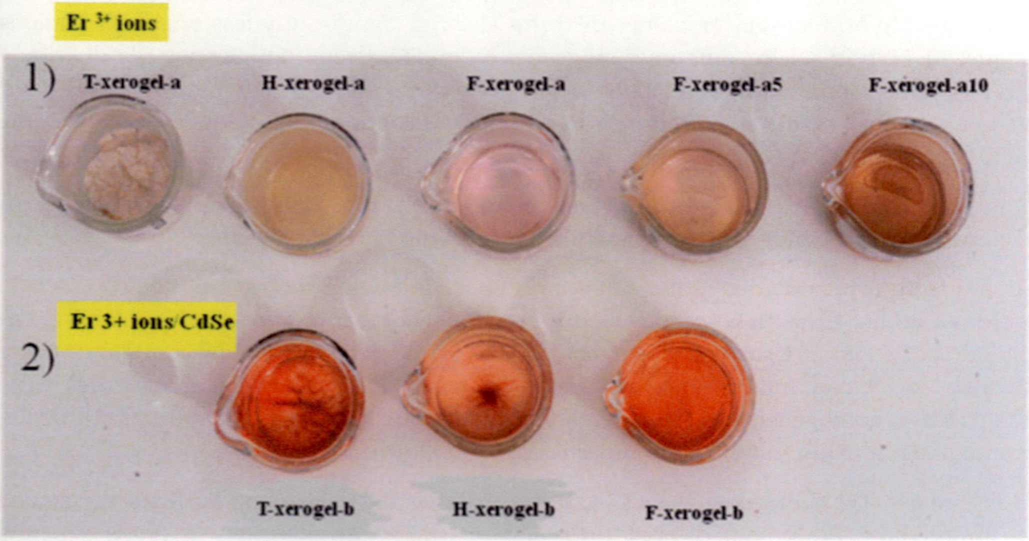

Figure 6. Mixing test.

2.3.5. *Sol–gel procedures*

Those sol–gel mixtures were then polymerized to produce the condensed xerogels under acidic condition using HCl as a sol-gel catalyst. Those xerogels were kept in a vacuum oven for 1–2 days to remove the remaining solvent and complete condensation.

2.3.6. *Solid state NMR experiments*

The ^{29}Si solid state NMR analysis of condensed xerogels was performed using a Varian Unity 400 solid state NMR spectrometer. The degree of condensation of undoped xerogels was computed from the single pulse magic angle spinning (SP/MAS) technique. A line fitting routine was also used in the analysis of the ^{29}Si NMR resonance in each spectrum to establish the siloxane ratio in the different structures.

2.3.7. *Fluorescence measurements*

Fluorescence study of Er^{3+}-ions doped into those glassy matrices was carried out by using the Ar+ ion laser (488 nm). Laser power densities ranging from 1.5 to 3 Wcm^{-2} were used for the measurements.

2.4. Results and discussions

2.4.1. *Mixing test*

During the mixing test (Figure 6), we observed that the TEOS-based mixtures revealed a substantial degree of undesirable phase separations after doping with the Er^{+3}-ion sources or Er^{+3}/CdSe nano-particles. For example, in Figure 6-1, the T-xerogel-a mixture shows a significant phase separation even at the lower erbium concentration.

In contrast, hybrid sol–gel monomer mixtures showed significantly less phase separations (Figure 6). Hybrid sol–gel monomer mixtures accommodate and homogeneously distribute the Er^{3+}/CdSe source without any significant phase separation. In mixing test with hybrid sol-gel monomer mixtures, we observed no momentous phase separation, especially in the highly fluorinated sol-gel mixtures.

It is apparent that the TEOS-based sol–gel mixture has a rather limited solubility of erbium-ions, and hence a limited capability for fluorescence enhancement.

Subsequently, we provided CdSe nano-particles by following the earlier method [44, 45], and then added CdSe nano-particles into those sol–gel monomer mixtures containing the the Er^{3+}/CdSe source. Usually, CdSe nano-particles synthesized in colloidal configuration are suitable for incorporation into a variety of hosts including sol–gel mixtures. The comparative homogeneity of the three sol-gel monomer mixtures containing both of the erbium isopropoxide and CdSe nano-particles is also shown in Figure 6-2.

In Figure 6-2, TEOS-based mixture (T-xerogel-b) shows the CdSe nano-particles segregated in the mixture; the orange-colored CdSe nano-particles are observed to phase separate within the mixture.

In contrast, a hybrid sol-gel monomer system (H-xerogel-b) shown in Figure 6-2, the CdSe nano-particles mixed better than the T-xerogel-b. In the H-xerogel-b mixture, most of CdSe particles were dissolved, except some of undissolved orange-colored CdSe residues toward the middle of the container. In fluorinated mixture (F-xerogel-b) shown in Figure 6-2, CdSe nano-particles were incorporated without phase separation. This result demonstrates that the fluoroalkelene-bridged sol-gel monomer has the capability of uniformly incorporating both types of dopants without any phase separation.

2.4.2. Solid state NMR analysis

The chemical composition and the degree of condensation for those condensed xerogels can be determined by solid state nuclear magnetic resonance, infrared, and Raman spectroscopies. [1]

We employed a solid state NMR analysis to determine the degree of condensation of hybrid glassy hosts. ^{29}Si solid state NMR was used to identify the Si–O–Si bonds in variety states of condensation for three matrices. Single pulse magic angle spinning NMR methods were employed for the characterization of T-xerogel, H-xerogel, and F-xerogel to calculate the degree of condensation.

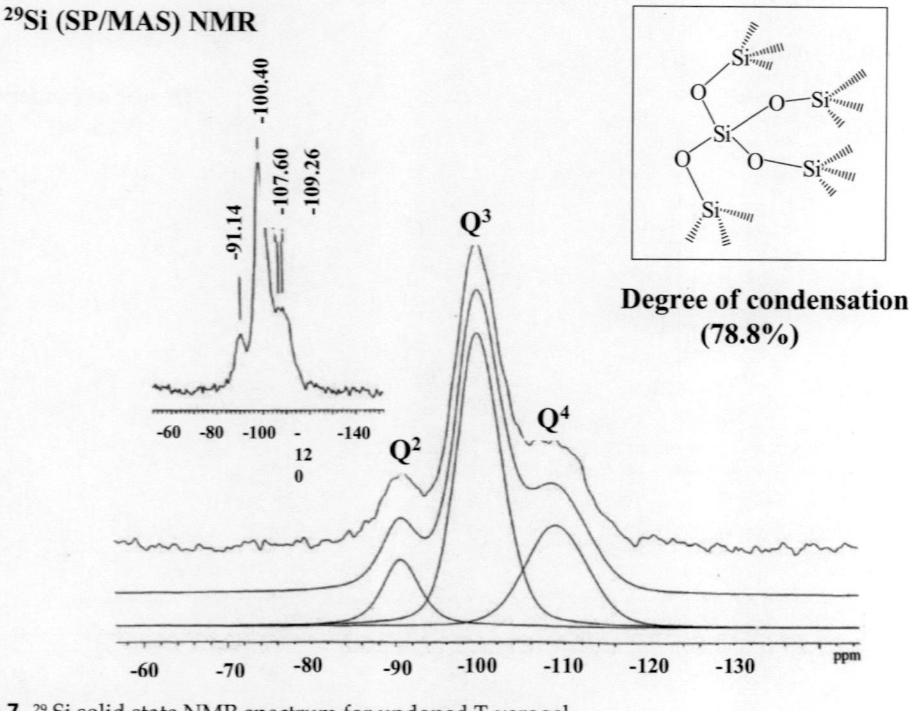

Figure 7. ^{29}Si solid state NMR spectrum for undoped T-xerogel.

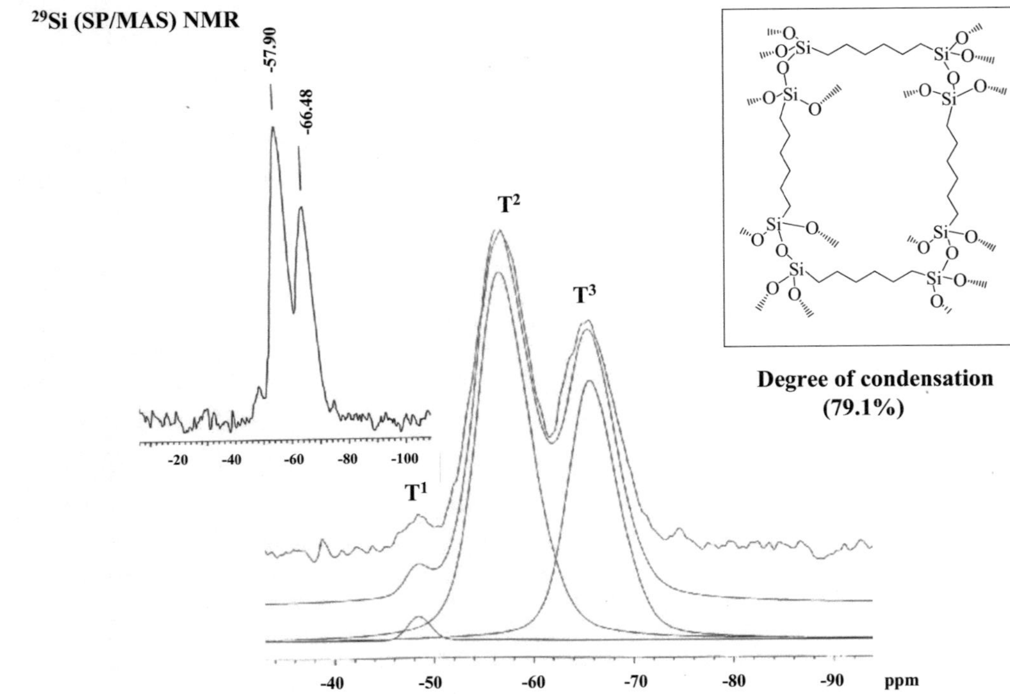

Figure 8. ^{29} Si solid state NMR spectrum for undoped H-xerogel.

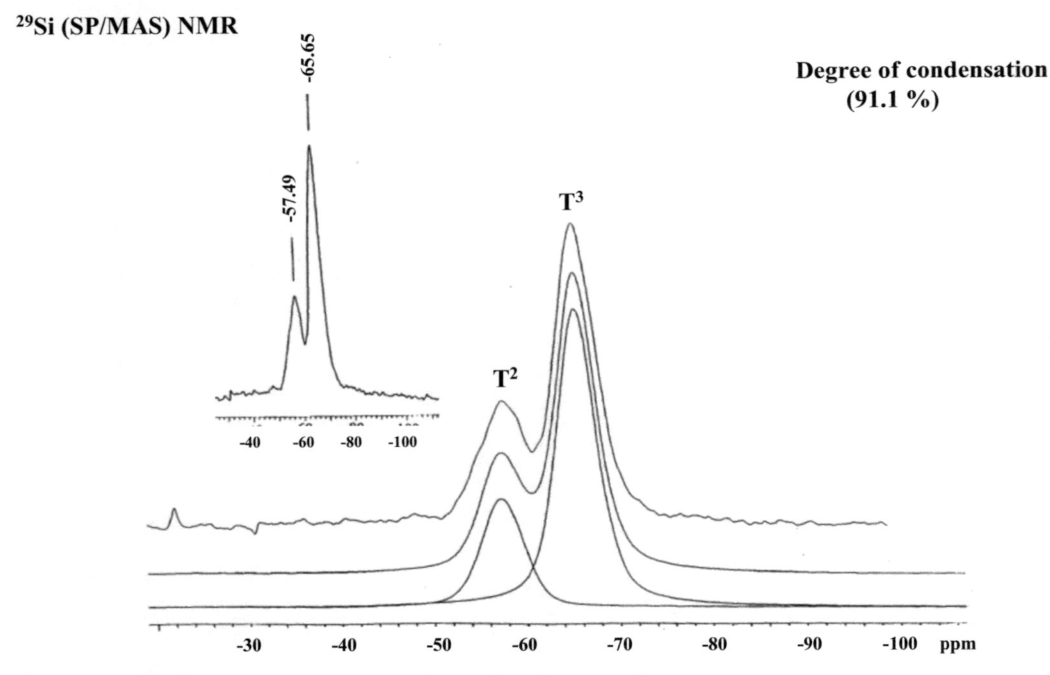

Figure 9. ^{29} Si solid state NMR spectrum for undoped F-xerogel.

Figures 7-9 show the result of ^{29}Si SP/MAS solid state NMR analyses and peak deconvolution lines of both normal and modified silicate systems.

Figure 7 shows the ^{29}Si SP/MAS solid state NMR spectrum of the undoped T-xerogel. As shown in Figure 7, it reveals three peaks, which correspond to the Q^2, Q^3, and Q^4. The degree of condensation for the T-xerogel was calculated to be 78.8%.

Figure 8 shows the ^{29}Si SP/MAS solid state NMR spectrum of the undoped H-xerogel. Three peaks shown in spectrum correspond to the T^1, T^2, and T^3. The degree of condensation of the H-xerogel was also computed to be 79.1%.

The ^{29}Si SP/MAS solid state NMR spectrum of the undoped F-xerogel is given in Figure 9. The T^1 peak is not observed in this system. Also, the T^3 peak intensity in Figure 9 is higher than that of T^2. We found that the degree of condensation was dramatically increased to 91.1% in this case.

The high degree of condensation in the undoped F-xerogel can be explained as a result of the electron-withdrawing effect of the fluorine, which causes a partial positive charge at the silicon facilitating nucleophilic attack in the sol–gel process thus accelerating hydrolysis and condensation.

The high degree of condensation, disappearance of T^1 peak, and enhanced T^3 peak in the F-xerogel indicate a low level of hydroxyl group content and a greater degree of condensation when fluorinated alkylene groups are present in the glassy hosts.

Si–OH groups show a high phonon energy (3000–3500 cm^{-1}) at 1540 nm. [43]

The reduction in hydroxyl content in the F-xerogel matrix decreases the phonon energy of the matrix. The exclusion of moisture from the high hydrophobicity of the fluoro-alkylene groups in the F-xerogel matrix may also contribute to reduce the absorption at 1540 nm with a concomitant increase the fluorescence intensity of Er^{3+}-ions.

All these effects contribute to the increased fluorescence from erbium-ions in the fluorinated hybrid glassy matrix.

2.4.3. XPS analysis

We also employed a XPS study to determine the chemical composition of F-xerogel-a. [46]

A full XPS scan was obtained in the 0–1100 eV range. Detail scan was also recorded for the Er (4d) region. Figure 10 shows a full spectrum for F-xerogel-a. The XPS spectrum of F-xerogel-a shows an erbium peak at ~169 eV, which corresponds to the presence of Er_2O_3.

The atomic compositions were evaluated in this study. The concentration of each element (atomic %) was calculated; O(1s)—22.74 atomic %, C(1s)—53.29 atomic %, Er(4d)—1.49 atomic %, F(1s)—9.02 atomic %, Si(2p)—10.91 atomic %, N(1s)—2.55 atomic %. From the XPS analysis, it was estimated that the F-xerogel-a contains ~1.49 atomic % of erbium.

2.4.4. Fluorescence measurements

Erbium-ions incorporated into glassy matrices exhibit well defined energy level transitions in 4f-shell electronic configurations.

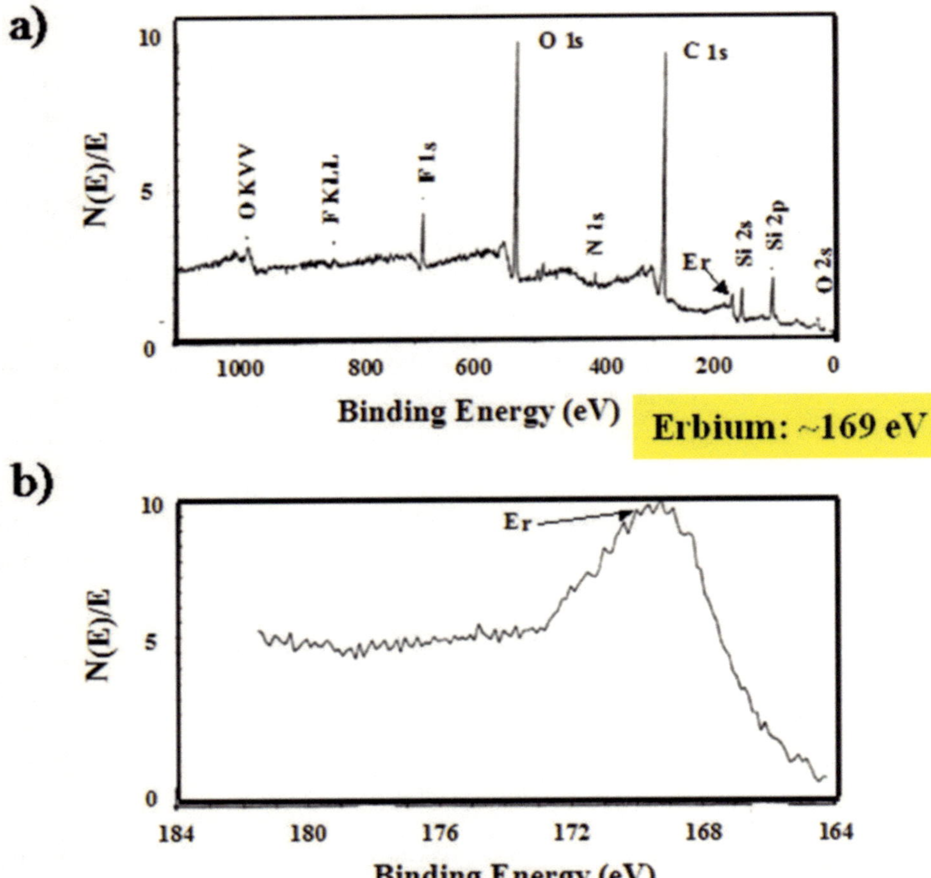

Figure 10. XPS Analysis of F-xerogel-a.

For erbium-ions, the $4I_{13/2}$ to $4I_{15/2}$ transition is important in optical communications; because, it results in fluorescence at 1540 nm, which is the most important wavelength regime for optical communication applications, especially in long-distance telecommunication networks. [47]

We examined how the fluorescence intensity of erbium-ions was dependent upon the matrix environment when fluorine and CdSe nano-particles were incorporated into hybrid glassy hosts. We carried out fluorescence analysis of erbium-ions around 1540 nm. The results are shown in Figures 11-13.

A comparison of fluorescence intensities from erbium-ions doped into different glassy hosts is shown in Figure 11 (H-xerogel-a and F-xerogel-a).

In Figure 11, fluorescence intensity of erbium-ions increased significantly more in F-xerogel-a than the other hybrid system of H-xerogel-a.

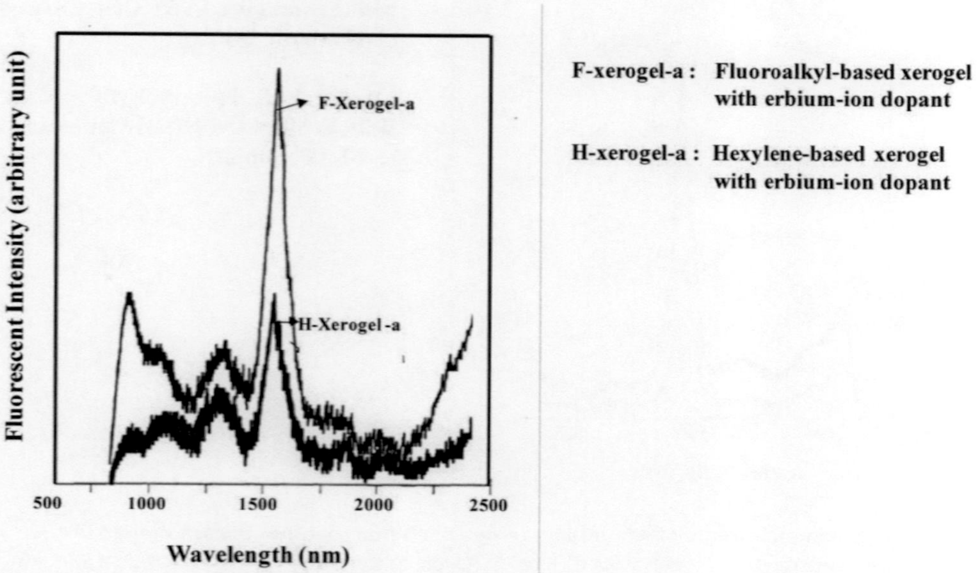

Figure 11. Photoluminescence of erbium-ions; different silicate matrices. Comparative fluorescence intensities of two different hybrid xerogels doped with Er^{+3} ions using a low power of 1.5 Wcm^{-2}.

The explanation of Figure 11 is that the enhanced fluorescence intensity of the fluorinated matrix is attributed to mainly its high hydrophobicity combined with the lower OH-group contents, which revealed in NMR study in Figure 9. Solid state silicon NMR analysis indicates an enhanced condensation in fluorinated xerogel compared to that of alkylene-bridged xerogel. The fluorinated hosts also showed an excellent chemical homogeneity in mixing test (Figure 6), which significantly affects the lasing performance.

Further investigations in fluorescence studies have been carried out. Since it is important to consider the erbium-ion concentration effect, fluoroalkylene-based glasses doped with two different levels of erbium concentrations were prepared for the determination of erbium-concentration effect (Figure 12).

Intensity of those xerogels (F-xerogel-a and F-xerogel-a5) was measured at a power density of 3 Wcm^{-2}. As shown in Figure 12, the fluorescence intensity increases (the upper curve) in the case of F-xerogel-a. It can be explained that the higher erbium concentration of F-xerogel-a5 caused a low lasing efficiency due to the "self-quenching effect."

Fluoroalkylene-bridged xerogel containing Er^{3+}-ions shows significantly reduced absorptions at the 1540 nm by reducing amounts of uncondensed hydroxyl groups.

The presence of CdSe nano-particles also significantly influences the fluorescence environment of Er^{3+}-ions in different glassy hosts, resulting in the increased fluorescence intensity. [43]

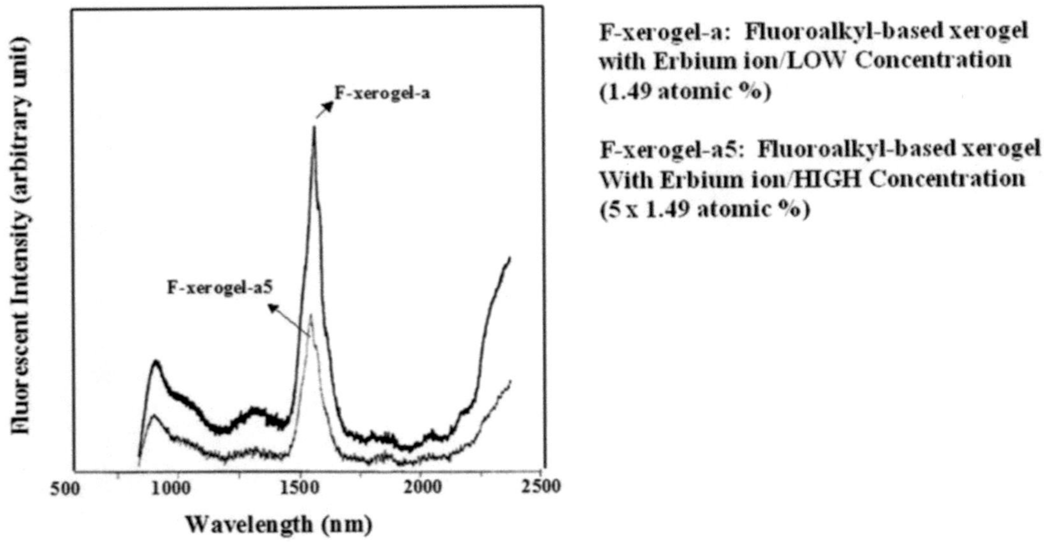

Figure 12. Photoluminescence of erbium-ions; different erbium concentrations. Comparative fluorescence intensities of xerogels with different Er^{3+}-ion concentrations (F-xerogels-a and -a5) using a power of 3.0 Wcm^{-2}.

CdSe nano-particles were used to modify the photochemical environments of erbium-ions in glassy hosts by taking advantage of a low phonon energy of CdSe phase (200 cm^{-1}) [43], since the incorporation of semiconductor nano-particles resulted in an enhancement of the semiconductor-to-erbium transfer when the quantum well and erbium-ion transition energies became close.

We thus examined the fluorescence of erbium-ions surrounded by CdSe nano-particles since it was anticipated that the presence of CdSe in the modified glassy hosts would affect the fluorescence performance (Figure 13). In order to test this, we prepared fluoroalkylene-bridged hybrid glasses doped without and with the CdSe nano-particles, F-xerogels-a5 and F-xerogel-b5, respectively.

As shown in Figure 13, the fluorescence intensity of F-xerogel-b5 is dramatically increased, which indicates an improvement in lasing efficiency by modifying photochemical environments of erbium-ions.

By taking advantage of the structural features and uniform doping capability in modified glassy matrices, we successfully demonstrate that the fluorescence environments of erbium-ions can be a key to improving the performances of optical devices like laser amplifiers to overcome the limitations in inorganic silica.

In conclusion, we have demonstrated here a promising result in laser amplifications by employing bridged polysilsesquioxanes doped with Er^{3+}-ions/CdSe nanoparticles.

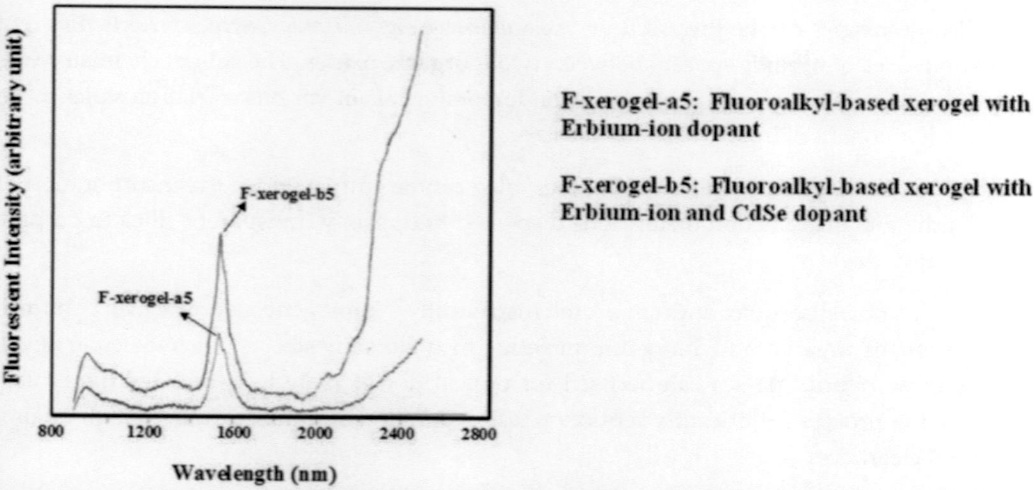

Figure 13. Photoluminescence of erbium-ions; CdSe nanoparticle effect. Comparative fluorescence intensities of fluoroalkylene-bridged xerogels doped without and with the CdSe nano-particles (F-xerogels-a5 and -b5), respectively, using a high power of $3Wcm^{-2}$.

To avoid the high phonon energy raised from OH-groups (3000–5300 cm^{-1}) at 1540 nm [43], we designed highly fluorinated hybrid glasses to shorten the fluorescence-level life times of dopants, which adversely affect optical device performance.

The presence of CdSe nano-particles, by virtue of its lowering of phonon energy, also appears to significantly influence the nature of the surrounding photochemical environment of Er^{3+}-ions in the fluorescence study.

From those study, we found that the control of such optical materials' properties affect the performance of optical devices via molecular tailoring strategy, which is molecular-level hybridization technique.

3. Novel optical device materials for acoustic wave [26]

3.1. Novel optical device materials based on polysilsesquioxanes

The preparation of semiconductors, metals, and ions in a variety of transparent materials has been actively pursued for optical devices. Scientists have taken a great attention to incorporate metal particles/ions in glassy hosts to develop high performance optical devices.

Chemists also have taken intensive challenges on how to achieve homogeneous incorporations of semiconductors or metal particles in glassy hosts, which are significantly influenced by diffusion of reagents, the number of nucleation sites, stabilization of growing particles by surface functionality, the boundary constraints of the growth matrix, and the opportunity for equilibration or "ripening" of particles formed under kinetic growth conditions. [2, 21, 48, 49]

Organically modified hybrid glasses, which were prepared by *'molecular-level hybridization'* (Figure 3), are a good candidate to develop laser device materials due to their easy processability and chemical modification.

Polysilsesquioxanes can be prepared by *'molecular-level hybridization'*, which inserts different types and sizes of organic spacers between two inorganic oxides. The sol-gel chemistry was employed to covalently incorporate semiconductors/metals in various oxidation states as an integral component of hybrid silicate matrices.

The 'molecular-level corporation techniques' also can be employed for the incorporation of semiconductors or transition metals/ions dispersed in optically transparent silica to prepare novel optical devices.

Especially, polysilsesquioxanes are microscopically homogeneous due to uniform distribution of organic and inorganic moieties in a domain size at the molecular level. Hence, these hybrid glasses can be used for optical device materials since the molecular-scale mixing process significantly reduces phase separation and thus produces high quality of optical clearance.

In this work, we molecularly designed a novel hybrid glass to develop alkylene-bridged polysilsesquioxanes doped with Cr^{o}/CrOx phases (Figure 14). The doped xerogel film effectively generates a HUGE ACOUSTIC WAVE.

3.2. Periodic alignment of alkylene-spacers to create molecular-scale optical grating for acoustic wave

Usually, scientists prepare 'periodic metal frames' to create 'diffraction grating' at the bulk-scales. For example, in a spectroscopic monochromator, 'optical grating effect' is used to split, and then diffract the light into several beams traveling in different directions.

In 'grating equation', the directions of these beams rely on the spacing/distances of the grating, which has a periodic structure of the media, and the wavelength of the light. Gratings, which modulate the phase rather than the amplitude of the incident light, can be also produced. A key idea of the periodic alignment of alkylene-spacers is to create 'a molecular-scale diffraction grating effect' and thus to generate a huge acoustic wave.

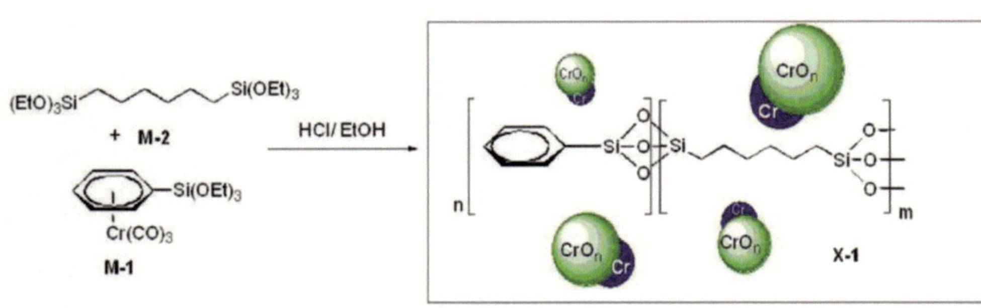

Figure 14. Hexylene-bridged polysilsesquioxane doped with Cr^{o}/CrOx.

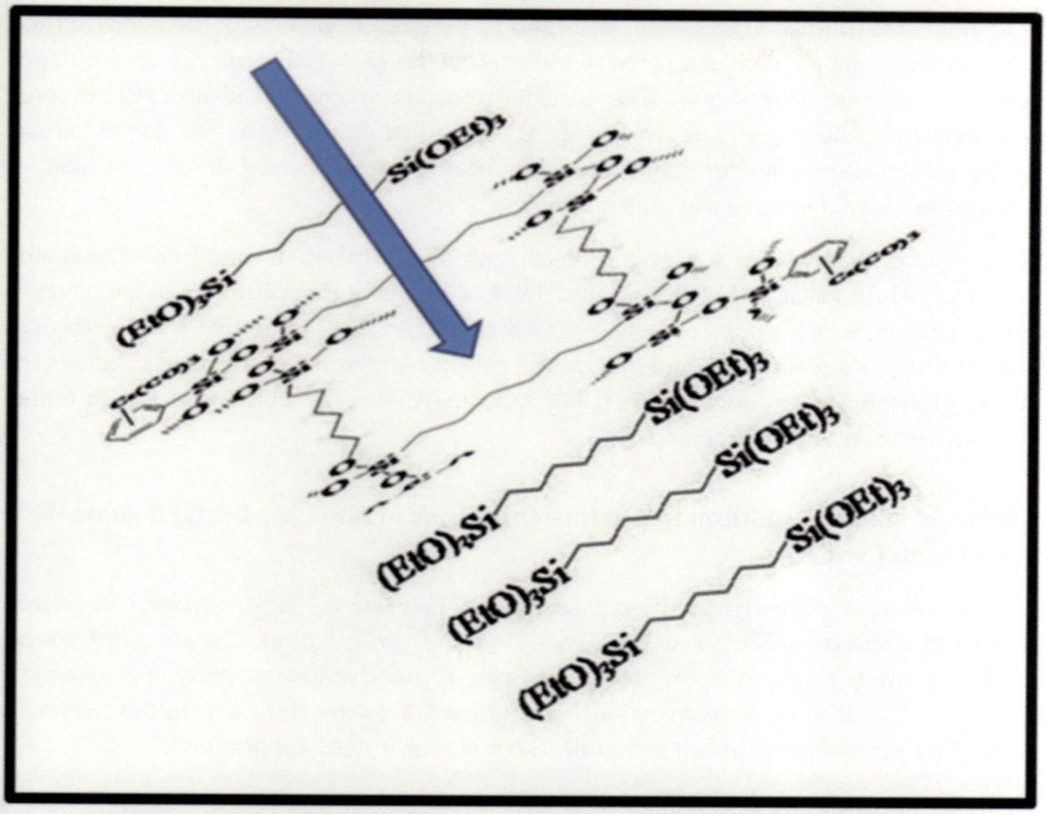

Figure 15. Periodic alignment of alkyl-chains to create "optical grating" at the molecular scales.

Figure 15 illustrates a light that passes through 'optical grating at the molecular-scales' designed in the hybrid silicate host. As you can see, the periodic alkylene-spacers create 'effective diffraction grating at the molecular scales' and thus produces a huge acoustic wave. This is a new optical phenomenon, which hitherto hasn't been discovered.

In addition, the distance between those alkyl-spacers can be also "controlled" by both choices of different organic spacers and sizes of sol-gel monomers. This is an effective method of developing new laser device materials based on organically modified silica, polysilsesquioxanes.

Usually, when the laser beam goes through a solid medium, the density wave is linear; because, in solid media, heat doesn't decay through the solid medium effectively.

Interestingly, the highly compressed xerogel, hexylene-bridged polysilsesquioxane doped with Cr^o/CrOx phases showed a strong 'acoustic response' as much as a liquid.

In our laser experiments, we calculated physical parameters of hexylene-bridged xerogel doped with Cr^o/CrOx phases. The coefficient of phonon diffraction (D) of the doped xerogel was FIVE times smaller than that of normal glass. Which means that the thermal conductivity of the doped xerogel, is FIVE times less than that of normal glass.

In addition, the diffraction efficiency, absorption light efficiency, (45%) of the doped xerogel is higher than that of methanol (25%), which means the COMPRESSIBILITY of the doped xerogel is as effective as liquid. The acoustic refractive intensity and acoustic response generated from the doped xerogel was as strong as liquid. Therefore, the doped xerogel serves as a 'heat generator,' and thus the heat gets transferred into expansion or compression wave (acoustic waves) effectively.

There were a lot of efforts to develop novel laser device materials based on organic and inorganic hybrid silica. [50-54] Especially, this is a unique approach to create an effective optical grating, which brings a HUGE ACOUSTIC RESPONSE by setting up a molecular-scale optical grating effect. By building up the periodic structure of long alkyl-chains in the hybrid glassy hosts, we obtained a HUGE ACOUSTIC WAVE, which hasn't been found. This is a new optical phenomenon.

3.3. Novel sol-gel condition to produce thin films of alkylene-bridged xerogel doped with Cr^o/CrOx

The new optical property partially rose from a new sol-gel mixing condition, which effectively produces HIGHLY COMPRESSED, THIN xerogel films. Conventional sol-gel conditions often result in poor optical transparency/mechanical property. For example, xerogels obtained from the conventional sol-gel conditions are thick and brittle materials. Those thick xerogels are difficult to handle, especially for optical applications.

The limitation motivated us to find a novel sol-gel mixing condition, which produces a highly compressed, thin sol-gel film with a smooth surface and excellent optical clarity.

We discovered a novel sol-gel mixing condition, which results in such a thin xerogel film of alkylene-bridged polysilsesquioxanes doped with Cr^o/CrOx with a low thermal conductivity and high compressibility (Figure 16). [26]

During the sol-gel polymerization, a volume of the sol-gel mixture was dramatically reduced, and then left a green sol-gel film with an excellent optical clarity (Figure 16).

For a source of chromium, we have synthesized a sol-gel processable chromium precursor [20, 26]; we prepared a green sol-gel film based on hexylene-bridged polysilsesquioxanes doped with Cr^o/CrOx phases using the chromium precursor. [20, 26]

In Figure 16 (top left), it shows a undoped hexylene-bridged polysilsesquioxane prepared from our novel sol-gel mixing condition; it was plastic-like a xerogel monolith. As you see in Figure 16 (top, left), it shows an excellent optical transparency.

Doped xerogels were also prepared in Figure 16 (top, Right and bottom). Those green xerogels were obtained as "plastic-like thin films" with an excellent optical compressibility and low thermal conductivity. It was prepared without any mechanical damage/cracking problem.

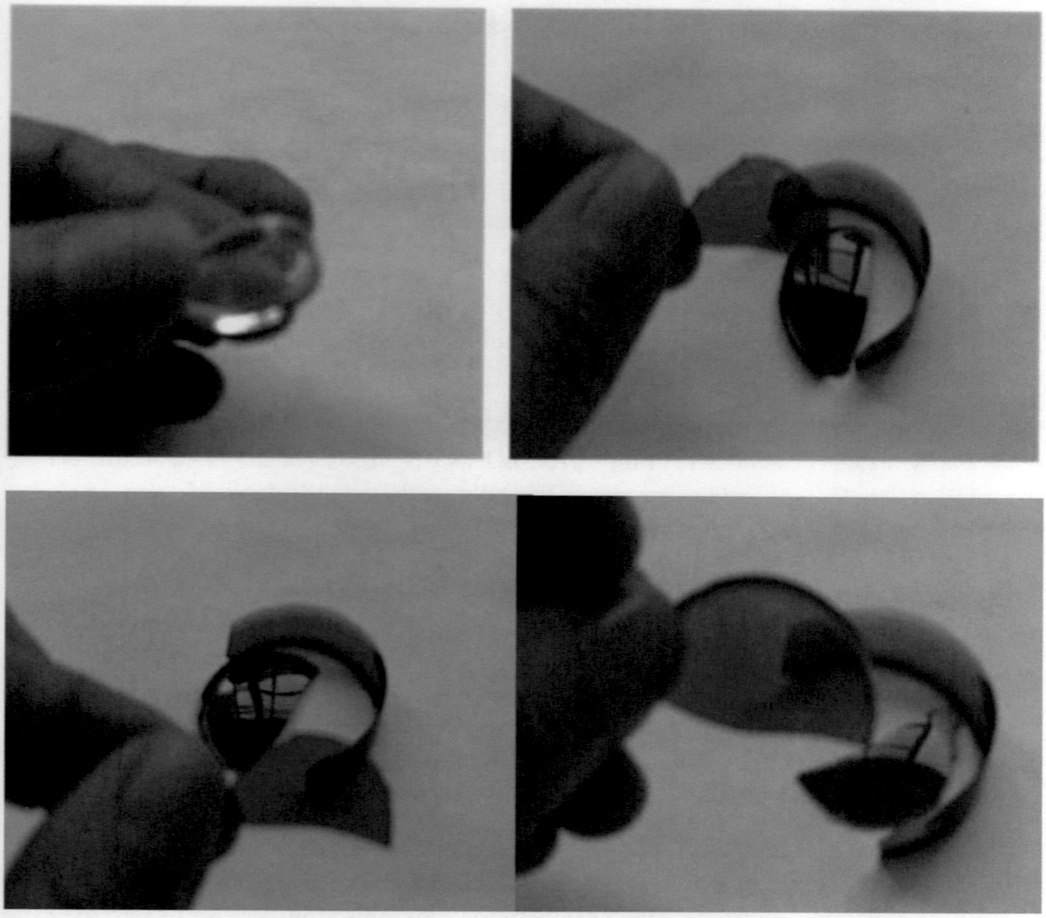

Figure 16. A comparative photos of undoped- (top left) and the Cr°/CrOx doped- (top right and bottom) xerogels based on hexylene-bridged polysilsesquioxane under an acidic condition.

3.4. Experiments

3.4.1. Preparation of doped xerogels

Syntheses of sol-gel monomers, 1,6-bis(triethoxsilyl)hexane and chromium precursor, η^6-chromium tricarbonyl(triethoxysilyl)benzene, were carried out by following the earlier procedures, respectively. [11, 20]

Subsequently, hexylene-bridged xerogel doped with Cr°/CrOx phases was prepared by a copolymerization of η^6-chromium tri-carbonyl(triethoxysilyl)benzene and 1,6-bis(triethoxsilyl)hexane under an acidic condition (Figure 17). A green glass was provided after the sol-gel polymerization (Figure 16).

Density of the doped xerogel was measured to be 1.2696 g/cm^3 using a He-gas pycnometer. From AAS analysis, the chromium amount was also analyzed to be 1.4% by weight.

3.4.2. TEM, EDAX, and electron diffraction analysis

A novel periodic alignment of alkylene-spacers in hybrid glass was specifically designed for creating 'a molecular-scale diffraction grating.' We employed TEM analysis to identify the molecular alignment built up in the hybrid silicate matrix.

The doped hybrid glass under an acidic condition (Figure 17) was ground to powders with a particle size (<150 μm), and deposited on a plasma-etched carbon substrate supported copper grid. TEM dark-field images were obtained using a Philips transmission electron microscope (TEM: CM 20/STEM, PW 6060).

The energy-dispersive X-ray diffraction (EDAX) pattern of the glassy particles was also obtained by an EDAX analyzer (Philips TEM-EDAX, PV 9800). For the electron diffraction, the camera length was calibrated experimentally with a gold standard, and an X-ray spectrum analyzer at 200 kV was used.

3.4.3. Laser analysis

To establish the optical properties of doped xerogels, we prepared a sample (<1mm thickness) fabricated with η^6-chromium tricarbonyl(triethoxysilyl)benzene loading of 2.4 % under an acidic sol-gel condition. The thin xerogel film exhibited a nonlinear property. The nonlinear optical (NLO) properties of doped xerogel films were measured by the degenerated into four wave mixing (DFWM) technique. [55] A quartz sample was used as a reference.

We used two types of lasers, a YAG laser with 50 ps pulse-width at 532 nm and a dye laser with 150 fs pulse-width at 640 nm. The output of either of lasers is split into two strong pump beams and a weak probe. The delay between two pump pulses is set to zero to create interference patterns in the doped sol-gel film. Variable delay line on the probe beam allows to measure the decay time of the diffracted beam.

3.5. Results and discussions

Figure 17 describes a sol-gel procedure for the preparations of hexylene-bridged polysilsesquioxanes doped with Cr^o/ CrOx.

The sol-gel process was carried out by a copolymerization of two sol-gel monomers, η6-chromium tricarbonyl(tri-ethoxysilyl)benzene (M-1) and 1,6-bis(triethoxsilyl)hexane (M-2) (Figure 14).

Those sol-gel monomers can be produced bridged Si-O-Si network under either acidic or basic condition, and thus processed into transparent glasses, glassy films, fibers, xerogels, and aerogels, and monoliths. [1,6] From the basic condition, it produced hybrid glasses containing chromium metal particles; because, the M-1 was stable under the basic sol-gel condition.

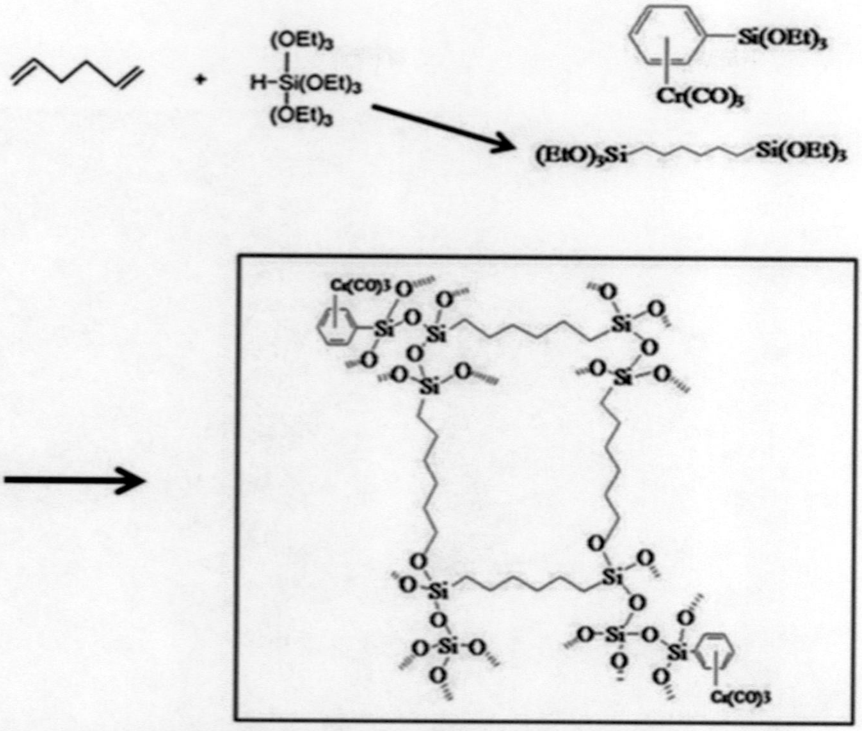

Figure 17. Synthesis of hexylene-bridged xerogel attached with the Cr° precursors.

In contrast, the acid-catalyzed system produced thinner sol-gel films than those of glasses obtained under base condition (Figure 16). Under the acidic sol-gel condition, decomposition of the M-1 competes with sol-gel copolymerization. The product of acid catalyzed decomposition reaction is "chromium oxides" and H_2 (Figure 17). [56-58]

In Figure 18, TEM images of alkylene-bridged silica doped with Cr°/CrOx phases reveal unusual nano-fringe patterns, which rose from the lattice fringes of the aligned alkyl-spacers in the silicate matrix. The novel molecular design results in 'a molecular level grating characteristic' when laser light passes by those periodic carbon-chains (Figure 15).

Based on those nano-fringe patterns, an effective optical grating was created in the hybrid silicate matrix. The TEM images of the doped hybrid glass reveal nano-fringe patterns, which are highly organized nano-periodicity (pointed with arrows in Figure 18). The nano-periodic patterns are sustained over substantial domains and appear to arise from lattice fringes.

In short, the formation mechanism of these nanoperiodic features observed in the TEM images arises from the highly arranged alkylene-spacers in the sol-gel monomer (Figure 15).

EDAX and electron diffraction analyses of these dark regions shown in the TEM images were also performed. In the EDAX spectrum, a Cr ($K\alpha$) peak was observed; thus, the dark contrast shown in the TEM images (Figure 18) was identified as a chromium phase spread over the hybrid glassy host by both of EDAX/electron diffraction analyses.

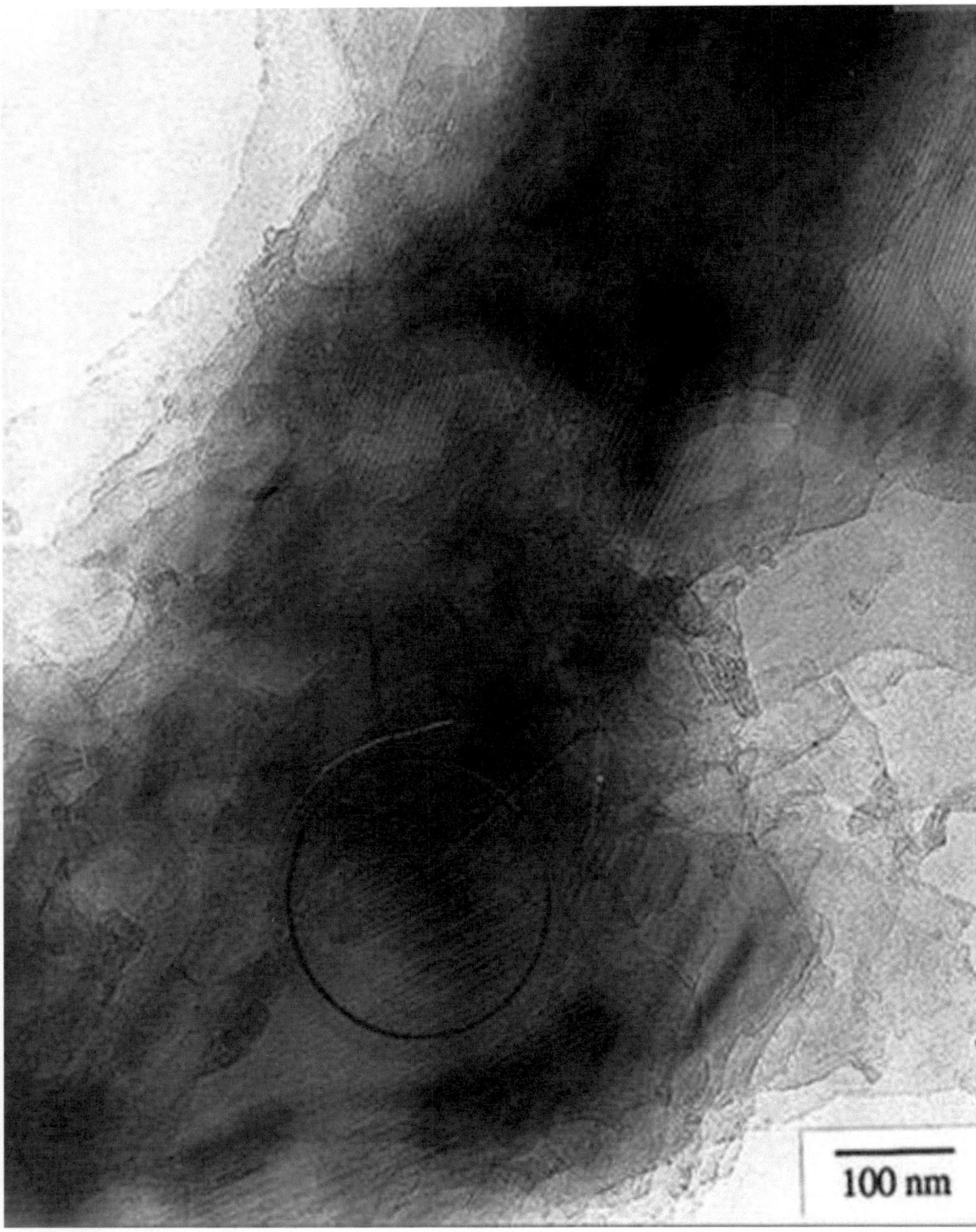

(a)

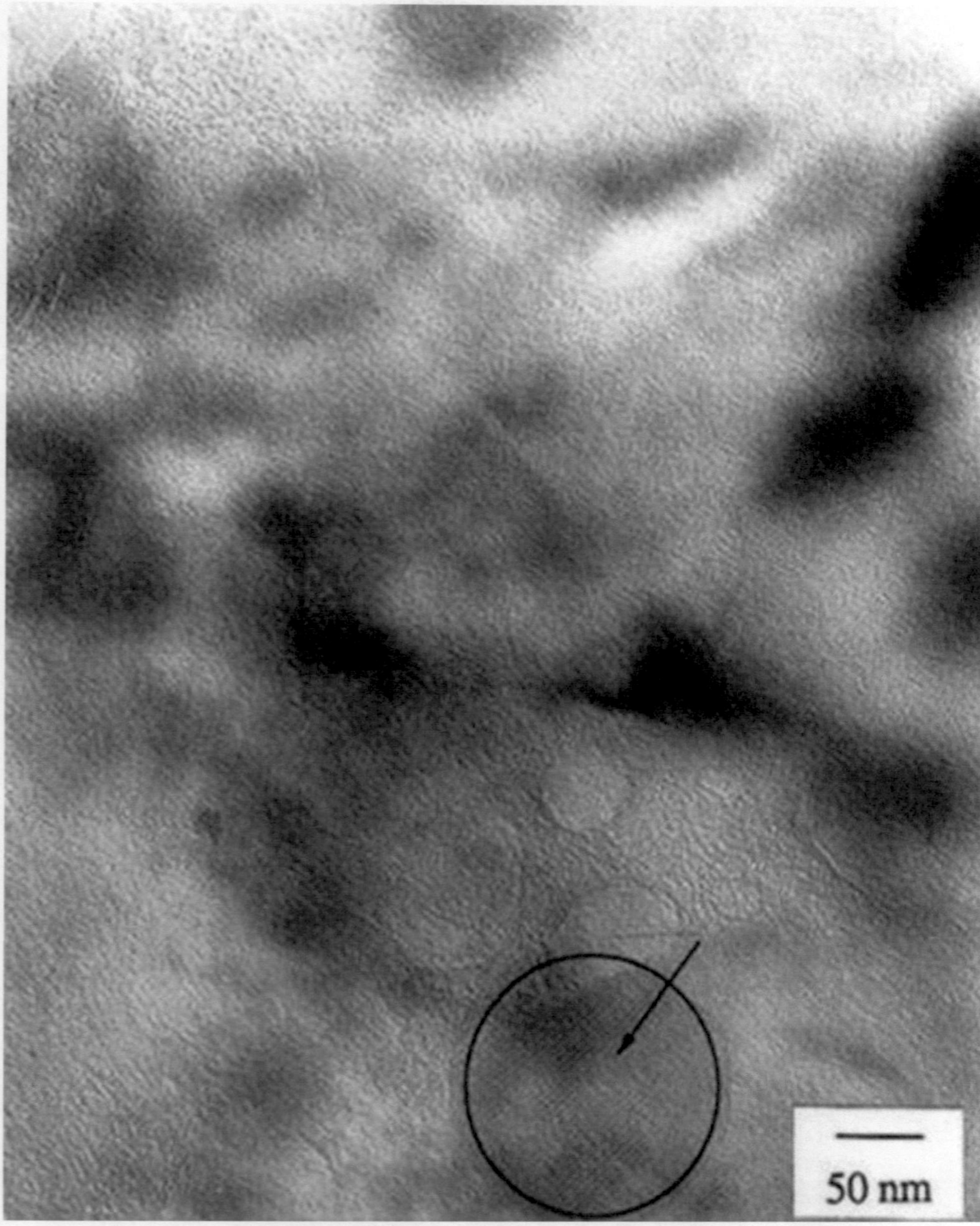

(b)

Figure 18. A and b. TEM Image of Cr°/CrOx-doped hexylene-bridged polysilsesquioxane under acidic condition; it reveals unusual nano-fringe patterns with alternating features of a lattice spacing of about 50 Å in different areas.

The $Cr^o/CrOx$ phases were produced by a chemical reaction, which is a simultaneous sol-gel copolymerization then decomposition of the chromium precursor under the acidic condition. The electron diffraction pattern of the dark areas in TEM images was also identified as a mixture of Cr^o/ CrOx phases.

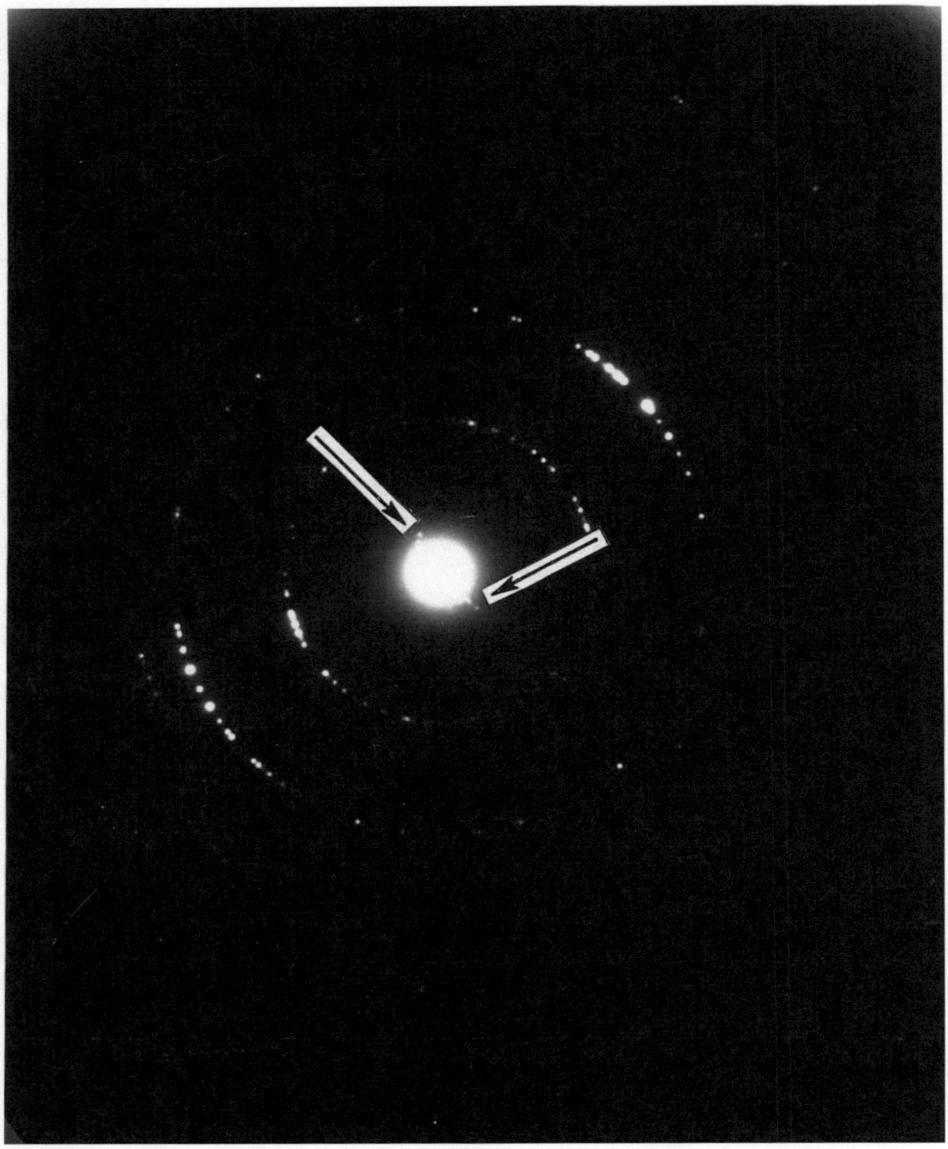

Figure 19. Electron diffraction pattern of doped hybrid glass; circled diffraction patterns correspond to the d-spacings of chromium metal. Highlighted features (arrows) correspond to the nanoperiodicity in the TEM images in Figure 18. From a distance between two diffraction spots point out each arrow a lattice spacing of the nanoscale fringes was calculated to be ~50 Å.

In order to calculate a lattice space of those highly organized nano-fringe patterns shown in the TEM images, we also carried out electron diffraction analysis of the doped sol-gel film. The result is shown in Figure 19. The nano-periodicity gives rise to features in the electron diffraction pattern.

Figure 19 shows the circled diffraction patterns that a rise from crystalline chromium metal and a set of diffractions near the center of the beam corresponded to the nanofringes observed in TEM images (Figure 18).

From the diffraction pattern corresponded to the nano-fringes, a lattice space of the nanostriped patterns observed in TEM images was calculated about 50 Å from a distance between two diffraction spots in two sets of diffraction patterns, which are pointed with arrows in Figure 19; each set consists of two diffraction spots.

The distance between alkyl-spacers significantly relies on the optical grating efficiency, and thus it can be "controlled" by inserting different types and sizes of organic spacers.

As shown in Figure 14, the reactions occurring during the formation of the sol-gel xerogels are complex and include simultaneous sol-gel copolymerization and decomposition of M-1 under the acidic condition. At present, the mechanism of formation of these nanoperiodic features observed in the TEM images may arise from the highly arranged alkyl-spacers in the sol-gel monomer (M-2).

In this study, the nonlinear optical (NLO) properties of the doped xerogel film were measured by the degenerated four wave mixing (DFWM) technique. [12, 13, 55]

In femto- and pico-second experiments, "electronic χ^3" and "population χ^3" of the doped xerogel film have been measured (Figure 20). The DFWM signals for both "pure electronic and population χ^3" are shown in Figure 20 as a small spike around t = 0. It is asymmetric and longer than the laser pulse. The trailing edge of the peak has decay, which is probably connected with population relaxation.

In thermal nonlinearity, the coefficient of phonon diffraction, which is proportional to the coefficient of thermal conductivity, has been calculated from the DFWM experiments as 1.9×10^{-3} cm^2 /sec using an equation, $D = \Lambda^2/4\pi\tau$ (where, Λ is the period of the diffraction grating and τ is the decay time of the thermal signal).

This number is about FIVE times lower than that of a normal glass. In other words, the thermal conductivity of the doped xerogel is FIVE times less than that of a normal glass. In acoustic study, the doped xerogel also shows an interesting new optical property, *an effective generation of a large acoustic wave.*

When the temperature of the doped xerogel at the maximum of the interference pattern goes up, the material expands then a counter propagating wave of expansion and compression start traveling inside the glassy host. Since the index of refraction depends on density of the material, on top of slow decaying thermal grating, we will have *a dynamic diffraction grating propagating with the sound velocity.*

By changing the delay on the laser probe beam, we measured *the period of acoustic grating and extract the sound velocity of the material* (Figure 20). We used YAG laser at 532 nm and 50 ps pulse-width for a laser analysis for the doped xerogel obtained by the novel sol-gel condition, which produces thin xerogel films with unusually high compressibility (higher density).

Small spike around t= 0 corresponds to the signal due to electronic nonlinearlity (Figure 20). The time required for the acoustic wave to travel from one interference maximum to another is twice the time between t=0 and the peak of the acoustic signal. The sound velocity (C) in the thin doped xerogel was calculated from the distance between two acoustic waves ($\Delta\tau$) as $C = 3.2 \times 10^{5}$ cm/sec.

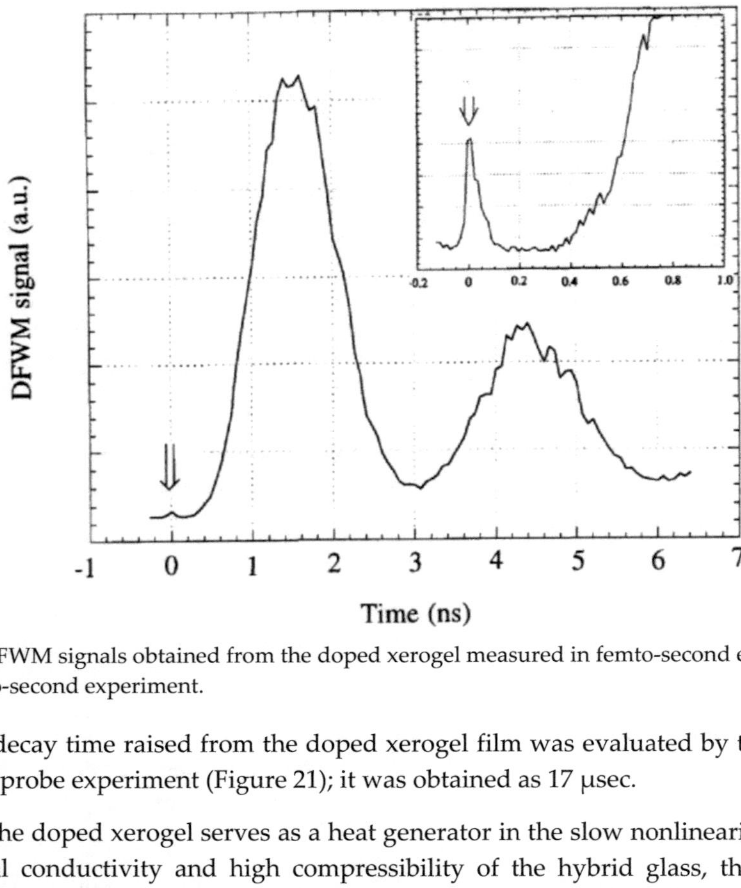

Figure 20. DFWM signals obtained from the doped xerogel measured in femto-second experiment (in a box) and pico-second experiment.

The signal decay time raised from the doped xerogel film was evaluated by the continuous wave (CW) probe experiment (Figure 21); it was obtained as 17 μsec.

Therefore, the doped xerogel serves as a heat generator in the slow nonlinearities due to the low thermal conductivity and high compressibility of the hybrid glass, thus the heat is transferred into expansion or compression wave (acoustic wave) very efficiently.

We also measured the diffraction efficiency of the probe beam at the delay time, corresponding to the peak of acoustic signal. At energy level about P = 0.47 J/cm^2, which is close to the optical damage threshold, the diffraction efficiency was 45 %.

The amplitude of the acoustic signal will depend upon how effectively the laser pulse energy is transferred to an expansion wave, which in turn depends on compressibility of the host materials. For a comparison we did the same measurement for a dye solution in methanol with the same optical density and the same energy density. The diffraction efficiency of methanol was 25 %, which means the compressibility of the doped xerogel film is as effective as liquid.

In a conclusion from the laser experiments, the doped xerogel film has a lower thermal conductivity than that of a normal glass. The compressibility of the doped xerogel is sufficiently high so the density grating formed in the doped xerogel could be effective to create high diffraction beam in the sound velocity.

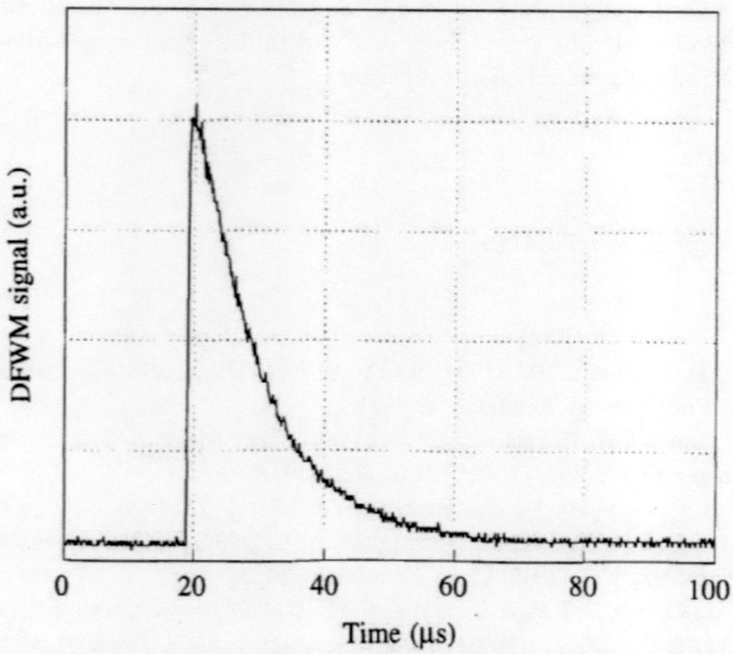

Figure 21. The decay of thermal signal for the doped xerogel measured in CW probe experiment.

We also believe that the nano-fringe patterns revealed in the TEM images (Figure 18), which rose from the lattice fringes of alkylene-bridged silicate matrix may result in an effective grating characteristics when the light passes by those long carbon-chains of alkylene-based polymeric networks.

The characteristics observed from the novel doped xerogel are new photonic properties, which hitherto have not been possible from simple physical mixing process of individual components at the bulk scale. In other words, those new optical properties are created by *the molecular-level hybridization*.

Based on these experiments, the (Cr^o/CrOx)-doped polysilsesquioxane with low thermal conductivity and high compressibility are suggested as a new type of optical device materials for optical applications, for example diffraction beam modulators.

We demonstrated here some examples of creating new optical properties by designing novel molecular building blocks at the molecular-scales via *the molecular-level hybridization.*

4. Conclusions

We introduce the *molecular-level hybridization* for the preparation of polysilsesquioxanes, which are hybrids of inorganic oxides and organic network polymers.

Optically transparent hybrid glasses are prepared by a molecular tailoring technique, which produces sol-gel processible monomers.

The resulting hybrid glasses show novel optical properties, which would not be expected from those individuals by the loss of individuals' identities after the molecular-level mixing process, thereby creating entirely new properties.

Author details

Kyung M. Choi
University of California at Irvine, USA

5. References

[1] Brinker, J. & Scherer, G. W. (1990) *Sol-Gel Science: The Physics and Chemistry of Sol-Gel Processing:* Published by Academic Press, Inc.
[2] Shea, K. J., Loy, D. A., & Webster, O. W. (1989) *Chem. Mater.,* 1, 572.; Shea, K. J., Loy, D. A., & Webster, O. W. (1992) *J. Am. Chem. Soc.,* 114, 6700.
[3] Corriu, R. J. P., Moreau, J. J. E., Thepot, P., & Wong, C. M. M. (1992) *Chem. Mater.,* 4, 1217; Corriu, R. J. P., Thepot, P., & Mang, M. W. C. (1994) *J. Mater. Chem.,* 4, 987.
[4] Loy, D. A. & Shea, K. J. (1995) *Chem. Rev., 95* (5), 1431.
[5] Cerveau, G., Corriu, R. J. P., & Lepeytre, C. (1997) *J. Orgmetal. Chem.,* 548 (1), 99.
[6] Choi, K. M. & Shea, K. J. (1998) *Photonic Polymer Systems, Fundamentals, Methods, and Application*: Edited by D. L. Wise et al. World Scientific Publishing Co. Pte. Ltd.
[7] Shea, K. J. & Loy, D. A. (2001) *MRS Bulletin,* 26, 368.
[8] Loy, D. A. & Rahimian, K. (May 2003), *Handbook of Organic-Inorganic Hybrid Materials and Nanocomposites*: Edited by Hari Singh Nalwa Formerly of Hitachi Research Laboratory, Hitachi Ltd.
[9] Shea, K. J., Moreau, J., Loy, D. A., Corriu, R.J.P., & Bour, B. (2003) *Hybrid Materials*: Published by Wiley Inter science, New York.
[10] Zhao, L., Vaupel, M., Loy, D. A., & Shea, K. J. (2008) *Chem. Mater.,* 20, 1870.
[11] Oviatt, H. W. Jr., Shea, K. J., & Small, J. H. (1993) *Chem. Mater.* , 5, 943.
[12] Oviatt, H. W., Shea, K. J., Kalluri, S., Shi, Y., Steier, W. H., & Dalton, L. R. (1995) *Chem. Mater.,* 7, 493.
[13] Dalton, L. R. (1998) *Polymers for Electro-Optic Modulator Waveguides in Electrical and Optical Polymer Systems; Fundamentals, Methods, and Applications*: Edited by Wise, D. L.,

Copper, T. M., Gresser, J. D., Trantolo, D. J., & Wnek, G. E., (World Scientific, Singapore) Chapter 18.
[14] Walcarius, A. (2001) *Chem. Mater., 13* (10), 3351.
[15] Matsubara, I. (2003) *AIST Today* , 3 (8), 32.
[16] Schaefer, D. W., Beaucage, G., Loy, D.A., Shea, K. J., & Lin, J.S. (2004) *Chem. Mater.*, 16, 1402.
[17] Pang, Y.X., Hodgson, S.N.B., Koniarek, J., & Weglinski, B. (2006) *J. of Magnetism and Magnetic Mater.*, 301(1), March 27, 83.
[18] Choi, K. M. & Shea, K. J. (1993) *Chem. Mater.*, 5, 1067.
[19] Choi, K. M. & Shea, K. J. (1994) *J. Phys. Chem.*, 98, 3207.
[20] Choi, K. M. & Shea, K. J. (1994) *J. Am. Chem. Soc.*, 116, 9052.
[21] Choi, K. M., Hemminger, J. C., & Shea, K. J. (1995) *J. Phys. Chem.*, 99, 4720.
[22] Choi, K. M. & Shea, K. J. (1995) *J. of Sol-Gel Sci. Tech.*, 5, 143.
[23] Choi, K. M. & Rogers, J. A. (2003) *J. Am. Chem. Soc.*, 125, 4060.
[24] Choi, K. M. (2005) *J. Phys. Chem. B*, 109, 21525.
[25] Choi, K. M. (2007) *Mater. Chem. Phys.*, 103, 176.
[26] Choi, K. M. & Shea, K. J. (2008) *J. Phy. Chem. C*, 112, 18173.
[27] Carpenter, J. P., Lukehart, C. M. , Stock, S. R., & Wittig, J. E. (1995) *Chem. Mater.*, 7, 201.
[28] Pocard, N. L., Alsmeyer, D. C., McCreery, R. L., Neenan, T. X., & Callstrom, M. R. (1992) *J. Am. Chem. Soc.*, 114, 769.
[29] Kim, J.H. & Holloway, P. H. (2005) *Adv. Mater.*, 17, 91.
[30] Xia, H. R., Lu, G. W., Zhao, P., Sun, S. Q., Meng, X. L., Cheng, X. F., Qin, L. J., Zhu, L., & Yang, Z. H. (2005) *J. Mater. Res.*, 20, 30.
[31] Mukhopadhyay, S., Ramesh, K. P., Kannan, R., & Ramakrishna, J. (2004) *Phys. Rev. B: Cond. Matter Mater. Phys.*, 70, 224202.
[32] Dantelle, G., Mortier, M., Vivien, D., & Patriarche, G. (2005) *J. Mater. Res.*, 20. 472.
[33] Xiao, S., Yang, X., Liu, Z., & Yan, X. H. (2004) *J. Appl. Phys.*, 96. 1360.
[34] Buscaglia, M. T., Buscaglia, V., Ghigna, P., Viviani, M., Spinolo, G., Testino, A., & Nanni, P. (2004) *Phys. Chem. Chem. Phys.*, 6, 3710.
[35] Stepanov, S., Hernandez, E., & Plata, M. (2004) *Opt. Lett.*, 29, 1327.
[36] Lin, H., Jing, S., Wu, J., Song, F., Peyghambarian, N., & Pun, E.Y.B. (2003) *J. Phys. D: Appl. Phys.*, 26, 812.
[37] Sohler, W. & Suche, H. (2000) *Opt. Eng.*, 66, 127.
[38] Neuman, W., Pennington, D., Dawson, J., Drobshoff, A., Beach, R., Jovanovic, I., Liao, Z., Payne, S., & Barty, C.P.J. (2005) *Proc. SPIE Int. Soc. Opt. Eng.* , 5653, 262.
[39] Huang, X., Liu, Y., Sui, Z., Li, M., Lin, H., Wang, J., Zhao, D., Wang, F., & Chen, J. (2005) *Proc. SPIE Int. Soc. Opt. Eng.*, 5623, 679.
[40] Sanchez, C., Lebeau, B., Chaput, F., & Boilot, J. P. (2003) *Adv. Mater.* , 15. 1969.
[41] Shea, K. J. & Loy, D. A. (2001) *Chem. Mater.* , 13, 3306.
[42] Schmidt, H. (1989) *Sol–Gel Science and Technology:* Published by World Scientific, Singapore.
[43] Urquhart, P. (1988) *IEEE Proc.-J.; Optoelectronics*, 135, 385.
[44] Murray, C. B., Kagan, C. R., & Bawendi, M. G. (1995) *Science*, 270, 1335.

[45] Empedocles, S. A., Norris, D. J., & Bawendi, M. G. (1996) *Phys. Rev. Lett.*, 77, 3873.

[46] Armelao, L., Gross, S., Obetti, G., & Tondello, E. (2004) *Surf. Coat. Technol.*, 109, 218.

[47] Di Giovanni, D. J. (1992) *Mater. Res. Soc. Proc.*, 244, 135.

[48] Rossetti, R., Hull, R., Gibson, J. M., & Brus, L. E. (1985) *J. Chem. Phys.*, 82 (l), 552; Zhang, Y., Raman, N., Bailey, J. K., Brinker, C. J., & Crooks, R. M. (1992) *J. Phys. Chem.*, 96, 9098.

[49] Schmid, G. (1992) *Chem. Rev.*, 92, 1709.

[50] Grate, J. W., Kaganove, S. N., Patrash, S. J., Craig, R., & Bliss, M. (1997) *Chem. Mater.*, *9* (5), 1201.

[51] Ponson, L., Boechler, N., Lai, Y. M., Porter, M. A., Kevrekidis, P. G. , & Daraio, C. (2010) *Physical Review E*, 82, 021301.

[52] Spadoni, A. & Daraio, C. (2010) *Proc. Natl. Acad. Sci. USA*, 107, 7230.

[53] Porter, M.A., Daraio, C., Herbold, E.B., Szelengowicz, I., & Kevrekidis, P.G. (2008) *Physical Review E*, 77, 015601(R).

[54] Sullivan, P. A., Olbricht, B. C., & Dalton, L. A. (2008) *Journal of Lightwave Technology*, 26 (15), 2345.

[55] Ma, H., Chen, B., Takafumi, S., Dalton, L. R., & Jen, A. K.-Y. (2001) *J. Am. Chem. Soc.* , 123, 986.

[56] η^6-Bisarene chromium complexes decompose in dilute acid. Oxidation and disproportion products (of both organic and Cr) are observed. [57] η^6-Arene $Cr(CO)_3$ complexes form adducts with Lewis and protonic acids but their decomposition products have not been investigated. [58]

[57] Gribov, B. G., Mozzhukhin, D. D., Kozyrkin, B. I., & Strizhkova, A. S. (1972) *J. Gen. Chem. USSR (Eng. Transl.)*, 42, 2521.

[58] Flood, T. C., Rosenberg, E., & Sarhangi, A. (1997) *J. Am. Chem. Soc.*, 99, 4334; Lillya, C. P. & Sahatjian, R. A. (1972) *Inorg. Chem.*, 11, 889.